New Directions in Agrarian Political Economy

T0331428

How relevant are the classic theories of agrarian change in the contemporary context? This volume explores this question by focusing upon the defining features of agrarian transformation in the 21st century: the financialization of food and agriculture, the blurring of rural and urban livelihoods through migration and other economic activities, forest transition, climate change, rural indebtedness, the co-evolution of social policy and moral economies, and changing property relations. Combined, the eleven contributions to this collection provide a broad overview of agrarian studies over the past four decades and identify the contemporary frontiers of agrarian political economy. In this path-breaking collection, the authors show how new iterations of long-evident processes continue to catch peasants and smallholders in the crosshairs of crises and how many manage to face these challenges, developing new sources and sites of livelihood production.

This volume was published as part one of the special double issue celebrating the 40th anniversary of *The Journal of Peasant Studies*.

Madeleine Fairbairn, Jonathan Fox, S. Ryan Isakson, Michael Levien, Nancy Lee Peluso, Shahra Razavi, Ian Scoones, Kalyanakrishnan "Shivi" Sivaramakrishnan

Critical Agrarian Studies
Series Editor: Saturnino M. Borras Jr.

Critical Agrarian Studies is the new accompanying book series to the *Journal of Peasant Studies*. It publishes selected special issues of the journal and, occasionally, books that offer major contributions in the field of critical agrarian studies. The book series builds on the long and rich history of the journal and its former accompanying book series, the Library of Peasant Studies (1973–2008) which had published several important monographs and special-issues-as-books.

New Directions in Agrarian Political Economy

Global agrarian transformations, volume I

Edited by
Madeleine Fairbairn, Jonathan Fox, S. Ryan Isakson, Michael Levien, Nancy Lee Peluso, Shahra Razavi, Ian Scoones, Kalyanakrishnan "Shivi" Sivaramakrishnan

Routledge
Taylor & Francis Group

LONDON AND NEW YORK

First published 2016
by Routledge
2 Park Square, Milton Park, Abingdon, Oxon, OX14 4RN, UK

and by Routledge
605 Third Avenue, New York, NY 10017, USA

First issued in paperback 2021

Routledge is an imprint of the Taylor & Francis Group, an informa business

British Library Cataloguing in Publication Data
A catalogue record for this book is available from the British Library

Typeset in Times New Roman
by RefineCatch Limited, Bungay, Suffolk

Publisher's Note
The publisher has gone to great lengths to ensure the quality of this reprint but
points out that some imperfections in the original copies may be apparent.

Disclaimer
Every effort has been made to contact copyright holders for their permission to
reprint material in this book. The publishers would be grateful to hear from any
copyright holder who is not here acknowledged and will undertake to rectify
any errors or omissions in future editions of this book.

ISBN 13: 978-1-138-12279-6 (pbk)
ISBN 13: 978-1-138-91639-5 (hbk)

Contents

Citation Information

The chapters in this book were originally published in *The Journal of Peasant Studies*, volume 41, issue 5 (November 2014). When citing this material, please use the original page numbering for each article, as follows:

Chapter 1
Introduction: New directions in agrarian political economy
Madeleine Fairbairn, Jonathan Fox, S. Ryan Isakson, Michael Levien, Nancy Peluso, Shahra Razavi, Ian Scoones and K. Sivaramakrishnan
Journal of Peasant Studies, volume 41, issue 5 (November 2014) pp. 653–666

Chapter 2
Cause and response: vulnerability and climate in the Anthropocene
Jesse Ribot
Journal of Peasant Studies, volume 41, issue 5 (November 2014) pp. 667–705

Chapter 3
The problem of property in industrial fisheries
Liam Campling and Elizabeth Havice
Journal of Peasant Studies, volume 41, issue 5 (November 2014) pp. 707–727

Chapter 4
The role of rural indebtedness in the evolution of capitalism
Julien-François Gerber
Journal of Peasant Studies, volume 41, issue 5 (November 2014) pp. 729–747

Chapter 5
Food and finance: the financial transformation of agro-food supply chains
S. Ryan Isakson
Journal of Peasant Studies, volume 41, issue 5 (November 2014) pp. 749–775

Chapter 6
'Like gold with yield': evolving intersections between farmland and finance
Madeleine Fairbairn
Journal of Peasant Studies, volume 41, issue 5 (November 2014) pp. 777–795

Chapter 7

Financialization, distance and global food politics
Jennifer Clapp
Journal of Peasant Studies, volume 41, issue 5 (November 2014) pp. 797–814

Chapter 8

Moral economy in a global era: the politics of provisions during contemporary food price spikes
Naomi Hossain and Devangana Kalita
Journal of Peasant Studies, volume 41, issue 5 (November 2014) pp. 815–831

Chapter 9

The government of poverty and the arts of survival: mobile and recombinant strategies at the margins of the South African economy
Andries du Toit and David Neves
Journal of Peasant Studies, volume 41, issue 5 (November 2014) pp. 833–853

Chapter 10

Rural-urban migration in Vietnam and China: gendered householding, production of space and the state
Minh T.N. Nguyen and Catherine Locke
Journal of Peasant Studies, volume 41, issue 5 (November 2014) pp. 855–876

Chapter 11

Forests lost and found in tropical Latin America: the woodland 'green revolution'
Susanna B. Hecht
Journal of Peasant Studies, volume 41, issue 5 (November 2014) pp. 877–909

Please direct any queries you may have about the citations to
clsuk.permissions@cengage.com

Notes on Editors

Madeleine Fairbairn is an Assistant Professor of Environmental Studies at the University of California, Santa Cruz. Her previous research examined the global food sovereignty movement and land grabbing in Mozambique. Her current work explores growing interest in farmland on the part of the financial sector, as well as the policy debate surrounding foreign farmland investment in Brazil.

Jonathan Fox is Professor in the School of International Service at American University. His publications include *Accountability Politics: Power and Voice in Rural Mexico* and *Subsidizing Inequality: Mexican Corn Policy Since NAFTA*. He currently works with the Open Government Partnership and serves on the boards of directors of Oxfam America and Fundar.

S. Ryan Isakson is an Assistant Professor of International Development Studies and Geography at the University of Toronto. His current research explores the financialization of food and agriculture, with a special focus upon rural vulnerability and the financialization of agricultural risk.

Michael Levien is Assistant Professor of Sociology at Johns Hopkins University. His current research focuses on the transformation of rural land dispossession in post-liberalization India.

Nancy Lee Peluso is Henry J. Vaux Professor of Forest Policy and Professor of Society and Environment in the Department of Environmental Science, Policy, and Management at the University of California, Berkeley. Her current project is an ethnography of globally connected, resource-based livelihoods and landscape transformations in Indonesia.

Shahra Razavi is Chief of Research and Data Section at UN Women, where she oversees the preparation of two flagship reports. She specializes in the gender dimensions of development, with a particular focus on livelihoods, agrarian issues, social policy, and care.

Ian Scoones is a Professorial Fellow at the Institute of Development Studies and Director of the ESRC STEPS Centre. He is a member of *The Journal of Peasant Studies* editorial collective.

Kalyanakrishnan "Shivi" Sivaramakrishnan is Dinakar Singh Professor of India & South Asian Studies, Professor of Anthropology, Professor of Forestry & Environmental Studies, Professor of International & Area Studies, and Co-Director, Program in Agrarian Studies at Yale University. His current research examines environmental litigation in India.

Preface

A fundamentally contested concept, food sovereignty has – as a political project and campaign, an alternative, and a social movement – barged into global agrarian discourse over the last two decades.[1] Since then, it has inspired and mobilized diverse publics: workers, scholars, and public intellectuals, farmers and peasant movements, NGOs, and human rights activists in the North and global South. The term has become a challenging subject for social science research, and has been interpreted and reinterpreted in a variety of ways by various groups and individuals. Indeed, it is a concept that is broadly defined as the right of peoples to democratically control or determine the shape of their food system, and to produce sufficient and healthy food in culturally appropriate and ecologically sustainable ways in and near their territory. As such it spans issues such as food politics, agroecology, land reform, biofuels, genetically modified organisms (GMOs), urban gardening, the patenting of life forms, labor migration, the feeding of volatile cities, ecological sustainability, and subsistence rights.

The idea and practice of food sovereignty have generated vibrant discussions and debates within and between various circles among social movement activists and academics. For this reason, several academic institutions decided to forge an alliance with key activist think tanks and NGOs to organize a critical dialogue around this topic. The Yale Agrarian Studies Program and the International Institute of Social Studies (ISS) in The Hague took the lead among the academic institutions, while the Transnational Institute (TNI), Institute for Food and Development Policy or Food First, and the Dutch development agency Inter-church Organization for Development Cooperation (ICCO) took the lead among non-academic research and advocacy organizations. Other collaborators played a critical role in constructing the space for dialogue: Yale Sustainable Food Project, Yale South Asian Studies Council, Initiatives in Critical Agrarian Studies (ICAS), Misereor, Kempf Fund, and *The Journal of Peasant Studies*.

Two successful events were held. The September 2013 Yale University event brought together close to 300 leading scholars and political activists who are advocates of and sympathetic to the idea of food sovereignty, as well as those who are skeptical about the concept of food sovereignty, to foster a critical and productive dialogue on the issue. The keynote addresses were given by Paul Nicholson (La Vía Campesina), Teodor Shanin (The Moscow School of Social and Economic Sciences), Bina Agarwal (Manchester University), and Olivier de Schutter (then the UN Special Rapporteur on the Right to Food). The January 2014 ISS event in The Hague brought together more than 300 scholars

---

[1] Part of the text of this Preface draws from the conference blurbs used at the "Food Sovereignty: a critical dialogue" colloquiums held in September 2013 at Yale University and in January 2014 at the International Institute of Social Studies (ISS) in The Hague, the Netherlands.

and social movement activists mostly from across Europe – key speakers included Tania Li and Bridget O'Laughlin. The two events were accompanied by close to a hundred conference papers. The conference organizers also produced high quality video clips of the plenary speakers' speeches. These video clips are available at the organizers' websites.

The exchanges and debates at the two conferences were collegial and comradely, tackling serious issues that unite and divide conference participants. Most of the conference papers have been transformed into journal articles. There were at least four journal special issues that came out from this critical dialogue: *The Journal of Peasant Studies* (*JPS*) vol. 41, no. 5, edited by Madeleine Fairbairn, Jonathan Fox, S. Ryan Isakson, Michael Levien, Nancy Peluso, Shahra Razavi, Ian Scoones, and Kalyanakrishnan ("Shivi") Sivaramakrishnan, explored new directions in agrarian political economy. This volume established the context for the other *JPS* special issue, vol. 41, no. 6 in 2014, which focused on food sovereignty and was edited by Marc Edelman, James C. Scott, Amita Baviskar, Saturnino M. Borras Jr., Deniz Kandiyoti, Eric Holt-Giménez, Tony Weis, and Wendy Wolford. A special issue of *Globalizations* in 2015 was guest edited by Annie Shattuck, Christina Schiavoni, and Zoe VanGelder. Finally, a special issue of *Third World Quarterly* in 2015 was guest edited by Eric Holt-Giménez, Alberto Alonso-Fradejas, Todd Holmes, and Martha Robbins.

Routledge decided to publish, together with the Transnational Institute (TNI), the book editions of the special issues by *JPS* edited by Fairbairn, *et. al.* and Edelman, *et. al.* These are the present collections being introduced through this Preface. By explaining the origin of these collections, we hope that the readers will be able to follow the spirit of critical dialogue that produced these volumes, and contribute to the continuing dialogue. Finally, we hope that readers of these volumes will also view the accompanying video clips and read the other two related journal special issues.

Saturnino ("Jun") M. Borras Jr.
2 March 2015, The Hague

Introduction: New directions in agrarian political economy

Madeleine Fairbairn, Jonathan Fox, S. Ryan Isakson, Michael Levien, Nancy Peluso, Shahra Razavi, Ian Scoones and K. Sivaramakrishnan

For four decades, *The Journal of Peasant Studies* (*JPS*) has served as a principal arena for the formation and dissemination of cutting-edge research and theory. It is globally renowned as a key site for documenting and analyzing variegated trajectories of agrarian change across space and time. Over the years, authors have taken new angles as they reinvigorated classic questions and debates about agrarian transition, resource access and rural livelihoods. This introductory essay highlights the four classic themes represented in Volume 1 of the *JPS* anniversary collection: land and resource dispossession, the financialization of food and agriculture, vulnerability and marginalization, and the blurring of the rural-urban relations through hybrid livelihoods. Contributors show both how new iterations of long-evident processes continue to catch peasants and smallholders in the crosshairs of crises and how many manage to face these challenges, developing new sources and sites of livelihood production.

Introduction

Forty years have passed since the first issue of *The Journal of Peasant Studies* (*JPS*) was launched. The 'broadly based but rigorous political economy' of peasant societies called for in its first editorial statement continues to flourish (Byres and Shanin 1973, 1). New trajectories of agrarian change generated by advanced capitalism, transformed political contexts and the emergence of new actors and social forces are prompting fresh research and theoretical reconstructions of classic questions of agrarian transition central to debates in *JPS* for many decades.

The contributions to this anniversary collection are testaments to the ongoing relevance of these questions to scholarly, activist and policy research. The essays in Volume 2 will focus on questions of agri-food systems and food sovereignty. In Volume 1, four themes are highlighted: land and resource dispossession, the financialization of food and agriculture, vulnerability and marginalization in an era of crisis, and the blurring of the rural/urban dichotomy in hybrid livelihoods. Many of these have been the subject of debate over the last 40 years, yet contemporary processes of agrarian change have generated fresh theoretical insights and important new lines of research.

Land and resource dispossession

Processes of land and resource dispossession, as well as their political drivers and consequences, have long been an emphasis of research on agrarian change. Yet in the last decade, diverse forms of land dispossession – often called 'land grabs' – have refocused scholarly research, debate and analysis. *JPS* has been a key site for documentation of the unprecedented extent of land and resource dispossession in the past few years, including many papers and several collections. This collection does not comprehensively address these contributions' arguments, but instead highlights several new dimensions of disposses-sion in relation to land and other resources.

Current research on land and resource dispossession builds on at least two previous waves of scholarly interest. The significance of European enclosures for capitalist develop-ment spurred fierce debates among Marxists and other social theorists for many decades (Dobb 1947; Hilton 1976; Aston and Philpin 1985), while their implications for agrarian transitions outside of the West were hotly contested in *JPS* and other journals in the 1970s and 1980s. In this first generation of dispossession scholarship, expropriation was largely seen as an historical stage in the development of capitalism – or so-called 'primitive accumulation' – although paths to capitalism varied (Lenin 1967; Bernstein 1982; Kautsky [1899] 1988; Byres 1991). At the time, expropriation was rarely viewed as an ongoing process recursively linked to advanced capitalism.

Agricultural land expropriations and other forms of dispossession nevertheless showed no sign of abating; scholars in every region of the world documented their various forms, causes and effects.[1] The 1980s and 1990s saw the emergence of political movements and both scholarly and activist documentations of 'development-induced displacement'. This period brought other kinds of contemporary and historical capitalist expropriations into critical view, including conflicts over dams, roads and other infrastructure projects (Hirsch 1990; Baviskar 1995), forestry and large-scale cattle production (Guha 1989; Hecht and Cockburn 1989; Peluso 1992; Fairhead and Leach 1996), and mega-mines and plantations (Mintz 1985; Stoler 1985). Many of these studies shifted attention from 'conflict in the factory and the field' to 'conflict around forests and rivers' (Baviskar 1995, 40). This period also saw the rise of political ecology as an approach to understanding environmental relationships and conflicts, including in agrarian settings (Blaikie 1985; Peet and Watts 1996; Neumann 1999; Carney 2001). This lineage of scholarship on displace-ment and the politics of resource access and control might be termed the second generation of dispossession studies, and it continues as a vibrant research program in the pages of *JPS* and other social science journals.

A third wave of dispossession studies has been prompted by the apparent acceleration of 'land grabs', mentioned above, as well as resistance to them in many parts of the world since roughly the mid-2000s. Though the precise extent of such grabs is difficult to quantify on the ground, the resurgence of publications and research on these latest, extensive dispos-sessions is not just an artifact of the academy. As conflicts around plantations, mines, dams, conservation projects, Special Economic Zones and other forms of urban-industrial devel-opment proliferate, new questions are being raised about land and property relations under advanced capitalism. This has generated renewed interest in Marx's theory of primitive accumulation and its reformulation by David Harvey as 'accumulation by dispossession' (De Angelis 2004; Kelly 2011; Benjaminsen and Bryceson 2012; Hall 2012; Levien 2012). To what extent and how, scholars ask, can such concepts be reconstructed to illuminate the regionally diverse land expropriations of globalized but locally articulated capitalisms?

---

[1]This literature, as well as many others alluded to in this introduction, is vast and impossible to do justice to here.

Recent research on regional political economies of dispossession has greatly enriched our understanding of the ever-shifting relationships between land, varied social formations and capitalism. New forms of agrarian politics and power relations, involving diverse actors and an array of natural resources, have opened up fresh avenues for debate, generated novel interpretations of the most classic works, and promise to inspire future generations of agrarian researchers. Two contributions in this collection cleave particularly closely to the traditional concerns of agrarian studies: in their contribution, Liam Campling and Elizabeth Havice explore the extraction of ground rents from an unexpected source, while in his paper, Julien-François Gerber analyzes the impacts of rural indebtedness.

Current debates about agrarian change extend beyond 'land' and landed property to encompass the air, climate and bodies of water. Of course, control over water resources has long been a concern of agrarian scholars, particularly the political economy of irrigation (e.g. Frankel 1971; Boyce 1987) and state-led river basin development (e.g. Barkin and King 1970). Yet the focus has broadened in recent years and *JPS* has featured many papers on new forms of water control, including its governance for consumption or production, or as a socio-natural context for the Earth's contested and stressed fish stocks (e.g. Gibbon 1997; Woodhouse 2012; Magee 2013; Tapela 2013). The latter focus is represented in this collection by the Campling and Havice contribution, which extends Marxian theories of ground rent to the property relations of marine fisheries. The authors argue that we need to understand 'exclusive economic zones' (EEZs), the fisheries between 12 and 200 nautical miles from land that contain 90 percent of global marine catch, as a form of modern landed property, challenging the claims of neo-classical economists who consider these fisheries as open access or non-property. Campling and Havice demonstrate that coastal states have had property rights in these fisheries since the 1982 United Nations Convention on the Law of the Sea (UNCLOS), allowing them to capture ground-rent through access payments from private capital. Following Marx's analysis of rent in Volume 3 of *Capital*, Campling and Havice argue that the neo-classical understanding of rent as a market-determined payment for accessing unevenly productive resources neglects the 'absolute' dimension of rent, which is based on historically contingent and politically contested claims to ownership. Their contribution underscores the potential extensions of agrarian political economy analysis to non-landed resources.

Like land and natural resource control, the role of debt in agrarian formations has been central to agrarian studies since its inception and has remained an important topic for scholars of agrarian societies over the past several decades. As a mechanism of exploitation, social differentiation, labor control and dispossession, debt continues to be a defining feature of rural lives and an important theme of scholarship, including sharp debates within *JPS*.[2] Gerber's contribution to this collection offers a panoramic survey of this literature, while advancing a broader conception of debt's numerous roles. Debt has been an important lever of accumulation, differentiation and labor control since early modern agrarian transitions. In addition, it has frequently enabled the diffusion of capitalist cultural practices and rationalities, dissolved social solidarity and contributed to ecological destruction (e.g. via subsidized credit for the expansion of livestock production). Gerber's fine-grained categorization of the forms and effects of debt thus consolidates the findings of agrarian studies over many decades and illuminates avenues for future research on rural livelihoods enmeshed in debt.

---

[2]See Banaji (1977); Hamid (1982); Islam (1985); Brass (1986); Srivastava (1989); Nazir (2000); Washbrook (2006); Shah (2012); Gray and Dowd-Uribe (2013).

If agrarian political economy continues to examine, as Kautsky ([1899] 1988, 12) put it, 'whether and how capital is seizing hold of agriculture', it has thus also expanded in the last few decades to examine the ways capital is seizing hold of land and natural resources more broadly, both destroying and reconfiguring social and socio-natural relations (e.g. Van der Ploeg 1993; Akram-Lodhi 1998; Hart 2002; Guthman 2004; McMichael 2012). As global capitalism penetrates agrarian formations and commons in new ways, agrarian scholars continue to reconstruct the theoretical heritage of the classic 'agrarian question', producing fresh insights on trajectories of agrarian change and politics.

The financialization of food and agriculture

One of the most important engines of transformation of the global agri-food system – and therefore of agrarian livelihoods – relates to the financial sector's recent rise in power and prominence. In the early 1970s, a suite of inter-related developments referred to generally as 'financialization' initiated sweeping changes across the global economic landscape. These changes have provided powerful new mechanisms of accumulation, and they intersect in numerous ways with the current wave of land and resource dispossession (McMichael 2012; Russi 2013). If early modern enclosures in western Europe and European powers' control of arable land in Asian, African and Latin American colonies were wedded to ideas of agricultural improvement and increased productivity, the early twenty-first century has unleashed international processes of land appropriation driven largely by speculative interests. Three contributions in this collection – by Ryan Isakson, Madeleine Fairbairn and Jennifer Clapp – explore the subtle (and not-so-subtle) ways in which finance is transforming agriculture and agri-food systems.

In seeing financialization as another dimension of ever-expanding capitalist processes, we recognize that its multi-faceted forms and effects have historical precedents, both in changing political economies and in scholarly analyses. Marxist scholars have, today and in the past, noted a growing reliance on financial profits in response to capitalist crises of accumulation caused by increasing international competition (Arrighi 1994; Harvey 2010) or by corporate concentration and the rise of monopoly capital (Magdoff and Sweezy 1987). On one level, therefore, contemporary mechanisms of financialization represent changing paths of capitalist accumulation (Arrighi 1994; Krippner 2011). The size and profitability of the financial sector in many national economies has ballooned relative to other sectors. Further, many non-financial corporations have expanded into financial service provision as a lucrative supplement to their regular productive activities (Krippner 2011). Buoyed by investor expectations, and only loosely tethered to real assets, financial profits can soar even when the productive economy is stagnant. The shift to finance, however, is a double-edged sword: some scholars argue that its rejuvenating effects have been coupled with increasingly frequent market bubbles and economic crises (Kindleberger and Aliber 2005; Parenteau 2005).

On another level, financialization has meant a reconfiguration of the institutional relationships between corporations and investors. Over the last three decades, the doctrine of 'shareholder value', which dictates that the purpose of a corporation is to create returns for shareholders above all else, has taken hold in corporate boardrooms everywhere (Useem 1996; Fligstein 2001). The shareholder value revolution has meant an increased alignment between the interests of stockholders and those of corporate managers via such mechanisms as manager compensation in stock options. Davis (2009, 5) compares these shifts to a Copernican revolution: 'from a social system orbiting around corporations and their imperatives, we have moved to a market-centered system in which the corporations themselves – along with households and governments – are guided by the gravitational pull of financial markets'.

Isakson's contribution to this collection provides a comprehensive overview of such transformations in the agri-food system, touching everything from agricultural inputs to the final products found on the supermarket shelf. Most evident has been the increasing role of financial actors in markets for agricultural commodity derivatives. Formerly a means for agricultural producers, traders and processors to hedge against possible price changes, these markets have recently seen major increases in the participation of financial speculators (Clapp 2012). The food crisis of 2008 brought this speculation into sharp relief, as commentators began to argue that the herd-like behavior of financial investors could be behind the increasing volatility of global grain markets (Masters 2008; De Schutter 2010b).

Beyond grain markets, financialization is reshaping the ways that food traders and retailers do business. Supermarket chains are subject to takeover and restructuring by private equity consortia, while they themselves increasingly offer financial services such as credit cards and check cashing to consumers (Burch and Lawrence 2009). Financial actors are increasingly investing in grain storage and transportation infrastructure, while the major grain trading companies have also begun selling commodity derivatives to third-party clients (Murphy, Burch, and Clapp 2012). In his essay, Isakson stresses this blurring boundary between financial and non-financial economic action. He further argues that financialization has exacerbated the already unequal power relations within agri-food supply chains. Retailers and traders, now beholden to financial backers, are finding new ways to produce shareholder value at the expense of the most vulnerable food system actors. In this way, food workers and small farmers find their positions further weakened in the drive for financial profits.

Madeleine Fairbairn addresses a critical finance-related development of interest to *JPS* readers and other scholars of agrarian change: the financial sector's growing interest in farmland since 2007. Students of the current land rush, mentioned above, have made clear that high agricultural commodity prices and the search for dependable investments in the wake of the 2008 financial crisis have made farmland an increasingly attractive investment, with many new farmland funds available. Following Harvey's (1982) early arguments about land as fictitious capital, Fairbairn argues that new farmland investors treat land as a financial asset in itself and that this drive to own real and productive assets is unusual for the sector. Fairbairn concludes that farmland financialization could increase land prices and volatility, ultimately fueling smallholder dispossession.

The question of potential paths to resistance in the food sector is primarily addressed in Volume 2 of this collection, but it is also taken up in this volume by Clapp, who uses the concept of 'distancing' (Princen 1997) to shed light on the political dynamics of the financialized food system. She contends that the immense geographic distances within agri-food supply chains are being compounded by another kind of distance that stems from increasing financial activity and the proliferation of agriculture-based financial investment tools. The growing number of financial intermediaries and the abstract nature of many financial investments, she argues, create a knowledge gap that obscures the negative social and environmental outcomes of financialization. This distancing has political ramifications; it opens up space for competing narratives that counter post-2008 calls to regulate financial investment. Taken together, the contributions by Clapp, Isakson and Fairbairn illuminate multiple dimensions of the financialization of food and agriculture, revealing developments that have produced new forms of vulnerability and precarity for rural people across the world.[3]

Vulnerability and marginalization in an era of crisis

From its inception, *JPS* has been a key forum for analyses of the marginalization and vulnerability of the rural poor within their broader historical, social and political-economic contexts (e.g. Byres and Shanin 1973; Bernstein and Byres 2001). These contexts have, however, changed dramatically in the four decades since the journal was launched. Two essays in this volume address important new stressors in the lives of marginal populations: climate change and rising food prices.

Increasingly, scholars predict that climate hazards will be crucial influences on agrarian livelihoods, particularly for subsistence and other small producers (Eakin 2006; Mercer, Perales, and Wainwright 2012). In this volume, Jesse Ribot argues that policy makers and scientists tend to depoliticize the anthropogenic origins of climate change, occluding the social processes that render certain populations more vulnerable than others to climate hazards. Ribot sees the causes of climate change and a group's relative vulnerability as inextricably linked. Reducing vulnerability is partially contingent on afflicted populations possessing the entitlements (i.e. rights and opportunities) to control resources that allow them to withstand climatic events (Sen 1981; Drèze and Sen 1989). Yet surplus extraction from the agrarian economy leaves certain people unable to act on their own behalf or to influence the provision of social protections. Reducing vulnerability is thus contingent on the ability of all population segments to influence the broader political economic processes shaping both entitlements and governance (Watts 1983, 1991; Appadurai 1984; Ribot and Peluso 2003).

Rising and increasingly volatile food prices are another contemporary stressor. Since 2007, wide swings in food prices have not only exacerbated the uncertainties faced by agricultural producers, but the record spikes punctuating that volatility have hurt the poorest food consumers (FAO 2009; De Janvry and Sadoulet 2010; Ghosh 2010; Spratt 2013). The corresponding erosion of food entitlements has generated food protests and political instability throughout the global South (Schneider 2008; Holt-Giménez and Patel 2009; Bush 2010; Moseley, Camey, and Becker 2010).

While dominant narratives attribute food protests and riots to population growth and Malthusian-style scarcities, critical analyses suggest that they are driven less by shortages than by gaps between expected and actual food entitlements (Hossain 2009; Patel and McMichael 2009; Sneyd, Legwegoh, and Fraser 2013). Much as the commodification of agricultural output unraveled the moral economies of 'pre-capitalist' societies (Scott 1976; Watts 1983; Swift 1989; Thompson 1991), neo-liberal policies restructuring agriculture and international commerce have exposed contemporary peasantries to a variety of interrelated threats, including: competition from subsidized industrial agricultural products from the North, streamlined supply chains that reduce the market power of small producers, rising input costs and reduced access to formal credit, and increased competition for farmland.

As Hossain and Kalita argue here, contemporary food protests are driven by beliefs about how food economies should function and the sense that governing authorities are not meeting their responsibilities. Drawing upon research in multiple sites in Asia and

---

[3]A trend noted by Friedmann (1993), much earlier.

Africa, they compare food politics during the 2011 global food price spike with the politics of 'moral economies' expressed during the eighteenth-century English food riots described by E.P. Thompson (1971, 1991). Hossain and Kalita document the widespread belief that access to food should not be governed by markets alone because markets can be manipulated by powerful actors. As in Thompson's examples, the poor and working classes protest to spur public authorities to act.

However, the feudal paternalism of eighteenth-century Europe has given way to *laissez faire* capitalism and representative democracy. The terms of negotiating moral economies have evolved accordingly (Scott 1976; Edelman 2005; Patel and McMichael 2009). As Hossain and Kalita argue, protesters no longer appeal to the reciprocal obligations inhering in patron-client relationships; instead, they must play to the electoral accountability of governing leaders. They also appeal to moral sentiments that basic subsistence needs should trump speculation and other forms of profiteering. Correspondingly, there has been a growing movement at multiple scales, both national and transnational, to formalize access to food from a moral claim to a codified right (Edelman 2005; De Schutter 2010a; Narula 2010). Attempts to ensure food rights include guaranteed access to wage labor (Argentina, Bangladesh, India), the public provisioning of agricultural inputs (Nicaragua), access to affordably priced food (India), and cash transfers (Brazil). Though their effectiveness varies, right-to-food programs and other social protection mechanisms contribute to the continued viability of many rural livelihoods.

Hybrid livelihoods and the blurring of the rural/urban dichotomy

As the literatures on rural livelihoods have highlighted, the dichotomies between 'rural' and 'urban' that are conventionally deployed to define peasantries and the poor in general do not hold in the contemporary context, a condition referred to as the 'new rurality' (Kay 2008; Hecht 2010). Rather, rural households are, in the words of Susana Hecht (this volume), 'largely semi-proletarianized, semi-globalized and increasingly semi-urban'. In response to the various encroachments of capitalist development, smallholders' agrarian livelihoods have not disappeared so much as diversified and hybridized, albeit with varying degrees of success. In many countries, agricultural income and rurality have been largely delinked, with various classes of labor remaining in rural areas but relying less and less on agricultural earnings (Wilson and Rigg 2003; Bernstein 2009; Fox and Haight 2010). Three of the contributions to this collection – by Andries du Toit and David Neves, Thi Nguyet Minh Nguyen and Catherine Locke, and Susana Hecht – evaluate the contemporary processes of agrarian change producing these hybrid livelihoods.

Du Toit and Neves analyze how the interplay between institutionalized inequality and economic agency give rise to the hybrid livelihoods of South Africa's poorest rural people. Specifically, they examine what they call the 'truncated agrarian transition' of rural South Africans who are dislodged from agricultural commodity production but lack opportunities in the non-farm economy. Du Toit and Neves' research challenges two dominant yet competing narratives – one which asserts that displaced populations will be seamlessly integrated into capitalist employment and a second in which they will be cast asunder as 'waste lives'. Instead, they describe the creative and resourceful manners by which poor South Africans cobble together complex livelihoods. These strategies muddle conventional dichotomies: households straddle rural and urban spaces, combine formal and informal sources of income, and draw upon the social care afforded by the state even as they generate revenue outside of publicly regulated spaces. Even within the context of de-agrarianization, agricultural self-provisioning is an important component of this 'recombinant bricolage' of complementary activities. While it does not allow for accumulation, subsistence agriculture helps to insure vulnerable South Africans against the vagaries of market-oriented activities (see also Lipton 1968; Annis 1987; Isakson 2009).

The complementarities between market and non-market, capitalist and non-capitalist, and multi-sited urban and rural economic activities are not unique to the South African context (e.g. Bebbington 1999; Barkin 2002; Isakson 2009). Indeed, several foundational texts in agrarian studies have documented similar relationships across space and time (Warman 1980; De Janvry 1981; Deere 1990; Bryceson, Kay, and Mooij 2000). De Janvry's (1981) 'functional dualism' thesis, for example, posited that the agricultural self-provisioning of semi-proletarianized farmers subsidized their workplace wages, thereby allowing their continued marginalization and the viability of capitalist production. Nguyen and Locke make a similar argument in their contribution to this volume, albeit with an important twist: rather than subsidizing capitalist activities, non-market reproductive labor in the countryside is essential to the functioning of market socialism (see also Deere 1990).

Analyzing rural-urban migration under market transformation in Vietnam and China, Nguyen and Locke show how the two socialist governments have manipulated gender norms and the complementarities between market and non-market activities to advance their economic and development goals. The dismantling of state support for household reproduction, combined with the enclosure of land and other assets, have catalyzed circular forms of rural-urban migration. At the same time, state control mechanisms, especially systems of household registration, have helped to separate the labor of migrant workers from their social reproduction. The resulting trans-local households, which are strained by evolving gender norms and expectations, provide a low cost and flexible labor force that has fueled economic growth in the two countries.

These new spatial complexities of contemporary rural livelihoods are reshaping environmental contexts. For the first time in human history, more people live today in urban areas than in rural, and more rural people than ever depend on remittances from urban and 'Northern' sites where family members have migrated to work (Padoch et al. 2008). Remittances constitute the largest international transfer of funds used for bolstering livelihoods, exceeding the amount of all international aid. Some scholars have suggested that various forms of permanent and circular migration and the concomitant transfer of wages have contributed to the de-agrarianization of the countryside (Wilson and Rigg 2003), forest resurgence (Hecht et al. 2005) and generally more complex urban-rural relations (Padoch et al. 2008), in addition to the long-recognized volatilities of commodity prices.

In her contribution to this volume, Susana Hecht explores the relationship between the surprising dynamic of 'forest transition' (i.e. declining rates of deforestation and, in some cases, reforestation) and the hybridization of rural livelihoods in Latin America. Through the critical lens of a political ecologist, she considers the potential linkages and impacts that various components of new ruralities, specifically migration and remittances, cash transfers, de-agrarianization and the emergence of green markets, have on tree cover. Much like earlier political ecological accounts that challenged the Malthusian notion that deforestation is the result of population growth (Hecht 1985; Blaikie and Brookfield 1987; Peluso 1992; Fairhead and Leach 1996), Hecht cautions against the simplistic narratives that attribute reforestation to urbanization and the exodus of humans from the countryside. She takes these arguments further by noting that many of Latin America's forests are inhabited but that those inhabitants' land use practices are changing in tandem with their livelihoods. Further, income from sources outside immediate agricultural production such as remittances, social care initiatives and non-farm employment have helped reconfigure rural livelihoods and, in some instances, underpinned the continued ability to engage in agricultural production. Of course, those livelihoods have followed a variety of trajectories, including some well-off actors engaging in activities that negatively impact forests, such as cattle-ranching, mineral extraction and plantation agriculture. Even so, however, the growth of tree plantations in the region also plays into forest transition, albeit in different ways than other forms of reforestation (Kroger 2014). In short, the agrarian roots of forest transition are complex and, as Hecht suggests, beg for a reconfiguration of the classic agrarian question, one that considers the contemporary hybridities of rural and urban landscapes and livelihoods. In so doing, she lays the foundation for a fascinating line of future research.

Conclusion

The authors in this fortieth anniversary collection document how the related pressures of financialization, transformed environmental contexts and diverse new forms of enclosure and exclusion are catching smallholders in the crosshairs of globalized food, energy, economic and environmental crises. They demonstrate the varied ways in which peasants and smallholders hold on, develop new sources of livelihoods and use mobilities to their advantage. The classic concerns around agrarian questions that filled the pages of *JPS* 40 years ago remain relevant in the radically changed political economic contexts of the present. Yet new contexts have also prompted new lines of empirical research and progressive theoretical reconstructions (e.g. Burawoy 1998). Forty years on, peasant and agrarian studies remains vibrant and relevant, offering new insights into the challenges of agrarian societies the world over.

References

Akram-Lodhi, A.H. 1998. The agrarian question, past and present. *Journal of Peasant Studies* 25, no. 4: 134–49.
Annis, S. 1987. *God and production in a Guatemalan town*. Austin: University of Texas Press.
Appadurai, A. 1984. How moral is South Asia's economy? – A review article. *Journal of Asian Studies* 43, no. 3: 481–97.
Arrighi, G. 1994. *The long twentieth century: Money, power, and the origins of our times*. New York: Verso.
Aston, T.H., and C.H.E. Philpin, eds. 1985. *The Brenner debate: Agrarian class structure and economic development in pre-industrial Europe*. Cambridge: Cambridge University Press.
Banaji, J. 1977. Capitalist domination and the small peasantry: Deccan districts in the late nineteenth century. *Economic and Political Weekly* 12, no. 33/34: 1375–404.
Barkin, D. 2002. The reconstruction of a modern Mexican peasantry. *Journal of Peasant Studies* 30, no. 1: 73–90.
Barkin, D., and T. King. 1970. *Regional economic development: The river basin approach in Mexico*. Cambridge: Cambridge University Press.
Baviskar, A. 1995. *In the belly of the river: Tribal conflicts over development in the Narmada Valley*. Oxford: Oxford University Press.
Bebbington, A. 1999. Capitals and capabilities: A framework for analyzing peasant viability, rural livelihoods and poverty. *World Development* 27, no. 12: 2021–44.
Benjaminsen, T.A., and I. Bryceson. 2012. Conservation, green/blue grabbing and accumulation by dispossession in Tanzania. *Journal of Peasant Studies* 39, no. 2: 335–55.
Bernstein, H. 1982. Notes on capital and peasantry. In *Rural development: Theories of peasant economy and agrarian change*, ed. J. Harriss. London: Hutchinson. [Originally published in *Review of African Political Economy*, 4(10), 60–73 (1977)].

Bernstein, H. 2009. Agrarian questions from transition to globalization. In *Peasants and globalization: political economy, rural transformation, and the agrarian question*, ed. H. Akram-Lodhi and C. Kay, 239–261. London: Routledge.

Bernstein, H., and T. J. Byres. 2001. From peasant studies to agrarian change. *Journal of Agrarian Change* 1, no. 1: 1–56.

Blaikie, P. 1985. *The political economy of soil erosion in developing countries*. London: Longman.

Blaikie, P., and H. Brookfield. 1987. *Land degradation and society*. London: Methuen.

Boyce, J.K. 1987. *Agrarian impasse in Bengal: institutional constraints to technological change*. Oxford: Oxford University Press.

Brass, T. 1986. Unfree labour and capitalist restructuring in the agrarian sector: Peru and India. *Journal of Peasant Studies* 14, no. 1: 50–77.

Bryceson, D.F., C. Kay, and J. Mooij, eds. 2000. *Disappearing peasantries? Rural labour in Africa, Asia and Latin America*. London: Intermediate Technology Publications.

Burawoy, M. 1998. The extended case method. *Sociological Theory* 16, no. 1: 4–33.

Burch, D., and G. Lawrence. 2009. Towards a third food regime: Behind the transformation. *Agriculture and Human Values* 26, no. 4: 267–79.

Bush, R. 2010. Food riots: Poverty, power, and protest. *Journal of Agrarian Change* 10, no. 1: 119–29.

Byres, T.J. 1991. The agrarian question and differing forms of capitalist agrarian transition: An essay with reference to Asia. In *Rural transformation in Asia*, eds. J. Breman and S. Mundle, pp.3–76. Delhi: Oxford University Press.

Byres, T.J., and T. Shanin. 1973. Editorial statement. *Journal of Peasant Studies* 1, no. 1: 1–2.

Carney, J. 2001. *Black rice: The African origins of rice cultivation in the Americas*. Berkeley: University of California Press.

Clapp, J. 2012. *Food*. Malden: Polity Press.

Davis, G. 2009. *Managed by the markets: How finance reshaped America*. Oxford: Oxford University Press.

De Angelis, M. 2004. Separating the doing and the deed: Capital and the continuous character of enclosures. *Historical Materialism* 12, no. 2: 57–87.

Deere, C.D. 1990. *Household and class relations: Peasants and landlords in northern Peru*. Berkeley: University of California Press.

De Janvry, A. 1981. *The agrarian question and reformism in Latin America*. Baltimore: Johns Hopkins University Press.

De Janvry, A., and E. Sadoulet. 2010. The global food crisis and Guatemala: What crisis and for whom?. *World Development* 38, no. 9: 1328–39.

De Schutter, O. 2010a. *Countries tackling hunger with a right to food approach*. FAO Briefing Note 01, May 2010.

De Schutter, O. 2010b. *Food commodities speculation and food price crises*. Geneva: United Nations Special Rapporteur on the Right to Food.

Dobb, M. 1947. *Studies in the development of capitalism*. New York: International Publishers.

Drèze, J., and A. Sen. 1989. *Hunger and public action*. Oxford: Clarendon Press.

Eakin, H. 2006. *Weathering risk in rural Mexico: Climatic, institutional, and economic change*. Tucson: University of Arizona Press.

Edelman, M. 2005. Bringing the moral economy back in … to the study of 21st century transnational peasant movements. *American Anthropologist* 107, no. 3: 331–45.

Fairhead, J., and M. Leach. 1996. *Misreading the African landscape: Society and ecology in a forest-savanna mosaic*. Cambridge: Cambridge University Press.

FAO. 2009. *The state of food insecurity in the world: Economic crises – impacts and lessons learned*. Rome: Food and Agricultural Organization of the United Nations.

Fligstein, N. 2001. *The architecture of markets: An economic sociology of twenty-first-century capitalist societies*. Princeton: Princeton University Press.

Fox, J., and L. Haight, eds. 2010. *Subsidizing inequality: Mexican corn policy since NAFTA*. Woodrow Wilson International Center for Scholars, Centro de Investigación y Docencia Económicas, University of California, Santa Cruz.

Frankel, F.R. 1971. *India's green revolution: economic gains and political costs*. Princeton: Princeton University Press.

Friedmann, H. 1993. The political economy of food: a global crisis. *New Left Review* 197, 29–57.

Ghosh, J. 2010. The unnatural coupling: Food and global finance. *Journal of Agrarian Change* 10, no. 1: 72–86.

Gibbon, P. 1997. Prawns and piranhas: The political economy of a Tanzanian private sector marketing chain. *Journal of Peasant Studies* 25, no. 1: 1–86.

Gray, L., and B. Dowd-Uribe. 2013. A political ecology of socio-economic differentiation: Debt, inputs and liberalization reforms in southwestern Burkina Faso. *Journal of Peasant Studies* 40, no. 4: 683–702.

Guha, R. 1989. *The unquiet woods: Ecological change and peasant resistance in the Himalaya.* New Delhi: Oxford University Press.

Guthman, J. 2004. *Agrarian dreams: The paradox of organic agriculture in California.* Berkeley: University of California Press.

Hall, D. 2012. Rethinking primitive accumulation: Theoretical tensions and rural Southeast Asian complexities. *Antipode* 44, no. 4: p 1188–1208.

Hamid, N. 1982. Dispossession and differentiation of the peasantry in the Punjab during colonial rule. *Journal of Peasant Studies* 10, no. 1: 52–72.

Hart, G. 2002. *Disabling globalization: Places of power in post-apartheid South Africa.* Berkeley: University of California Press.

Harvey, D. 1982. *The limits to capital.* Oxford: Blackwell.

Harvey, D. 2010. *The enigma of capital: And the crises of capitalism.* Oxford: Oxford University Press.

Hecht, S. 1985. Environment, development and politics: Capital accumulation and the livestock sector in eastern Amazonia. *World Development* 13, no. 6: 663–84.

Hecht, S. 2010. The new rurality: Globalization, peasants, and the paradoxes of landscapes. *Land Use Policy* 27: 161–9.

Hecht, S., and A. Cockburn. 1989. *The fate of the forest: Developers, destroyers, and defenders of the Amazon.* New York: Verso.

Hecht, S. B., S. Kandel, I. Gomes, N. Cuellar, and H. Rosa. 2005. Globalization, forest resurgence, and environmental politics in El Salvador. *World Development* 34, no. 2: 308–23.

Hilton, R., ed. 1976. *The transition from feudalism to capitalism.* London: Verso.

Hirsch, Philip. 1990. *Development dilemmas in rural Thailand.* Singapore: Oxford University Press.

Holt-Giménez, E., and R. Patel. 2009. *Food rebellions! Crisis and the hunger for justice.* Cape Town: Pambazuka Press.

Hossain, N. 2009. Reading political responses to food, fuel, and financial crises: The return of the moral economy?. *Development* 52, no. 3: 329–33.

Isakson, S.R. 2009. No hay ganancia en la *milpa*: The agrarian question, food sovereignty, and the on-farm conservation of agrobiodiversity in the Guatemalan highlands. *Journal of Peasant Studies* 36, no. 4: 725–59.

Islam, M.M. 1985. M.L. Darling and the Punjab peasant in prosperity and debt: A fresh look. *Journal of Peasant Studies* 13, no. 1: 83–98.

Kautsky, K. [1899] 1988. *The agrarian question.* London: Zwan Publications.

Kay, C. 2008. Reflections of Latin American rural studies in the neoliberal globalization period: A new rurality?. *Development and Change* 39, no. 6: 915–43.

Kelly, A.B. 2011. Conservation practice as primitive accumulation. *Journal of Peasant Studies* 38, no. 4: 683–701.

Kindleberger, C., and R. Aliber. 2005. *Manias, panics, and crashes: A history of financial crises.* 5th ed. Hoboken: Wiley and Sons.

Krippner, G. 2011. *Capitalizing on crisis: The political origins of the rise of finance.* Cambridge, MA: Harvard University Press.

Kroger, M. 2014. The political economy of the global tree plantation expansion: A review. *Journal of Peasant Studies* 41, no. 2: 235–61.

Lenin, V.I. 1967. *The development of capitalism in Russia.* Moscow: Progress Publishers.

Levien, M. 2012. The land question: Special economic zones and the political economy of dispossession in India. *Journal of Peasant Studies* 39, no. 3: 933–69.

Lipton, M. 1968. The theory of the optimising peasant. *Journal of Development Studies* 4, no. 3: 323–51.

Magdoff, H., and P. Sweezy. 1987. *Stagnation and the financial explosion.* New York: Monthly Review Press.

Magee, D. 2013. The politics of water in rural China: A review of English-language scholarship. *Journal of Peasant Studies* 40, no. 6: 1189–208.

Masters, M. 2008. Testimony of Michael W. Masters before the Committee on Homeland Security and Governmental Affairs, United States Senate.www.hsgac.senate.gov//imo/media/doc/052008Masters.pdf?attempt=2 (accessed April 11, 2014).

McMichael, P. 2012. The land grab and corporate food regime restructuring. *Journal of Peasant Studies* 39, no. 3–4: 681–701.

Mercer, K.L., H.R. Perales, and J.D. Wainwright. 2012. Climate change and the transgenic adaptation strategy: Smallholder livelihoods, climate justice, and maize landraces in Mexico. *Global Environmental Change* 22, 495–504.

Mintz, S. 1985. *Sweetness and power: The place of sugar in modern History*. New York: Penguin.

Moseley, W.G., J. Carney, and L. Becker. 2010. Neoliberal policy, rural livelihoods, and urban food security in West Africa: A comparative study of The Gambia, Côte d'Ivoire, and Mali. *Proceedings of the National Academy of Sciences* 107, no. 13: 5774–9.

Murphy, S., D. Burch, and J. Clapp. 2012. *Cereal secrets*. Oxford: Oxfam.

Narula, S. 2010. Reclaiming the right to food as a normative response to the global food crisis. *Yale Human Rights and Development Law Journal* 13, no. 2: 403–20.

Nazir, P. 2000. Origins of debt, mortgage and alienation of land in early modern Punjab. *Journal of Peasant Studies* 27, no. 3: 55–91.

Neumann, R. 1999. *Imposing wilderness: Struggles over livelihood and nature preservation in Africa*. Berkeley: University of California Press.

Padoch, C., E. Brondizio, S. Costa, M. Pinedo-Vasquez, R. Sears, and A. Sequeira. 2008. Urban forest and rural cities: Multi-sited households, consumption patterns, and forest resources in Amazonia. *Ecology and Society* 13, no. 2: 2. [online: available at: http://www.ecologyandsociety.org/vol13/iss2/.]

Parenteau, R. 2005. The late 1990s' bubble: Financialization in the extreme. In *Financialization and the world economy*, ed. G. Epstein, 111–148. Cheltenham, UK: Edward Elgar Publishing.

Patel, R., and P. McMichael. 2009. A political economy of the food riot. *Review (Fernand Braudel Center)* 32, no. 1: 9–36.

Peet, R., and M. Watts, eds. 1996. *Liberation ecologies: environment, development, social movements*. London: Routledge.

Peluso, N.L. 1992. *Rich forests, poor people: Resource control and resistance in Java*. Berkeley: University of California Press.

Princen, T. 1997. The shading and distancing of commerce: When internalization is not enough. *Ecological Economics* 20, no. 3: 235–53.

Ribot, J. C., and N.L. Peluso. 2003. A theory of access. *Rural Sociology* 68, no. 2: 153–81.

Russi, Luigi. 2013. *Hungry capital: The financialization of food*. Winchester: Zero Books.

Schneider, M. 2008. 'We are hungry!' A summary report of food riots, government responses, and states of democracy in 2008. Unpublished manuscript.

Scott, J. 1976. *The moral economy of the peasant*. New Haven: Yale University Press.

Sen, A. 1981. *Poverty and famines: An essay on entitlement and deprivation*. Oxford: Oxford University Press.

Shah, E. 2012. 'A life wasted making dust': Affective histories of dearth, death, debt and farmer's suicides in India. *Journal of Peasant Studies* 39, no. 5: 1159–79.

Sneyd, L.Q., A. Legwegoh, and E.D.G. Fraser. 2013. Food riots: Media perspectives on the causes of food protest in Africa. *Food Security* 5, 485–97.

Spratt, S. 2013. Food price volatility and financial speculation. Future Agricultures Consortium, Working Paper 047.

Srivastava, R. 1989. Interlinked modes of exploitation in Indian agriculture during transition: A case study. *Journal of Peasant Studies* 16, no. 4: 493–522.

Stoler, A. 1985. *Capitalism and confrontation in Sumatra's plantation belt*. Berkeley: University of California Press.

Swift, J. 1989. Why are rural people vulnerable to famine?. *IDS Bulletin* 20, no. 2: 8–15.

Tapela, B. 2013. Conflicts over land and water in Africa. *Journal of Peasant Studies* 40, no. 1: 319–21.

Thompson, E.P. 1971. The moral economy of the English crowd in the eighteenth century. *Past & Present* 50: 76–136.

Thompson, E.P. 1991. *Customs in common*. London: Penguin.

Useem, M. 1996. *Investor capitalism: How money managers are changing the face of corporate America*. New York: Basic Books.

Van der Ploeg, J. 1993. Rural sociology and the new agrarian question: A perspective from the Netherlands. *Sociologia Ruralis* 33, no. 2: 240–60.

Warman, A. 1980. *'We come to object': The peasants of Morelos and the national state*. Trans. S.K. Ault. Baltimore: Johns Hopkins University Press.

Washbrook, S. 2006. 'Una eslavitud simulada': Debt peonage in the state of Chiapas, Mexico, 1876–1911. *Journal of Peasant Studies* 33, no. 3: 367–412.

Watts, M.J. 1983. *Silent violence: Food, famine, and peasantry in northern Nigeria*. Berkeley: University of California Press.

Watts, M.J. 1991. Entitlements or empowerment? Famine and starvation in Africa. *Review of African Political Economy* 18: 9–26.

Wilson, G. A., and J. Rigg. 2003. Post-productivist agricultural regimes and the South: Discordant concepts?. *Progress in Human Geography* 27, no. 6: 681–707.

Woodhouse, P. 2012. New investment, old challenges: Land deals and the water constraint in African agriculture. *Journal of Peasant Studies* 39, no. 3–4: 777–94.

Cause and response: vulnerability and climate in the Anthropocene

Jesse Ribot

Causal analysis of vulnerability aims to identify root causes of crises so that transformative solutions might be found. Yet root-cause analysis is absent from most climate response assessments. Framings for climate-change risk analysis often locate causality in hazards while attributing some causal weight to proximate social variables such as poverty or lack of capacity. They rarely ask why capacity is lacking, assets are inadequate or social protections are absent or fail. This contribution frames vulnerability and security as matters of access to assets and social protections. Assets and social protections each have their own context-contingent causal chains. A key recursive element in those causal chains is the ability – means and powers – of vulnerable people to influence the political economy that shapes their assets and social protections. Vulnerability is, as Sen rightly observed, linked to the lack of freedom – the freedom to influence the political economy that shapes these entitlements. In the Anthropocene, human causes of climate hazard must also now be accounted for in etiologies of disaster. However, attention to anthropogenic climate change should not occlude social causes of (and responsibility for) vulnerability – vulnerability is still produced in and by society.

Introduction

> Efforts to reduce suffering have habitually focused on control and repair of individual bodies. The social origins of suffering and distress, including poverty and discrimination, even if fleetingly recognized, are set aside. (Margaret Lock, 'Displacing suffering', in Fassin 2012, 21)

In the Anthropocene, climate events and associated suffering can no longer be cast as acts of God or nature. They are now at least partly linked to human agency and responsibility. Of course, causes of climate-related disaster have always been social. Vulnerability is, by definition, the social precarity found on the ground when hazards arrive. It does not fall from the sky. While there is no disaster without hazard, without vulnerability, hazard is nil (Blaikie *et al.* 1994). The conditions of precarity have first to be in place. Vulnerability analysis identifies the causes of this precarity.

Sincere thanks to friends and colleagues Tom Bassett, Trevor Birkenholtz, Vasudha Chhotray, Erin Collins, Tim Forsyth, Samantha Frost, Zsuzsa Gille, Lauren Goodlad, Roger Kasperson, Colleen Murphy, Nancy Peluso, Vijay Ramprasad, Malini Ranganathan, Lisa Schipper, Sheona Shackelton and Ben Wisner for your incisive comments on drafts of this article. A special thanks goes to Christian Lund for early discussions while developing my arguments. Thanks also to two fabulous anonymous reviewers for their guidance.

As we enter the Anthropocene, climate disasters are being attributed to anthropogenic climate change (Hayes 2009, IPCC 2013, Myers and Kulish 2013, 1, 10). Yet to what degree is this attribution appropriate or complete? What generates the pre-existing vulnerabilities? Ironically, while some responsibility for stressors may now travel through the sky, the renewed focus on climate hazard is clouding attention to the grounded social causes of precarity that expose and sensitize people to hazard. Both vulnerability and at least part of climate are now anthropogenic. A bifurcated analysis of social cause is needed that keeps underlying generative structures of vulnerability in frame.

Peasant studies has a long history of explaining the marginality and flexibility, vulnerability and security of peasant households through their embeddedness, as an economy within an economy, in layered social and political-economic relations (e.g. Shanin 1971, Scott 1976, Wolf 1981, Watts 1983a, Deere and deJanvry 1984, Blaikie 1985, Bernstein 1996). Understanding rural vulnerability – including food insecurity – requires and has used the same kind of multi-scale analytics (O'Keefe *et al.* 1976, Wisner 1976, Chambers 1989, Swift 1989, Agarwal 1993, Watts and Bohle 1993, Blaikie *et al.* 1994). Such analyses explain why peasants have limited assets and inadequate protections, as well as what enables them to cope with stress under conditions of exploitation, subordination to landlords, dependencies, skewed market access and policies ranging from conscription and corvée to taxation, unequal exchange or skewed access to social services.

Grounded social-science research does not explain the precarity of the peasant household or its security and ability to withdraw into subsistence as a mere proximate relation between a household and the environment or hazard. Precarity and security are explained by locating the individual in the household, community, polity, market, nation and a differentiated global political economy. They are explained by people's political leverage to shape these contexts. This applies to any social analysis of precarity – of the peasant, the young, the old, the disenfranchised – including climate-related vulnerability analysis (Sen 1981, Drèze and Sen 1989, Watts and Bohle 1993, Blaikie *et al.* 1994).

In the Anthropocene, some causal analysis must trace stressors to greenhouse gas effluents, explaining how these effluents are enabled and how their regulation and mitigation are products of a complex social and political-economic history. These are the causes of stressors in the sky. They are distinct from underlying vulnerability. This contribution focuses on the vulnerability tine of the now bifurcated analysis of social cause. The paper links vulnerability analysis with the burgeoning adaptation and resilience approaches to climate response through 'capacity' by including causal explanation of capacity, ability and capability. As Downing and Patwardhan (2005), Cardona *et al.* (2012) and Manyena (2006) suggest, vulnerability analysis is a necessary complement to adaptation planning. Capacity and its causes are part of a complete evaluation of vulnerability.

The framing in this contribution starts with entitlements theory (Sen 1981, Drèze and Sen 1989), places it in a broader political economy and empowerment approach (Watts 1983a, Watts and Bohle 1993) and links it to more recent capacities and capabilities thinking (Sen 1984, 1999, Bebbington 1999, Yohe and Tol 2002). Bringing together various readings of causality, the framing also outlines recursive[1] elements of vulnerability analysis by exploring ways in which those at risk shape the political economy that shapes their precarity and security (Watts 1991). Emancipatory recursive elements, which are most in need of development, include representation (Sen 1981, 1999, Watts and Bohle 1993, Appadurai 1984, Lappé 2013), structural relations (Polanyi 2001 [1944], Scott 1976, Swift 1989,

---

[1]By 'recursive' I mean looping back, iterative or producing feedback.

Watts 1991, Moore 1997, Pelling 2003) and discursive effects (à la Beck 1992, Rose 1996, Butler, 1997, 2009, Fraser 2000, Luhmann 2002, Agrawal 2005, Wolford 2007, Wilkinson 2010, Connolly 2013). Causal chains (Blaikie 1985) and access theory (Ribot and Peluso 2003) frame the empirical starting point for explaining assets, social protections and relations of emancipation.

The ability of vulnerable people to shape the political economy that shapes their securities and vulnerabilities – that is, emancipation – remains under-researched. In calls for a recursive relation in causal structures of vulnerability, Appadurai (1984) brought 'enfranchisement' into famine studies, Drèze and Sen (1989) evoked the role of a free media in supporting food entitlements (the food people could obtain), and Watts (1991) brought in empowerment, where empowerment is the ability to influence the political economy that shapes those entitlements. Lappé (2013) sees populist democracy as the path to security. These are calls for democracy – public means to discipline government to respond to demands (à la Manin *et al.* 1999). Democracy must be integrated into any full analysis of causality. Indeed, to equip public debate and demands to undertake 'transformative solutions', democracy must be informed by analyses of cause – that reveal underlying 'generative frameworks' (Fraser 2008, 28). An informed polity brings government back in, letting God and natural hazards take a rest.

The next section of the paper, 'Hazards of occlusion', explores the links between cause, responsibility and the visibility of vulnerability as a social-historical product. 'Climate and society' outlines tensions between hazards analysis and socially rooted analysis within human-climate theory while exploring capacity as a point of integration. 'Vulnerability analysis' sketches the causal theories of climate-related vulnerability in two parts – foundational framings of causality, and recursive elements of representation, structure and discourse. Before concluding, the last section outlines an agenda for 'Causal research'. The review and framing are neither attempting to be complete nor theoretically consistent. The objective is to evoke a range of models to start building a repertoire of potential causal relations that any researcher should be consciously attempting to identify and test.

Hazards of occlusion: cause and blame in the Anthropocene

> ... no one person suffers a lack of shelter without a social failure to organize shelter in such a way that it is accessible to each and every person. And no one person suffers unemployment without a system or a political economy that fails to safeguard against that possibility. (Judith Butler, 'For and against precarity' 2011)

> Blaming nature can, of course, be very consoling and comforting. It can be of great use especially to those in positions of power and responsibility. Comfortable inaction is, however, typically purchased at a very heavy price – a price that is paid by others, often with their lives. (Drèze and Sen, *Hunger and public action* 1989, 47)

The vast majority of policy-oriented and scholarly publications on climate-related vulnerability and adaptation attend to response rather than causality (Bassett and Fogelman 2013, 47). They seek to identify *who* is vulnerable rather than *why*, indicators rather than explanation, fixes rather than causes – as if cause were not part of redressing vulnerability and its production. Some occlude causes of vulnerability and crisis behind hazards, nature or God – as acts that need no further explanation.[2] Many stop with convenient proximate explanations

---

[2]On roles of God, including distraction, see Schipper 2010.

such as assets or poverty – without asking how these are produced. Others, from adaptation and resilience schools, cordon off causality in capacities – like adaptive capacity or the capacity to bounce back (Manyena 2006). These approaches focus attention on 'innate' characteristics of the individual, household or group – the unit at risk (Gaillard 2010, 220).[3] Capacity is now a common explanatory factor in most risk and vulnerability frameworks (Yohe and Tol 2002, Manyena 2006, Folke et al. 2010, Cardona et al. 2012, 72).[4] But capacity as cause is not enough – it begs us to ask what shapes capacity.[5] So, even while these analyses point inward, it is still hard to escape that causality ultimately points outward to a broader set of social, political-economic and structural variables. Indeed, all that enables or disables people's abilities to maintain their security is part of vulnerability's causal structure.

Occlusion of cause is no surprise. Causality is theoretically, ethically and politically contentious – as are the transformative solutions to which causal analysis may point (Pelling 2011, O'Brien 2012, 670–1). We must take a structured look back to evaluate how and why societies place and leave certain categories of people at risk (e.g. O'Keefe et al 1976, Wisner 1976, Watts 1983a, Swift 1989, Watts and Bohle 1993, Blaikie et al. 1994, Wisner et al. 2004, Somers 2008). Yet, while such understanding of causality is a necessary element of response (see Somers 2008, Miller et al. 2010), explanation quickly generates conflict – of theory, method, interpretation, but also, and more fundamentally, over implication and interest. Causality is a contentious category of mind. Cause indicates blame, responsibility and liability, linking damages to social organization and human agency. Causal analyses and the transformations they imply present deep challenges to the status quo (also see O'Brien 2012, 668). The tracing of causality from any instance of crisis is a threat to those who might have played a role – of ignorance, negligence, intent, hubris or greed – in the production of pain. It is a threat to those who benefit, passively or actively, from unacceptable but everyday relations of production, exchange and consumption. It is no surprise indeed that analyses of climate focus on *who* is vulnerable rather than *why*. *Why* is socially and politically contentious. Yet contention should not stop us. It should be fodder for public debate – enabling democratic process around risk and response.

Thirty years ago, Appadurai (1984, 491) divided explanations of famine into ' … evolutionary approaches which stress adaptation and function, and historical approaches, which stress causality and contingencies'. Similar divisions persist in climate risk studies (Adger 2006, Füssel and Klein 2006, 305). Yet these views are not so separable. As Somers (2008, 10) observes in her analysis of post-Katrina New Orleans, 'we cannot look forward until we look

---

[3]This is akin to Rose's (1996) point that states produce risk subjects pointing cause inward toward the individual and group so as to make them responsible for their own sins – blaming the victim and asking the victim to blame her or himself. While there are such 'internal' characteristics (à la Chambers 1989), they still must be understood in the broader political economy that produces them.
[4]Folke (2006, 253) takes the social into account in his model of interplay as including 'social processes like, social learning and social memory, mental models and knowledge-system integration, visioning and scenario building, leadership, agents and actor groups, social networks, institutional and organizational inertia and change, adaptive capacity, transformability and systems of adaptive governance'. These processes are all proximate (except the ill-defined 'governance').
[5]For Cardona et al. (2012, 76), 'Drivers of capacity include: an integrated economy; urbanization; information technology; attention to human rights; agricultural capacity; strong international institutions; access to insurance; class structure; life expectancy, health, and well-being; degree of urbanization; access to public health facilities; community organizations; existing planning regulations at national and local levels; institutional and decisionmaking frameworks; existing warning and protection from natural hazards; and good governance'. Through these variables capacity can be understood as an outcome of social stratification in a broader political economy.

back to learn how we came to be who we are and until we know what we have lost, or gained'. Social-historical vulnerability analysis is a necessary complement of adaptation planning (Downing and Patwardhan 2005). Rather than looking back in time, however, most practitioners of adaptation, resilience and disaster relief still start by attributing climate-related disasters to acts of nature, or, in the Anthropocene, to anthropogenic climate change (see Gaillard 2010, Bassett and Fogelman 2013). In so doing, rather than seeking causality in social history, they continue to locate risk within the hazard to which people adjust, implicitly attributing pain and suffering to droughts, floods, and storms (Gaillard 2010, 223 – who observes this hazards frame 'regaining ground'). Rather than explaining vulnerability, they continue to frame disaster as a direct impact of climatic events. Nevertheless, many climate-risk theorists and analysts bring social causality into integrated models, locating it in the 'capacity' to adjust, withstand or re-establish. But such snapshot proximate analyses tell only part of the story.

Climate disasters, by definition, occur at the intersection of hazard and vulnerability. Without climate hazards there is no risk of climate disaster, and there are no disasters without vulnerability (Blaikie *et al.* 1994, 49). Blaikie *et al.* (1994), working on disasters before and since climate change, placed analyses of causality and of options for redress entirely on the vulnerability side of the risk equation. For them, the hazard side of the problem is to model and predict probability – to produce imaginary risk futures (à la Beck 1992). Hazards are probabilistic events that, while expected, are not subject to local manipulation. Hence, Blaikie *et al.*'s analysis of causality and disaster prevention focuses on vulnerability assessment and reduction while taking hazard to be a condition outside of the equation of redress. In this frame, the hazard cannot be removed but the vulnerability can be reduced. Today, while mitigation could change climate hazards, it is still true that climate is beyond local manipulation. Against this backdrop of external hazard, Wisner (1976), O'Keefe *et al.* (1976), Chambers (1989), Swift (1989), Watts and Bohle (1993), and Blaikie *et al.* (1994 reprinted in Wisner *et al.* 2004) brought social causal analysis of vulnerability to the center of the social study of disaster risk reduction.

While triggered by climate stress, it is well documented that crises are historical, social and political-economic products. The 1943 West Bengal famine was caused by well-functioning markets, not drought or absolute shortage (Sen 1981). The 1959–1960 famine in China was produced by administration, not drought (Jisheng 2012). The 2011 Somali famine was produced by 'interplay of livelihoods, clan and politics', not drought (Majid and McDowell 2012, 37). The 1300 fatalities in New Orleans in 2005 resulted from government negligence, not hurricane Katrina (White House 2006, Hayes 2009, also see Somers 2008). Conversely, the 150-fold reduction in fatalities in Bangladesh (from over 500,000 to 3406 deaths) between the comparable cyclones Bohla and Sidr was due to planning reforms (Bern *et al* 1993, CEDMHA 2007, Batha 2008, Government of Bangladesh 2008, Ministry of Food and Disaster Management of Bangladesh 2008). The inability to sustain stresses is produced by on-the-ground processes of social differentiation, unequal access to resources, poverty, poor infrastructure, lack of representation and inadequate systems of social security, early warning and planning (Ribot 1995, 2010). These factors translate climate vagaries into suffering and loss.

Today, in the Anthropocene, we face a new dilemma in explaining causes of climate-related disaster. Nature has become more-recognizably cultured. Some part of climate is anthropogenic. The hazards themselves, climate events, are no longer natural and blameless.[6] It seems 'natural' that cause and blame be turned back toward the hazard, that

---

[6]Cardona *et al.* (2012, 69) call these 'socio-natural hazards'.

disasters be attributed to climate change – and traced to the perpetrators driving SUVs in New Jersey. Such blame and responsibility has long been debated in climate negotiations.[7] More and more global institutions, through agreements with the United Nations Framework Convention on Climate Change (UNFCCC), are, at least implicitly, taking responsibility for climate change by aiming adaptation funds to support people to avoid the 'additional' stress that climate change is projected to produce. Developing countries are also calling for redress, as occurred after super-cyclone Haiyan in the Philippines (Khan and Roberts 2013, Myers and Kulish 2013, 1, 10).

Yet, under UNFCCC, adaptation funds are earmarked to redress only the damages of the *additional stress* that climate change might cause. Article 4.4 'refers to assistance to be given by developed country Parties in meeting the costs of adaptation that arises from climate change impacts' (Khan and Roberts 2013, 182). This additionality stance, along with calls for 'polluter-pays' positions and the UNFCCC 'agenda of loss and damage', implicitly acknowledges that climate change is anthropogenic and that the responsible parties should fund adaptation (Khan and Roberts 2013).[8] But additionality also implies non-responsibility for the preexisting precarity of those at risk – most of whom were vulnerable in the face of climate stress well before climate change was on the horizon (Khan and Roberts 2013, 182). The additional stance is laying down a cut-off for vulnerability redress. It only acknowledges the increment of suffering associated with added stress – despite that suffering is still attributable to the underlying conditions that turn *any* climate stress, anthropogenic or not, into crisis.[9]

The targeting of adaptation funds toward the anthropogenic increment of climate change accepts that nature has been cultured, but, paradoxically, requires that the pre-existing misery of precarity be naturalized – as a background condition. Disaster management schools share this tendency to aim to return disaster-stricken groups to 'normal', naturalizing their pre-disaster state (Manyena 2006, 438). The populations most affected or made worse off by climate change, however, are already the most vulnerable in the face of ordinary climate extremes (Drèze and Sen 1989, 60, Ribot *et al.* 1996, Cannon *et al.* 2004, 5, Anderson *et al.* 2010, Heltberg *et al.* 2010, Figueiredo and Perkins 2012, 192, IPCC 2012, 76, European Commission 2013, 5). Pre-existing poverty remains the most salient of the conditions that shape climate-related vulnerability (Sen 1999, 171–2, Yohe and Tol 2002, 29, Prowse 2003, 3, Pelling 2003, 52, Cannon *et al.* 2004, 5, Anderson *et al.* 2010, Heltberg *et al.* 2010, Cardona *et al.* 2012, 67). The poor, least able to buffer themselves against and rebound from stress, live in a state of precarity. But their pre-existing precarity in the additionality frame is the 'normal' condition – no longer framed as

---

[7]The Alliance of Small Island States, China and the Group of 77 pointed to liability and compensation for climate change as early as the 1990s (Khan and Roberts 2013, 175).

[8]Framing of adaptation as restitution was supported by the G77 but rejected by the Annex I countries, but then, later, the inclusion of an 'agenda of loss and damage' in COP (Conference of the Parties) 16 in 2010 and in Doha at COP18 in 2012 seems to show some acceptance by the developed nations (Khan and Roberts 2013, 184).

[9]Khan and Roberts (2013, 182) make the point that 'this global premise of adaptation as an additional burden for development in the particularly vulnerable countries presents "risks" from climate change due to a biophysical change in the atmosphere, rather than factors that make people vulnerable to these changes'. They connect these factors to 'existing development needs and contexts', continuing, 'on the basis of this consideration, developed countries argue that their responsibility in supporting adaptation should be limited to the problem itself, i.e., adaptation action in addition to the baseline, that the developing countries would undertake in absence of climate change; so the responsibility part for the wealthy nations relates only to damages attributable to human-caused climate change'.

anthropogenic. Paradoxically, then, in its welcomed emphasis on human agency, response to anthropogenic climate change has the effect of naturalizing and thus obscuring pre-existing anthropogenic vulnerability.

This pre-existing precarity that climate change finds in place is produced within the same global political economy that enables climate change – Rodney (1973), Wallerstein (1974) and many others have long since established these global connections. In the face of the anthropogenic increment, the international community appears to be mobilizing anew to take responsibility – yet they are simultaneously transmuting pre-existing vulnerability to natural – blaming no one. Such aid requires a natural baseline beyond which the producers of climate change are no longer responsible. It cordons off liability. Together, additionality and proximate adaptation analyses occlude social cause, erase history and extractive relations, masking the structural violence that created the poor's systemic climate- and non-climate-related vulnerabilities, across multiple axes, geographies and histories.[10]

While naturalizing misery, the logic of additionality also turns attention back to hazards. Social grounded causality is doubly obscured, framing hazards as culpable and existing precarity as natural. How do we square cultured nature with un-natural socially generated vulnerability? Now that nature is cultured, we can indeed trace the social causes of stress through climate change. Yet despite that hazards are now socially produced, an anthropogenic climate does not mean that the cause of vulnerability shifts to the hazard. Because the biophysical events are partly anthropogenic, the causal explanation of the hazard must, of course, now account for human will, intentionality, negligence and interest, and, of course, people, courts and governments are appropriately blaming society for climate events (Hayes 2009 on Katrina, Myers and Kulish 2013, 1, 10 on Haiyan). Social attribution becomes even more acute with the advent of intentional monkeying with climate through geo-engineering (Klein 2012). Even if disasters were never acts of God or nature (O'Keefe *et al.* 1976, Drèze and Sen 1989, 46, Smith 2006), climate events, which were viewed as external to the social world, are now traceable to social systems and agents (Arthur 2002, Jones and Edwards 2009). These new liabilities still, however, reside in society, not in the sky. They don't add to or erase the causes of vulnerability. Rather, they add to the hazardscape, which, when combined with vulnerability, is responsible for disaster. Causal chains behind hazard and vulnerability remain distinct – while also overlapping and interacting (Leichenko and O'Brien 2008),[11] and perhaps sharing root causes.

Responsibility in the Anthropocene is now bifurcated. Hazards and vulnerabilities have social cause. God and nature can no longer absolve us. Of course, it is not as if society could ever – with or without anthropogenic climate change, with or without God – have washed its hands of the production of vulnerability. Vulnerability on the ground is (and always has been) as much a product of far-away social forces as are the changes we now see in the skies. Stress articulates through climate events due to protected actions of real people in real places who, without direct liability through the rules, structures and subjectivities of

---

[10]I owe Erin Collins for this insight and wording.

[11]O'Brien and Leichenko (2003) and Leichenko and O'Brien (2008) speak of 'double exposure' – to climate and globalization. This split is not analogous. They look at both as parallel and interacting stresses to which people are exposed. In a climate-vulnerability analysis, climate is the stress people are exposed to. Vulnerability remains in society, and globalization is, of course, a part of its the causal structure, explaining why people are exposed. Also see Shackleton and Shackleton (2012) for a model that places social factors as parallel to climate stressors, casting the vulnerable populations as 'exposed' to these multiple stressors – among which climate is one.

differentiation, shape patterns of inclusion and exclusion that externalize the cost of their desires and their profit on others far away. The structure of vulnerability remains social. The differentiated causes of vulnerability in a given place must still be traced from that place through the social relations of production, exchange, domination, subordination, governance and subjectivity. They still have to be analyzed and understood starting from the instance of crisis in a real place and real time. While acknowledged anthropogenic climate change provides a new pathway for attributing social causality, and therefore responsibility and blame – as well as claims for redress and compensation (Jones and Edwards 2009, Hyvarinen 2012) – vulnerability remains a social condition of the exposed.

Being anthropogenic profoundly changes the meaning of climate events. Humans are now demonstrably (to non-deniers) responsible – not only for the vulnerability on the ground, but also for the stressors that arc across the sky. Indeed, anthropogenesis adds a new dimension to a global connectivity that has long been apparent to historians and to social and political-economic theorists (e.g. Wallerstein 1974, Wolf 1981). Social causes of place-based vulnerability *and* of stressors in the sky – the two chains of cause and blame – are interlinked. Unequal access to the opportunities that produce climate-changing greenhouse gasses is partly responsible for the poverty and marginality that places some people in secure standing and others at risk. Those who can consume well beyond subsistence are less vulnerable than those who cannot (see Sen 1981, Watts 1983b, Agarwal and Narain 1991). Unfettered access to resources and goods – protected through a differentiated global political economy with rules and social relations that protect some actors and subordinate others – enables the excess consumption that is changing the climate and increasing the stresses on those at risk. The social stratifications that create unequal patterns of vulnerability on the ground simultaneously contribute to stress articulated through a changing climate system. This is one direction human-environment integrated modeling needs to go – away from the myopically direct human-nature interface where people meet the elements.

The remainder of this contribution explores elements of a grounded causal analysis of vulnerability and insecurity – so that anthropogenic climate change cannot be added to the repertoire of obfuscations already occluding the multi-scale causes and responsibilities for climate disasters. While causal analysis may now be bifurcated, the analysis in this essay is not. The paper does not examine causal structures of anthropogenic climate change. These are already partly covered, in a most-proximate causal sense, as functions of effluents (see IPCC 2013 for a proximate technical analysis). Of course, a full generative social and political-economic analysis of those effluents and attribution of responsibility is needed (à la Agarwal and Narain 1991, Sachs 2008). A multi-scale, multi-stranded causal analysis of specific vulnerabilities can point to the multiple social scales at which solutions may reside. Responses must then be forged in the crucible of politics.

Climate and society: theories of vulnerability

... To call the frame into question is to show that the frame never quite contained the scene it was meant to limn, that something was already outside, which made the very sense of the inside possible, recognizable. (Judith Butler, *Frames of war* 2009, 9)

Vulnerability is driven by inadvertent or deliberate human action that reinforces self-interest and the distribution of power in addition to interacting with physical and ecological systems. (Adger 2006, 270)

Vulnerability analysis is often polarized into what are called risk-hazard and social constructivist frameworks (Füssel and Klein 2006, 305, also see Adger 2006, O'Brien *et al* 2007, 76, Cardona *et al.* 2012, 70).[12] Risk-hazard is characterized as the *positivist* (or realist) school while the entitlements and livelihoods approaches are lumped together as *constructivist*. This 'social constructivist' label is a misnomer as neither entitlements nor livelihoods approaches are founded on constructivist tenets.[13]

For the positivists, 'risk … is a tangible by-product of actually occurring natural and social processes. It can be mapped and measured by knowledgeable experts, and within limits, controlled' (Jasanoff 1999, 137). 'In social constructivist views, risks do not directly reflect natural reality but are refracted in every society through lenses shaped by history, politics and culture' (Jasanoff 1999, 139). The climate-vulnerability literature falsely contrasts the positivism or 'realism' of the natural sciences to the social constructivism of the social sciences. Moreover, many social scientists focused on climate seem to have accepted these misleading categories (e.g. Ribot 1995, Adger 2006, Fussel and Klein 2006, O'Brien *et al* 2007, 76).

It should be evident to any social scientists, however, that both the risk-hazards and the entitlements and livelihoods approaches can be positivist (à la Jasanoff 1999). Both kinds of analysis can also be subject to, or can incorporate, a social constructivist view. If one distinguishes between a constructivist *ontology*, referring to the nature of things, and a constructivist *methodology*, a way of understanding situated knowledge, constructivism need not suggest that the conditions and causes of vulnerability are not 'real' (Leach 2008, 7). Such a methodology would respect the phenomenology of vulnerability – understanding its material and affective effects. Further, it is perfectly positivist to assert that the socially constructed meanings that emerge from differently positioned actors shape causality (see Rebotier 2012). In short, we need to discard this false dichotomy, which serves only to discredit social-science analyses by contrasting them with the 'real'.[14] Forsyth's (2001) critical realist view, for example, accomplishes this goal by acknowledging the possibility of using constructivist approaches to produce better and more broadly relevant science (also see Beck 2007, 89 on 'reflexive realism').

There still remain two distinct primary schools of thought concerning climate's relation to risk. The ostensibly 'natural-science' risk-hazards model tends to evaluate the multiple outcomes (or 'impacts') of a single climate event (see Figure 1), while the entitlements

---

[12]Prowse and Scott (2008) label these behaviorist and structuralist approaches. Miller *et al.* (2010) places resilience into the 'systems' (meaning systems theory) camp and puts vulnerability with adaptation into an 'actors' category. They seem to have no place for 'structure'.

[13]Cardona *et al.* (2012) use the term differently, simply referring to the idea that vulnerability is a social 'product', hence it is 'constructed' in the sense of manufactured or produced. But this is not the social science use of the term 'constructed' (see Douglas and Wildavsky 1982).

[14]There is no positivist reasoning that would prevent analysis of interpretation and positionality as being part of the analytics of causality – since difference and struggles over meaning and interpretation are part and parcel of causality. In addition, discourse is no less 'real' than a tree or a storm system. The causes of decisions that shape security and damage are the results of discursive battles for domination, for authority, for decision-making power and ultimately for policy and practice. Positionality shapes people's behavior and is therefore part of the material political-economic structure of causality. These are not trivial observations of categorization. The very placing of the social-science analyses into 'social constructivist' and non-'realist' categories is a means of delegitimizing these perspectives as if social, discursive, constructivist factors are not part of the 'real' causal structure of vulnerability. Indeed, they are essential to it. Of course, any 'realist' who does not understand that interpretation is multi-faceted and meaning is attributed misses the point that these observations do not deny the materiality of their 'science'.

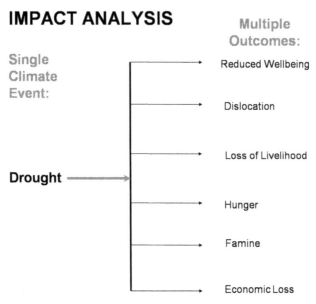

Figure 1.   Impact analysis.

and livelihoods approaches characterize the multiple causes of single outcomes (Figure 2) (Ribot 1995, Adger 2006). Both approaches, of course, could be conducted using positivist, constructivist or, more powerfully, combined analytics. A key difference between them is that the risk-hazards approach traces a linear causal relation back to the environmental hazard itself while the entitlements and livelihoods approaches tend to trace cause to multiple social and political-economic factors. The entitlements-livelihoods approach locates

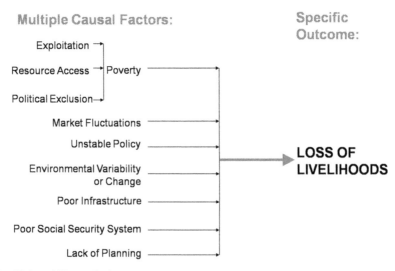

Figure 2.   Vulnerability analysis.

causality in society and hence tends to see natural phenomena as playing a role but not as having 'caused' the risk or damage in the face of an event.[15]

The risk-hazard approach, which defines vulnerability as a 'dose-response relation between an exogenous hazard to a system and its adverse effects' (Füssel and Klein 2006, 305; also see Bassett and Fogelman 2013), is concerned with predicting the 'impact' of a given climate event or stress, and estimating the increment of damage caused by an intensification from 'normal' climatic conditions to the conditions expected under climate-change scenarios.[16] This approach, which views people as vulnerable to *hazards*, locates risk in the hazard itself and, as scholars have argued, inadequately incorporates the social dimensions of risk (Watts and Bohle 1993, Ribot 1995, Adger 2006, 270, also see Cannon 2000). This approach also enables the 'additionality' views common in climate policy circles, which are based on the imaginary (or 'constructed') notion that the effects of climate change are separable from underlying social conditions.

By contrast, the entitlements and livelihoods schools are concerned with the chains of events that lead to vulnerability. They consider people to be vulnerable to *undesirable outcomes* such as the loss of a valued asset. They are also concerned with the likely aftermath of a climate event or trend. While these approaches view the risk of disaster and suffering as social, they view climate itself as an external phenomenon. Externalizing hazards places the burden of explanation – for vulnerability and ensuing disaster – within the social system. Adger (2006, 270) has described this school as depicting 'vulnerability as lack of entitlements' or a lack of sufficient means to protect or sustain oneself in the face of climate events, where risk is shaped by society's provision of food, productive assets and social protection arrangements. While some scholars have suggested that entitlements and livelihoods approaches ignore biophysical factors, as Blaikie and others argue, this school of thought does explicitly link vulnerability to biophysical hazards by acknowledging that hazards change the resources available to a household and can thereby intensify vulnerability (Blaikie 1985, 110, cf. Blaikie *et al.* 1994, 21–2, Shackelton and Shackelton 2012, 275). This is a simple but strong analytical relation between biophysical events and social vulnerability. This framing still attends to the question of why that household is so close to the thresholds of risk, the condition that renders a resource change so critical.

Integrative frameworks link the two views. They tend to be risk-hazard-based with borrowings from entitlements and livelihoods models. The social-biophysical integration remains uncomfortable and runs into boundary problems. Integrative frameworks view vulnerability as depending on both biophysical and human factors. Many characterize vulnerability as having 'an external dimension, which is represented ... by the "exposure" of a system to climate variations, as well as an internal dimension, which comprises its "sensitivity" and its "adaptive capacity" to these stressors' (Füssel and Klein 2006, 306; see

---

[15]By locating environment, including climate, within a social framework, the environment may appear to become marginalized – set as one among many factors affecting and affected by production, reproduction and development (also see Brooks 2003, 8). This seems to be one more reason for resistance to vulnerability analyses from the climate science side. But this does not diminish the importance of environmental variability and change. Indeed, it strengthens environmental arguments by making it clear *how* important – in degree and manner – the quality of and access to natural resources or natural insults are to social wellbeing. Of course, it also points us to the least costly ways to reduce the problem we are ultimately trying to solve – damage.

[16]O'Brien *et al.* (2007, 75–6) label the entitlement-livelihoods perspective as 'starting-point' and risk-hazards as 'end-point' vulnerabilities: the prior starting with the threatened attribute of value and the latter starting at the end point of the analysis, with the hazard itself. I find this language confusing.

Chambers 1989, 1). These notions of external and internal, however, are entirely contingent on how one draws the boundaries of the system under analysis. Linked approaches also draw on resilience theories (Gunderson and Holling 2002). These all tend to integrate the social into systems theory approaches in ways that do not account for social theory of political or economic change and history. Resilience and systems theories are still struggling to integrate social theory and to expand their boundary conditions (Brooks 2003, Manyena 2006, Duit *et al.* 2010, Miller *et al.* 2010, Beymer-Farris and Bassett 2012). To date most (save Beymer-Farris and Bassett 2012) come across as hopped-up hazards models – with extra social bells and whistles.

The analytic approach outlined in this contribution builds on Blaikie 1985 and Watts and Bohle 1993 (cf. Turner *et al.* 2003a, 2003b, 8074–5) to get around these boundary problems.[17] These authors have adopted a socially-rooted entitlements-livelihoods approach that empirically traces the causes of vulnerability from specific instances of crisis – explaining why a given individual, household, group, nation or region was at risk of a particular set of damages (see Figure 2). By tracing causality out from each affected unit, their model views the entire system as a whole. It accounts for all factors – biophysical and social – that lead to crisis of the unit of concern (Kasperson *et al.* 2005, 159–61).[18] It allows for multi-scale, multi-factor analyses of vulnerability. It traces causal chains outward toward whatever factors – material, social, political, discursive – are found to shape vulnerability. Where climate is part of that constellation, say as a trigger of crisis, it emerges as being significant – a significance that still requires social explanation of vulnerability. This sociological and political-economic 'progressive contextualization' (coined by Vayda 1983, who applied it in an a-theoretical manner) or the political ecology framing (Watts and Bohle 1993, Ribot 1995) or new 'network political-ecology' approach to vulnerability analysis (Birkenholtz 2011) focuses attention on the full array of causes, thus enabling the analyst to identify the multiple causes of undesirable outcomes.[19]

Two other schools are commonly summoned to assess climate-society relations. The climate 'adaptation' and 'resilience' literatures are primarily concerned with vulnerability reduction – engineering and managing of change. They often proceed, however, without applying a broad vulnerability analytic to understand root causes of risk. They often take

---

[17]Blaikie *et al.*'s (1994) 'pressure-release model' also adopts this causal-chains approach. They view at-risk people as being pressed between hazard and vulnerability, with vulnerability as the locus of cause and therefore treatment – pressure is relived by reducing vulnerability (updated in Wisner *et al.* 2004). A similar approach, differentiated by the new language of 'network political ecology', is described by Birkenholtz (2011, 10), who explains: 'For network political ecology, this means a focus on both vertical (hierarchical) and horizontal (non-hierarchical) connectivity in places experiencing a common effect of climate change, understood through their connections to other processes'. In practice, this 'network' approach, despite the appearance of a new Latourian approach, works much like the methods other theorists have been using for decades.
[18]This unbounded view does not collapse the categories of 'internal' and 'external' vulnerability often used in the literature (see Chambers 1989, Brooks 2003, Füssel 2007, 158). Rather, it leads us to acknowledge the hazard as a stress or trigger of events, which are still explained as outcomes of the social (or internal) aspects of the system. We can see the roles hazards play, but those roles remain socially shaped (as does the very term 'hazard' which is meaningless without vulnerability).
[19]Birkenholtz (2011) uses language of 'effect of climate change'. This looks like a kind of slippage into hazards language. It may also be a result of his use of Latour – whose flat ontology makes objects into actors (with troubling anti-humanist implications). The flat-ontology approach contrasts with the one used in this contribution. However, Birkenholtz's article, despite Latour, retains the focus on the social and uses 'effect' to indicate that climate produces action and re-action recursively – both of which are within society. The action then remains with humans, so Latour can be dumped.

a hazards approach while parking vulnerability in the term 'capacity' – resulting in anodyne prescriptions for 'capacity building'.[20] Nevertheless, this term 'capacity' – the capacity to adapt, resist or rebound – provides an opening for causal analysis.[21] Some approaches, such as 'adaptive capacity' as framed by Yohe and Tol (2002), provide a partial converse of vulnerability (Brooks 2003, Ribot 2011).[22] The question of what causes vulnerability is partly answered by asking why a given individual or household does not have adaptive, resistive or rebounding capacities. The causes of this lack are among the causes of vulnerability. A thorough analysis of capacity, however, would require a causal-chain, progressive-contextualization analysis of causes of capacity – rather than just viewing capacity as explanation. This analysis of causes of capacity would then also have to be complemented by an analysis of the social and political-economic context of such variables as social protections, which may fall outside of the definition of capacities. Social protections can form a tradeoff with capacities if, for example, security is framed as capacities (abilities, capabilities) of individuals and groups plus the social security arrangements that can complement and supplement individual and group resources.

Capacity, then, still needs to be placed in the broader structural political economy that stratifies society, enabling some while disabling others, providing for some while excluding others. As will be developed later, capacity analysis would also need to be completed by the recursive relations made up of the very capacity, or ability, to influence the governing political system – shaping both the basis of capacity, e.g. endowments and assets, as well as social protections (this link is made by Sen 1984, 1999, Bebbington 1999, Prowse and Scott 2008).[23] Here, vulnerability analysis, which has interrogated and theorized micro-macro relations of differentiation and access to resources and power, can inform adaptation and resilience studies to steer their analyses toward generative structures, preventing them from remaining internally oriented, proximate, ahistorical or atheoretical. Tracing out the chains that cause and disable capacity is then one productive entry point into a fuller analysis of vulnerability – and a necessary foundation for any adaptation or resilience program.

---

[20]Dill (2013, 13–5) argues that capacity is often a dead-end explanation of development failures – pointing toward analytics that deal with the institutional and cultural arenas, while pointing out that these explanations often fail to have historical depth. The same is true when applied to climate adaptation and resilience. Ribot (2004) also makes the point that lack of 'capacity' is often evoked as an excuse not to empower people, despite that they would gain capacity were they empowered to act (a kind of 'catch 22'). Without capacity they cannot be entrusted with power. Without power they cannot gain capacity.

[21]Resilience schools focus on capacity to resist or bounce back (Manyena 2006, Gaillard 2010, or Beymer-Farris et al. 2012 who add differentiated valuation and access to resilience models).

[22]Yohe and Tol (2002, 27–8) frame capacity as having causes that are location-specific and path-dependent, with micro and macro determinants. They then trace causality to (1) availability of technology plus (2) a set of factors, including resources, institutions, human capital, risk-spreading processes, information management and 'attributing signals of change to their source'. Their model can capture some causality of the 'capacity' to adjust, but it still fails to explain the factors or 'determinants' they identify. It does not provide a root-causal analytic. They look at variables that correlate with disaster without explaining causal pathways or genesis. Their approach does not provide a map for generative solutions; rather, it indicates technical fixes and obstacles to their feasibility and implementation – a universal problem in this literature. In a more sophisticated approach, capacities (akin to abilities and capabilities) include the ability to influence politics.

[23]Prowse and Scott (2008, 43) lump tangible and intangible 'capitals' into assets – physical and financial as well as human and social capitals. They attribute the ability to influence those who govern to only the intangible social capital. They bring in politics, but only as a product of social capita.

Causal analyses of vulnerability, including the entitlement and livelihoods approaches, are often classified as social constructivist, too complex or not relevant, and ignored by most climate-society modelers (Wisner *et al.* 2004, 61). These modelers continue to take the hazards stance while including the 'social factors' only as they interact directly with the landscape and hazard. The hazards stance, even where 'social factors' are included, continuously bends causality back to the interface between people and nature. But a truly integrated social model asks what caused people to be exposed to and damaged from the presence of hazard. In essence, a truly integrated analysis asks what made the hazard a hazard versus a mere event. The ability to transform hazards into events or events into hazards is social – without vulnerability there is no hazard (just as there is no disaster). For policy this is critical since it turns attention to the agencies (in all senses) and social configurations that generate crisis. Vulnerability analysis has to work hard to pull attention from the hazard back onto society. In the Anthropocene, this includes the physical production of events – the agency behind effluents and their effects on nature – as well as the social production of vulnerability that turns those events into hazard. Both need progressively contextualized explanation.

This section sketched some of the main models that relate climate events to the loss of valued attributes. The section highlights models that explain the root causes of loss – with an eye to stemming the production of risk and developing protections. A critical realist framing of vulnerability can build on positivist models, recognizing that science itself is socially constructed, while also acknowledging that framing and discourse shape action and have material consequence. We need to understand causes that can also lead to redress – by identifying causes that can be changed as well as the responsible structures and institutions. These causes, in a complete analysis, can be material, discursive and recursive. The next section expands an unbounded framing for vulnerability analysis. It is followed by a section outlining some of the recursive relations that shape the ability of people to shape the political economy that shapes wellbeing. These are relations of emancipation.

Vulnerability analysis – an unbounded access framing

> No one would say that a lack of money in the world is the reason there are poor people; yet, many blithely suggest that a lack of food is the reason a billion go hungry. (Lappé 2013, 227)

> Climate, 'over-population' and war, while potentially significant as proximate or trigger factors, have been substantially discredited as primary factors. (Watts 1991, 15)

> The challenge today is to integrate agency and structure in examinations of the production of vulnerability, in specific places, whilst also acknowledging the importance of physical systems in generating hazard that can trigger disaster. (Pelling 2003, 47)

Two objectives of any policy-oriented vulnerability analysis for climate action are to identify who is vulnerable and how to assist them. Analysts need to ask: *where* should we spend public funds earmarked for vulnerability reduction or climate adaptation, and to preserve *what* values. The first question, how to target expenditures, requires identifying which regions and social groups (who) and the things of value to them that are at risk (what). The question of what we need to invest in, projects and policy reforms, requires an understanding of the characteristics of their vulnerability to understand the reasons (why) these people, and things are at risk, so we can assess the full range of means for reducing that vulnerability. *Who* and *what* are very different questions than *why*. Knowing *who* and *what* tells us how to target expenditures – this is the world of vulnerability indicators.

Knowing *why* tells us what to modify or improve in these targeted places and communities – this is the world of causal structure of vulnerability. *Why* also indicates the complexity and cost of short- and long-term solutions to vulnerabilities associated with climate variability and change – this is the world of causal analysis. Climate action should be guided by who, what and why. Much attention has been given to impact assessment, indicators, and mapping for targeting (see Downing 1991, Adger *et al.* 2004, Kasperson *et al.* 2005, 150, Deressa *et al.* 2008). This section trains attention on the material, recursive and discursive elements of an analysis of causal structures of vulnerability – elements of why. This is a starting point for generative reform.

Framing causality

Sen (1981, 1984; also see Drèze and Sen 1989) laid the groundwork for analyzing causes of vulnerability to hunger and famine with what he calls 'entitlements'. Entitlements are the total set of rights and opportunities with which a household can command – or through which they are legally 'entitled' to obtain – different bundles of commodities. For example, a household's food entitlement delineates the food that the household can command or obtain through production, exchange, or extra-legal legitimate conventions, such as reciprocal relations or kinship obligations (Drèze and Sen 1989). A household may have an endowment or set of assets that include productive means, stores of food or cash, and claims they can make on other households, patrons, chiefs or government (Swift 1989, 11; cf. Drèze and Sen 1989, Bebbington 1999). Assets buffer people against food shortage (Swift 1989, 11). They may be stocks of food or things people can use to make or obtain food. A person's entitlements can fail 'either because of a fall in her endowment (e.g. alienation of land, or loss of labour power due to ill health), or because of an unfavorable shift in her exchange entitlement (e.g. loss of employment, fall in wages, rise in food prices, drop in the price of goods or services she sells, decline in self-employed production)' (Drèze and Sen 1989, 23).

The concept of 'entitlements', as I think Sen (1984, 1999) recognizes, is also part of the problem of exclusion – it is predicated on a strong justification of private property following Locke, Kant and Nozick (see Nozick 1974).[24] Entitlements are viewed as properties that are just if obtained in a 'just acquisition' – in ways recognized by a particular legal system. Sen (1981, 1984, 312) clearly shows that famines unfold in the face of legitimate (legally sanctioned) ownership and exchange. His observation brings the moral basis of the legal regime into question. If the procedure is legal-legitimate but the outcome unacceptable on moral grounds, then 'rights'-based approaches are clearly inadequate. Sen (1984, 312), to move beyond entitlements, advocates for a limited 'consequentialism' or 'consequence-sensitivity' in which outcomes can be used to judge and to justify an override of legal procedure. Sen (1984, 313) notes, 'Since it is implausible – indeed I believe incredible – to claim [by privileging just acquisition] that consequences in the form of life or death, starvation or nourishment, indeed pleasure or pain, are intrinsically matters of moral indifference, or have only very weak intrinsic moral relevance, it is not easy to see why history-based rules of procedure should be so invulnerable to the facts of their consequences'.

Here is where consequentialism brings us back to causal analysis. It requires an analysis of cause and effect – consequence is a consequence of something – that can include as cause legally legitimate rules and procedures, among other things. My objective is to flip Sen's

---

[24]See also Gasper (1993) for an analysis of limits of Sen's entitlements approach.

brilliant insight into an empirical inquiry – to ask what causes different observed outcomes. This accomplished, one can interrogate the moral tradeoffs between procedure and consequence with the causal chain weighing into the moral judgment. By exposing causalities, the moral judgment can be a matter of public debate. Further, by understanding cause, such public debate can be extended to the weighing of the role of individual or public agency or interest behind cause. In this manner responsibility can also be attributed.[25] The consequence Sen is most concerned with is 'freedom' – freedom from and freedom to; in particular the freedom from poverty and the freedom to 'be' and 'do', to 'function', to achieve desirable outcomes (Sen 1984). These freedoms require food entitlements as a necessary condition – as laws and markets shape access to these necessities. Broader desired outcomes also, of course, require freedom from oppressive regimes, social stigmas, skewed cultural norms and petty theft, as well as state and structural violence.

The notion of 'entitlements', if taken at face value (without consequence sensitivity), implicitly legitimizes any existing distribution of property while occluding non-legal, extra-legal and illegal forces. Nevertheless, the framing and process of analysis of entitlements failure gives us a strong basis for accounting for hunger. A broader empirical view would frame assets as depending on the 'ability' (as supported by rights *and* other structures or powers), rather than just the 'right', of the household to produce a surplus that it can store, invest in productive capacity and markets, and use in the maintenance of social relations (see Scott 1976, Berry 1993, Ribot 1998, Ribot and Peluso 2003). Rights-based approaches and rule of law are not everything. Access theory – explaining the ability of people to benefit from things – provides broader empirical (rather than just legal) analytic of what people are able to obtain and use (Ribot and Peluso 2003). Further, Sen's model does not account for moral economy – the basis of expectations that people have on those who govern (Scott 1976, Swift 1989). Ability is also broader than capacity insofar as it is not about innate characteristics of those at risk, but rather about all that enables and disables – since access theory focuses on the ability to benefit from things tangible and intangible: including material assets, knowledge, ideologies, discourses, doxas, habitus, social relations, social status, social structures, legal and political structures, stealth and violence (Ribot and Peluso 2003).

Vulnerability in an entitlements framework is the risk that the household's alternative commodity bundles will fail to buffer them against hunger, famine, dislocation or other losses. This is risk of 'entitlements failure' to Sen or 'access failure' (failure to access, or enjoy benefits of, these alternative commodity bundles) in an access framing. Vulnerability is a relative measure of the household's proneness to such failure (Downing 1991; also see Downing 1992, Watts and Bohle 1993, 46, and Chambers 1989, 1). By starting with the 'entitlement' framing's components (that is, production, investments, stores and claims) of what enables households to maintain food consumption, an access framework allows us to analyze the causes of food crises. By analyzing chains of factors that produce household crises, a whole range of causes are revealed – hopefully signaling potential policies to reduce vulnerability (Blaikie 1985, Turner *et al.* 2003a, 2003b). This social model applied to instances where climate events are associated with food crisis replaces eco-centric models of natural hazards and environmental change (Watts 1983b). By showing a range of causes, legal, extra-legal or illegal environmental stresses are located among, and their

---

[25]The reason to evaluate agency is to attribute responsibility. Similarly with interest. With or without agency, there are winners and losers for any outcome. These winners or losers may apply their agency to maintaining or changing that outcome (and procedures that created it), depending on where they stand in the stream of damage or benefit.

role explained within, other material and social conditions that shape household wellbeing. Hunger, for example, may occur during a drought because of privatization policies that limit pastoral mobility making pastoralists dependent on precarious rain-fed agriculture (Smucker and Wisner 2008; also see Leichenko and O'Brien 2008).

Household-based social models also illustrate how important it is that assets are sufficient to cope with or adjust to (as in buffer against) environmental variations and changes so that land-based production activities are not undermined by and do not undermine the natural resources they depend on (Blaikie 1985). Household models, however, often fail to account for intra-household gender and age differences in production, consumption and reproduction; so internal household struggles must be made an explicit part of any complete analysis. Gender-differentiated access to household food and assets, and to natural resources, jobs, markets, services and representation, is foundational to household and individual wellbeing (see Guyer and Peters 1987, Vaughan 1987, Carney 1988, Hart 1992, Agarwal 1993, Schroeder 1999, Turner 2000). If not fully theorized, the household models may also miss broader structural relations of production and exchange within markets and a globalized system that shape broader-scale distributions (Polanyi 1944, Leichenko and O'Brien 2008, Butler 2009, Fraser 2011). But all of these should be linked into chains of causality by a full causal analysis of access failures.

Leach et al.'s (1999) 'environmental entitlements' framework introduced the notion of a sub-component entitlement, a set of utilities that a particular resource or sector contributes to wellbeing – e.g. environment – allowing the analysis of inputs from particular livelihood arenas. Their framing, building on Swift (1989, 10, also see Chambers 1989), modulates Sen's concept of entitlements from the household to any social unit (or exposure unit in the case of climate related analyses), such as individuals, households, women, ethnic groups, fishers, farmers, organizations, communities, nations or regions. (Drèze and Sen 1989, 30 also indicate groupings as units of study, with a focus on occupational groups, emphasizing a more disaggregated view than offered by class analysis.) Leach et al. (1999) also expand the idea of rights such that things may be 'claimed' rather than just legally 'owned' (on claims, also see Swift 1989). In this framing, claims may be contested – something Sen fails to capture by missing non-legal forms of contestation and legal pluralisms. Hence, endowments such as natural resources can still be accessed through social relations that may introduce cooperation, competition or conflict mediated by systems of legitimization other than or contesting state law (see Lund 2008, 2013, Lund and Boone 2013). They introduce a plural notion of rights (à la von Benda-Beckman 1981, Griffiths 1986), which Sen takes as singular and static. Claims in this framing are based on multiple, potentially conflicting, social and political-economic relations of access (à la Blaikie 1985, Ribot and Peluso 2003).

Many causal chains have been identified that shape household assets and entitlements beyond Sen (1980, 1984, 1999) (also see Moser's 2007 neoclassical approach to assets).[26]

---

[26]Moser shifts the focus from entitlements and livelihoods toward accumulation and attention to means of accumulation (Moser 2007). Instead of maintaining livelihoods or social protections, her model calls for social policy that promotes asset accumulation (see Prowse and Scott 2008, 48–9). Her model's objective is to improve returns to assets through improving infrastructure and competition within markets – as if well-functioning markets were not part of the problem (Prowse and Scott 2008, 48 – citing Moser and Dani 2008; also see Sen 1981). This neoclassical accumulation approach does not take into account that under a certain income threshold, called the Micawber threshold, many households cannot limit consumption sufficiently to accumulate (Prowse and Scott 2008, 48–9; for a nuanced threshold analysis, see Luers et al. 2003). Prowse and Scott (2008, 49)

Deere and deJanvry (1984) identify extractive mechanisms that market and institutional models do not attend to – to explain why households have insufficient surplus to invest in their own wellbeing and development. These include tax in cash, kind and labor (corvée), labor exploitation and unequal terms of trade. These processes siphon off household wealth – with the systematic support of governments and policies. Scott (1976, also see Alavi 1965, Berry 1993) also shows how peasant households' assets are drained through sharecropping and corvée in exchange for uncertain security. Isakson (2013) shows how financialization of agricultural markets is moving further down the supply and demand chains, diminishing profits retained by farmers. Ribot and Peluso (2003) show how complementarities of factors of production and exchange shape people's ability to form assets. Land without labor or labor without land or products without market access – contingent on identity or access to government to get permits or licenses for market access – can add up to failed asset formation or destitution. Controlling both elements of profit – property in land and labor – makes the land rights more lucrative (deJanvry *et al.* 2001, 5). Blaikie (1985) shows how assets are a function of people's identities within a larger political economy. These identities shape access to resources, credit, markets, jobs, rights and social services that are the basis of asset formation. Many of these factors are products of policy – that is, like markets, they are products of, or they can be enabled or disabled by, policy. These are not 'natural' self-regulating social systems with invisible hands (see Polanyi 1944).

Agrawal (2010) provides another important causal pathway to vulnerability and security by showing how rural institutions (public, civic and private organizations) enable or disable collective action in risk pooling. Rural populations protect themselves by risk pooling via storage (over time), migration (over space), sharing assets (among households) and diversification (across assets). In his model, exchange (via markets) can substitute for any of these risk-pooling responses. Rural institutions/organizations play different roles in enabling each of these risk-reducing practices. Building on Agrawal's research, a fuller causal analysis would also ask three other important questions: what in addition to institutions (e.g. moral economy, assets, the political and regulatory environment) enables (or even necessitates) people to engage in risk pooling; what resources and legitimation regimes enable these institutions to play supportive roles; and what shapes people's access to various institutions? Cardona *et al.* (2012, 85) adds, 'expanding the institutional domain to include political economy … and different modes of production – feudal, capitalist, socialist … – raises questions about the vulnerability *of* institutions and the vulnerability caused *by* institutions (including government)'. The individual and household ability to engage with institutions is partly addressed by other authors above. The landscape of institutions, the mix of institutions in it, the forces that enable and disable institutions, are not well explained (for a framing see Ribot *et al.* 2008).

Policy, policy making and the politics of influence, are, of course, always part of the causal structure of vulnerability and security – and part of many of the political, economic, institutional and structural relations discussed above. Policies, including the polities that create, foster or undermine local institutional landscapes, shape people's freedoms and

---

then argue that social services are required; acknowledging a need for an assets approach, a need to understand quality and role of assets required to affect a pro-poor adaptation strategy and a need to offer a floor (the Micawber threshold) below which 'pro-poor adaptation' must support households – reminiscent of 'means-tested entitlements'. Their very neoclassical accumulation model depends entirely on access to assets and the decisions they make about asset use (Pelling 2003, 58). Its author, Moser (2007), takes a conservative negative view of government intervention – claiming is it ineffective.

people's assets, capabilities and freedoms are always part of their ability to shape those pol-icies. Understanding the authorities that produce them as well as their effects on authorities at different scales is part of vulnerability's causal chain. Policies for reducing vulnerability, as well as other laws and regulations, will have some intended effects while also having damaging effects and ancillary benefits (Burton *et al.* 2002, Turner *et al.* 2003b, Moser and Satterthwaite 2010, Beymer-Farris and Bassett 2012, Marino and Ribot 2012). Policies can also be damaging or emancipatory by reshaping the authorities they work through, thus reconfiguring representation and rights (Ribot *et al.* 2008, Osborne 2011, 875, Poteete and Ribot 2011, Beymer-Farris and Bassett 2012). Policy is the formal expression of the intended structuring of the larger political economy. Its formation and effects are key elements of causal analysis.

A full analysis of the causes of access must follow all chains outward to explain the state of assets and entitlements – including the landscape of extended entitlements or social pro-tections. It must include a full analysis of the structure of people's access to goods and ser-vices. Starting with a limited entitlements framing, theorists have outlined a larger set of material and discursive factors that shape people's assets, social protections and entitle-ments sets. Concepts such as capacity or policy must find their place in causal chains, rather than being mere endpoints of explanation. The next sub-section explores the recur-sive relations between those at risk and the authorities that govern – channels through which broader relations of policy and practice are shaped.

Recursive elements – representation and emancipation

> If famine is the socially differentiated lack of command over food, it is naturally about power, politics and rights broadly understood, all of which are embedded in a multiplicity of areas from the domestic (patriarchal politics) to the national/state (how ruling classes and subaltern groups acquire and defend certain rights). (Watts, 'Entitlements or empowerment?' 1991, 21)

> From the standpoint of any sophisticated economic theory, an individual's command over public resources forms part of his [sic] private resources. Someone who has power to influence public decisions about the quality of the air he or she breathes, for example, is richer than someone who does not. So an overall theory of equality must find a means of integrating private resources and political power. (Dworkin, 'What is equality?' 2013, 283)

Representation and citizenship

While providing a buffer against stress and shocks (à la Swift 1989), assets are also a necessary ingredient for influencing the rights, recourse and representation required to shape the political economy that in turn shapes both accumulation and social protections (Watts and Bohle 1993, Bebbington 1999, 2022).[27,28] Surplus assets – time and resources – enable people to make demands and apply pressures to the systems that govern them. Freedom from risk, then, is enabled by surplus in its double role of enabling people to adjust on their own and enabling them to turn society toward transformative

---

[27]As Leach *et al.* (1998, in Bebbington 1999, 2033) suggest, it is 'important to invest in people's capa-bilities to control and defend assets'.

[28]Moser and Norton (2001, xi), in their market-oriented approach, argue for accumulation as a central means for security. They also argue that democracy and human rights frameworks are a resource that empowers people to make claims for government accountability in providing basic necessities and social securities. In the context of vulnerability, they (2001, x) view mobilization to claim basic rights as an important means for poor people to shape the larger political economy.

restructurings that serve their needs and aspirations. This emancipatory power is a central part of the recursive relation between individuals, households or communities and the regimes that govern them. The recursive relation with the larger political economy links back to wellbeing by shaping both private assets and public social protections – together assets and social protections form the basis of security or insecurity (Sen 1981, Drèze and Sen 1989).

Means by which people shape the political economy that shapes their entitlements, or by which they influence those who govern them, have long been seen as a critical part of durable wellbeing, food security and policy formation. Appadurai (1984, 481), building on Sen's (1981) treatment of 'entitlements', called for attention to 'enfranchisement', 'the degree to which an individual or group can legitimately participate in the decisions of a given society *about* entitlement'.[29] Drèze and Sen (1989, 263) observed the role of certain types of political enfranchisement in reducing vulnerability, specifically the role of media in creating crises of legitimacy in democracies. Adding 'empowerment', Watts (1991), Watts and Bohle (1993) placed vulnerability in a multi-scale political economy, arguing that vulnerability is configured by the mutually constituted triad of entitlements, empowerment and political economy – where empowerment is the ability to influence the political economy that shapes entitlements. Sen (1999) later included public dialogue in the context of elections and party politics. Watts and Bohle's (1993) empowerment stance, however, more broadly implies that protests, resistance, class struggle, social movements, union, civil-society pressures and direct representation all shape policy and political processes or the broader political economy that shapes household entitlements. Their framing incorporates the politics of the production as well as the contestation of marginalization processes (also see Swift 1989 based on Watts 1983a).

Taking another approach to wellbeing, Sen (1999, 75) defines capabilities as the set of 'functionings' that an individual has the freedoms to achieve (also see Sen 1984). Functionings are made up of 'beings' and 'doings' that individuals have reason to value. Hence, capabilities are the set of desired outcomes people are *able* to achieve – whether or not they choose to. Food entitlements, discussed earlier, describe the alternative bundles of food people can obtain. Capabilities describe the alternative bundles of achievements people are able to attain. Being consequentialist, capabilities are broader than food entitlements, since they include all outcomes (not just access to food) to which a person is entitled. Capabilities expand 'assets' by taking into account the individual's characteristics in addition to the assets they command. Entitlements and capabilities are similar insofar as they are based on legal legitimate forms of production and exchange. Among the functionings that Sen recognizes are 'their direct relevance to well-being and freedom of people', 'their indirect role through influencing economic production' and 'their indirect role through influencing social change' (Sen 1984). Capabilities are in effect the ensemble of outcomes to which individuals have access given their personal characteristics plus the ensemble of resources and rights that enable them to act. Capabilities outline the substantive freedoms of the individual within a social-political-administrative-legal regime – including the ability to shape that regime.

In essence, capabilities are the alternative set of outcomes to which a person or group has access (the ability to enjoy) – whether or not they use that access. Hence, given the emphasis on what people 'can' do rather than what they 'do', Sen (1984, 1997, 1999) characterizes these potential outcomes as their substantive freedoms. The capabilities

---

[29]Italics in original.

approach helps Sen (1984) move from a pure procedural-utilitarian framing toward one that is consequence-sensitive that can account for the moral acceptability of outcomes. This move enables a judgment of laws when those laws produce unacceptable results – hunger, famine or any form of deprivation. In his framing, legally legitimate 'just exchange' does not justify outcomes whereby some people have insufficient assets to survive. In this sense, capabilities theory, without dismantling the notion of procedural justice, provides a moral basis for expanding freedoms such that everyone has at least the capability to survive, and hopefully to also lead desired lives.

Sen (1999) defines development in terms of the freedom capabilities enable. Sen (1999, 152–3,178) attributes increased capabilities and freedoms partly to democracy, which he defines as free media, regular elections and opposition parties.[30] These must be complemented by public dialogue/deliberation and by citizen engagement predicated on civic virtue. Political and civil rights include the allowing and encouraging of 'open discussion and debates, participatory politic and unpersecuted opposition' (Sen 1999, 158). Norms and values that mobilize people to engage are formed by such public discussion. These debates and discussions are enabled by political freedoms and civil rights and they shape the very values expressed when there are such political freedoms (Sen 1999, 158). Sen's capabilities represent a brilliant move into a world where freedoms that enable wellbeing – including the ability to influence social change – become a matter of human rights.

One of the powers of Sen's capabilities framing is its attention to consequence – the potential achievements that add up to freedoms. The framework, however, needs to be complemented by an empirical explanatory method that traces causality from outcomes. In particular, when attending to climate-related vulnerability, we are interested in explaining two phenomena: (1) entitlement (or access) security or failure (i.e. sufficiency of assets and social protections) and (2) the ability to influence those who govern and the broader political economy. How these are attained is an empirical question to which access theory applies – by asking what enables or disables access to assets and influence? Like entitlements theory, Sen's capabilities framing remains legalistic even while bringing law into question. Bringing attention back to assets, the material basis of both subsistence security and influence, and to social protections which shape assets required for survival and freedom, access theory would ask how access to security (assets and social protections) and influence are structured. Sen attributes this structuring to law and the ways law is influenced – by moral argument and public opinion within democracy. Access theory attends to a broader set of factors, including legal, illegal, extra-legal, identitary, structural and discursive causal relations (Ribot and Peluso 2003). It asks: what enables the freedoms needed to avoid risk and to influence those who govern and the broader political economic system? To what means must individuals and groups have access? Access-based causal analysis starts with entitlement theory and traces the causal chains outward to and beyond law and market operations, including structural and discursive spheres.

Whether one buys that Sen's idea of liberal democracy is the only means of influence, his capabilities framing provides democracy as a loop back to possible changes in the broader political economy. However, this loop can exist whether or not there are legal rights or elections, whether or not there is real electoral competition, whether or not

---

[30]Sen's approach to capabilities is limited by the same legalist approach used in entitlements theory and critiqued by legal pluralists and access theory (Leach *et al.* 1999, Ribot and Peluso 2003).

protest and criticism are permitted, whether or not government has the means to respond. People struggle for change using their surplus assets and other public means even in non-democratic settings – surplus assets and public means are two key elements in the capability to engage. Access to surplus assets is one element that must be evaluated to understand engagement. The legal-political enabling environment (whether procedural democracy à la Sen, or otherwise) must also be evaluated, but separately. While imbricated, their analysis needs to be disaggregated – being merged in capabilities theory. So, capabilities, like Watts and Bohle's (1993) empowerment, expand the vocabulary around security's link to engagement. The means to act is a basic element of freedom and influence. Separating the legal environment from assets opens us to a broader set of analytic pathways – this way we can take the assets part of capabilities as a starting point (as do Sen 1981, Watts and Bohle 1993, Bebbington 1999), but without limiting the enabling context to Sen's idealized democracy. This is not to say that liberal democracy is not the ideal means of influencing those who govern and the larger political economy, but to say for the completeness of analysis that it is not the only one.

Building on Sen, Bebbington (1999, 2022) places assets at the center of the link between capabilities and emancipation, arguing that in addition to allowing people to survive, adapt and escape poverty, assets ' … are also the basis of agent's *power* to act and reproduce, challenge or change the rules that govern the control, use and transformation of resources … '. He sees assets as essential to 'making a living', 'making living meaningful', and to 'emancipatory action (challenging the structures under which one makes a living)'. Bebbington (1999) casts assets as 'capitals' and influence as a matter of investment in social capital. Access theory would place influence in a larger material and structural context. The capabilities framing, including that based on 'capitals', roots explanation of assets and influence in law and legal production and exchange. In lieu of social capitals, Ribot and Peluso (2003) use the language of identities, social status and social relations as a key part of 'access' to government or the ability to derive benefits (including changes in laws and practices) from influence of the state. These work in conjunction with material resources, finance, knowledge, ideology, voice, collective action, sabotage, protest, stealth and violence as means that are used to shape conditions that shape an access regime. The access approach provides an empirical method for explaining what enables gain or loss by mapping the causal chains in any instance where a benefit is attained or lost – including changes in assets, social protections or in the regime of asset and protection formation.

The critical relation between people and government is recognized as important but still vague in the vulnerability literature (Appadurai 1984, Drèze and Sen 1989, Watts and Bohle 1993, Leach *et al.* 1998, Bebbington 1999, Sen 1999, Lappé 2013). Representation can be substantively defined as responsiveness of authority to people's needs and aspirations. It is called democratic when driven by means of sanction or accountability (Manin *et al.* 1999). Mirroring representation, citizenship can be substantively defined as the ability to influence those who govern – an ability to hold government accountable via sanction (Isin and Turner 2002, 4, Ribot *et al.* 2008, cf. Somers 2008). This definition of citizenship is the substantive opposite of 'subject' – a condition intentionally produced by those who dominate (see Mamdani 1996). Substantive citizenship is the part of capabilities that constitutes the ability to affect structural change – representation, democratic or not, is often part of the structure being changed and producing change. Surplus assets enable valued functionings beyond subsistence, including citizenship. Frameworks of capabilities and empowerment need to expand from notions of political accountability through media or social movements to a much broader array of accountability means or what Agrawal and

Ribot (1999) called 'counter-powers' (also see Agrawal and Ribot 2012, Ribot 2001, 2004).

Climate interventions, like all policy interventions, shape representation and citizenship through the local institutions they support (Ribot 2001, Marino and Ribot 2012).[31] Different institutions have different forms of belonging (residency, identity or interest based; inclusive or exclusive) and of accountability (upward, downward; narrow or broad based) (Ribot et al. 2008). These characteristics imply different degrees of representation and democracy, hence it matters which institutions policy interventions support. Agrawal (2010) found that civic organizations support risk pooling more often than local governments. The implication seems to be that supporting civics would be preferable for vulnerability reduction. But why are civics able while elected authorities are unable to serve local needs, and what are the implications for long-term representation and security? Manor (2005) shows that central governments and aid institutions overfunded self-appointed or externally created non-governmental organizations (NGOs) while underfunding elected local governments – making formal local representation less able and less relevant. Similarly, participatory processes or indigenous leaders are often favored by central governments and external donors over elected local government in ways that undermine representation (Mansuri and Rao 2003, 2012, Swyngedouw 2005, Ribot and Mearns 2008, Burga Cahuana 2013, Mbeche forthcoming). Such favoring of non-representative local institutions at the expense of formal representation are examples of what Swyngedouw (2005) critiques as horizontally organized 'governance-beyond-the-state' – which undermines formal representation and democracy.

The social actions and protections that local institutions can and do support are part of a larger political economy behind the strategic production of institutions (Bates 1981, Ribot 2007, Ribot et al. 2008). As Majid and McDowell (2012) show, famine relief to Ethiopia in 2011 was withheld in order to prevent it from supporting or legitimating insurgent organizations. Institutions are not just there to be chosen by local risk-poolers. They do not just organically or 'naturally' emerge from the polycentric ether (à la Ostrom 2009). Institutions and the forms of representation or service that they afford are products of local histories embedded in higher-level political decisions (Ribot 2007). As Bates (1981) argued, governments choose policy options based on political utility. Governments and international organizations cultivate local authorities and institutions along similar lines, creating or fostering local authorities to support their external objectives (Ribot 2007, Ribot et al. 2008). When higher-level institutions shape local institutions and authority structures, they are shaping the ability of people to sustain themselves, to be represented, and to shape the policies they are subject to.

In sum, two causal chains are required for vulnerability analysis – one concerning what shapes access to assets and another concerning access to influence within the political economy that shapes entitlements. These chains are recursively linked. Assets enable engagement and are a productive starting point for the analysis of freedom.

---

[31]Note that responsiveness is based on those who are governing having the powers to respond. Without power government cannot respond. Hence, structural adjustment programs that weaken governments can undermine their ability to respond even if accountability mechanisms are in place. Accountability without empowered government is tantamount to giving people the ability to beat a dead horse – not an uncommon phenomenon in local government (Ribot 2004). It is not democracy. Also see Gaillard and Mercer (2013, 108) who call for accountability in climate adaptation.

Structural recursive – Scott, Polanyi, Fraser and social protections

Other recursive relations of influence between people and authority, beyond formal democracy, also need to be taken into account to explain the political economy that shapes well-being. Scott (1976) shows that peasants rebel and assert demands on a moral basis when patrons fail to provide food. This ethic of reciprocity and moral expectation can be eroded away by commodification (Watts 1983a, 1991, 22, McElwee 2007, 81–96). There are parallel dynamics shaping urban demands and riots (e.g. Harvey's 2008 'right to the city'). Swift (1989, 12) remarks that 'the growth of commodity production and market relations has strengthened food security in some aspects, but has also undermined the redistributive guarantees [moral economy] of the pre-colonial economy, replacing them with an uncertain market mechanism' (also see Polanyi 2001[1944], Stiglitz 2001, and Pelling 2003, 53). Swift (1989, 12) also notes reciprocal expectations between government and people in which taxation generates the moral expectation of support (also see Moore 1997).

Polanyi (1944, 187–200) addresses how social protections (as well as environmental protections) are mobilized by society. He argues that social protections are an artifact of a double movement of capitalism. The first movement is capitalism's tendency to destroy both labor (life itself) and land (writ nature or environment), because, as 'fictive commodities' that are neither produced by nor for the market, labor and land are undervalued and overexploited.[32] In response, a second movement emerges within society to protect labor and land. Such protections find support in the enlightened self-interest of capitalists. Hence capitalism's destructive forces provoke a protective social response. Fraser (2011) sees both movements as mediated by a third movement of emancipation. Markets in Fraser's view are emancipatory and destructive. So are protective policies (see Marino and Ribot 2012). Hence, society demands that both markets and protections be subject to public scrutiny. This scrutiny is called for and, in her sense of right, must be subject to the criteria of participatory parity – a judgment based on equal access to representation. These movements form another set of potential loops constituting people's relation to government, connecting security to a wider political economy.

Polanyi's (1944) fictive commodities offer another chain of relations linking the larger political economy and vulnerability. Risk – as in the probability of stressors – is a product of nature (storms, quakes) and a byproduct of markets (effluents, toxins). It is not intentionally generated nor produced for the market. Risk appears to be another fictive commodity – whose commodification causes a dysfunctional production of risk itself through quantifying, packaging and sale of this abstract derivative of circumstance. In this case, rather than destroying a positive market input (land or labor), the market enhances this destructive force as a source of profit – e.g. moral hazard generated by insurance and disaster aid, which rely on risk as their object of intervention. Indeed, the profitability of risk is complemented by what Rose (1996) called the production of risk subjects – where through governmentality, individuals internalize the explanations of risk as if it were produced by their behavior and not by broader social and political-economic forces. In this manner, risk, and the demand for protection, is turned into demand for insurance. By blaming themselves for the risk, risk subjects take on the burden of self-protection rather than seeking social protection. This is a causal link that dampens government accountability and demand for

---

[32]His third fictive commodity is finance, not addressed here as it is tangential to micro-notions of security. At a macro level, of course, such fictive commodities as derivatives undermine everyone's security.

response (dampening representation à la Manin *et al.* 1999) – making the commodification and packaging of risk a causal link in the recursive relation of government and subjects.

A further Polanyian twist on risk production comes from viewing nature through the lens that Polanyi (1944) viewed the social history of markets. Markets served people during the period of merchant capital. In the transition to industrial capitalism, however, people were transformed into labor to serve markets (generating the industrial revolution's 'Satanic mills'). As labor, people became inputs to markets, rather than the market being an input to human production and reproduction – an input to people's lives. Nature, similarly, once served people. More and more people now serve nature – as a way of serving markets to which nature is subordinated as an ever-scarcer input. Sato (2013) shows that 'governance of environment goes hand in hand with the governance of people'. He shows how, in Thailand, natural resources play a role in state-society formation at societal margins – integrating hill people and others into society through extractive resource relations. With commodification of nature, people are subordinated to serving nature's extraction and reproduction (through exploitation and management). In the process nature and labor are commodified with no mechanisms to reflect the costs of their reproduction. As margins are incorporated, people transition from using nature to live to being used for (subordinated to) nature's economic production. This great human-environment transformation inverts agrarian people's relation to nature – from being served to serving. In the process they are also excluded from its bounty while their labor is commodified and exploited. This transformation is part and parcel of marginalization. This marginality is a foundation of vulnerability. It begs the Polanyian-Frasierian question: under what conditions does such vulnerability foment a second movement demanding social protection and a third toward emancipation, and how are these movements seen in practice?

Discursive recursive in representation and emancipation

Discursive relations form another recursive link between government and people. Discourse is a different form of 'representation' that systematically affects – is a causal element in – material insecurity (see earlier governmentality example from Rose 1996). Discourse shapes individual and group expectations and behaviors and all scales of politics (shaping virtue à la Sen 1999, 158; public sphere of Sen 1999, 158 and Habermas 1991; the governmentalities/environmentalities of Agrawal 2005 and Rose 1996, à la Foucault; the doxa and habitus of Bourdieu 1977). Rebotier (2012) develops a risk analysis framework for understanding the iterative biophysical and social production of risk. He examines how discourse, the naming of a place, a community, a geographical area of a city as 'risky', creates its own outcomes and can have the effect of a self-fulfilling prophecy. Rebotier also shows that interpretation of risk is always stratified by the differentiated relation of individuals and groups to physical risks and to the discourses about it. In this sense, he shows how risk is also always political – its interpretations imply actions that differently serve people with different social identities and means.

Rebotier also shows that once risk is identified and translated into meaning – that is, interpreted – it becomes performative and instrumental. The identification of risk, the words we use to describe it and its inscription in place imply actions and interventions with consequences for the control and use of spaces. He observes, 'territories are spaces in which meanings are inscribed, and in addition to the physical transformation of territories that risk may imply, risk is itself one of the meanings inscribed within these spaces, shaping the relationships as well as the actions carried out by their occupants, including those who govern' (2012, 392). In this sense, Rebotier's 'territorialization-of-risk framework' requires us to take a holistic view that bridges the gap between material fact and representations – placing both in the political space of risk apprehension and assessment. Here, through its performative nature, insult becomes injury – deepening material marginality through its perlocutionary effects (Butler, 1997).

In another perlocutionary manner, risk is inscribed in land reforms in Brazil. Wolford (2007) shows how land insecurity can be traced to common beliefs of both the right and left. While the neoliberals blame the state and populists blame the market for land inefficiencies, they both presume that rights to property are rooted in labor investments (à la Locke 1823). The result is that from both sides the farmer is pushed to demonstrate evidence of productivity in order to secure and maintain their property rights. Land reform beneficiaries who have won access to land based on a labor theory of property find it difficult to feel secure in their ownership – unless they use land in ways that are consistent with collective social norms regarding pro-ductivity and productive-ness. (2007, 552). She shows that those norms corral farmers into self- and mutual surveillance of land use, producing ownership insecurities that lead to land-use conforming with government programs – whether or not those produce greater land-use efficiency. The analysis calls for attention toward the framing of land reforms, forms of land title and the community norms of land use as means to soften the insecurities and peer pressures that reform discourses are producing.

Vulnerability is also established discursively at a much higher scale of social organ-ization. The very framing of the 'third' or 'developing' world as far away and other pro-duces otherness. As Butler (2009, 25) states, ' ... those whose lives are not 'regarded' as potentially grievable, and hence valuable, are made to bear the burden of starvation, underemployment, legal disenfranchisement, and differential exposure to violence and death'. Butler goes on to note that it is impossible to distinguish whether the 'regard' leads to the 'material reality' or it is the material differences that shape ill regard (cf. Taylor 1994, Fraser 2000, and Kymlicka, 2002 on directionality in effects of recog-nition and redistribution). The key point is that the categories themselves are perpetually crafting the material world. In short, perception has a material effect. Framings matter.[33]

More work has been done in science and technology studies, sociology and anthro-pology on the politicization of risk (including the opening or closing of debate), its defi-nitions, its identification, its communication, its perception, its judgment, its discursive effects on individuals and politics, its very nature as deviation from normal (e.g. Beck 1992, Demeritt 2001, Luhmann 2002, Wilkinson 2010, Fassin 2012, Connolly 2013). More work needs to be done on how science and politics occlude causal ana-lyses, favoring hazard framings and turning blame and responsibility away from society. These are beyond the scope of this contribution, which focuses on causal analy-sis of vulnerability – despite that such vulnerability and its causes may not even be visible or treatable in certain discursive, scientized and governmentalized circumstances. In short, a full analysis of vulnerability traces out chains of causality and the recursive relations between those chains, those at risk, and those analyzing and governing cause. When cause is occluded, those responsible and able to respond are shielded from blame

---

[33]See also Forsyth (2003) and O'Brien (2011) on the importance of framing in environmental and climate analytics. See Jasanoff (2010) on how the view of earth and climate framed scientifically reconfigures people's relations to nature their sensibilities concerning justice and rights to and around resources.

and responsibility, and without informed public discourse, representation is truncated and injustice facilitated. Any turn away from causality is a turn away from redress while serving to support the legitimacy and legal protection of regimes of theodicized, naturalized, quick, slow and silent injustice.

Recursive integration

While vulnerability and hazard need separate analytics, climate change does reconfigure vulnerability. The causes of that change trace back to the social origins of effluents. The first level of integration of climate-human relations is where the causal chains of climate change and vulnerability converge in a global political economy that is partly responsible for the production of both. A second locus of integration is in the co-formation of hazard and crisis. Neither exists without the other. Increased climate stressors shift the threshold of vulnerability, increasing the required assets and protections to maintain security (see Luers *et al.* 2003 for an excellent threshold model). This threshold shift is the additionality for which international adaptation policy would like responsibility. It is perhaps measurable. But it is not the line that determines responsibility since the convergence of causes shows that there is no distinct line distinguishing the production of (or responsibility for) hazard from that of vulnerability. By tracing out causality for both hazard and vulnerability, we can see how these two elements are integrated at origin and in their co-production of crisis.

Climate change has a meta-recursive relation to democracy. The shift in the vulnerability threshold is also a shift in surplus deployment – a shift from its service of *freedom to* be and do to *freedom from* risk and depravation. In this sense it represents a diminishing of positive freedoms, including the emancipatory freedom to shape the political economy that shapes assets and protections. These are the freedoms to shape the political economy in which causes of climate change and vulnerability converge. Observing the long arc of history, Chakrabarty (2009, 208) observed that cheap energy freed society, yet its effluents are diminishing that freedom. 'In no discussion of freedom in the period since the Enlightenment was there ever any awareness of the geological agency that human beings were acquiring at the same time as and through processes closely linked to their acquisition of freedom'. If development is freedom (à la Sen 1999) then climate change raises the bar/threshold for its achievement. Freedom is at risk in the Anthropocene.

This section on the unbounded framing of vulnerability analysis provides some elements of a causal-chain analysis of instances of vulnerability. A causal analysis aims to explain why people do or do not have access to the essentials of security. Two key elements that need to be explained are assets and social protections along with the recursive relations by which individuals or groups shape the political economy that shapes these foundations of wellbeing. The next section briefly outlines a comparative research agenda on causal structures of vulnerability.

Causal research: toward reduced vulnerability

> Overall, the promotion of resilient and adaptive societies requires a paradigm shift away from the primary focus on natural hazards and extreme weather events toward the identification, assessment, and ranking of vulnerability … . Therefore, understanding vulnerability is a prerequisite for understanding risk and the development of risk reduction and adaptation strategies to extreme events in the light of climate change … . Cardona *et al.* (2012, 72)

> Vulnerability research generally seeks to understand the underlying causes of vulnerability …
> … resilience approaches aimed at securing future sustainability cannot be realized without understanding the socio-political processes that underpin the foundations of vulnerability. (Miller *et al.* 2010, 6–7)

Empirical analysis of climate-related vulnerability (1) starts by identifying exposure units that have lost valued attributes and the distribution of those losses across individuals, households and groups (who lost what); (2) links losses to specific asset and protection failures; (3) assesses the immediate causes of failed access to adequate assets and protections, and (4) maps these immediate causes to capacities, knowledge, identities, intra-household relations, local social and political hierarchies and production and exchange relations, and to the larger physical, social and political-economic relations in which the exposed unit is located (see Turner *et al.* 2003a, 8075, Blaikie 1985, Downing 1991, Watts and Bohle 1993, Füssel 2007). A full analysis then (5) evaluates the means and mechanisms by which exposed units can influence or are prevented from influencing structures they operate within and those who govern (Sen 1981, 1999, Watts and Bohle 1993, Bebbington 1999).

Analyzing the 'chains of causality' (Blaikie 1985) behind chronic deprivation or crisis, by showing how outcomes are produced by proximate factors that are in turn shaped by more distant events and processes,[34] can tell us what kinds of promotions or protections (Drèze and Sen 1989, 60) – to which I would add emancipations, restructurings and redistributions – might stem the production of vulnerability at what scales; and, where relevant, who should pay the costs of vulnerability reduction. As Drèze and Sen (1989, 15) note, the important thing is to 'examine all causal influences on these matters' (also see Sen 1999, 161, Miller *et al.* 2010, 6–7, Cardona *et al.* 2012, 72). These are the kinds of analyses that should precede and complement adaptation and mitigation planning. To make this kind of research robust, a large number of in-depth case studies would need to be developed and compared to identify the most salient causal factors at different social, political-economic and institutional scales (see Ribot 2013 for further discussion of comparative vulnerability research).

Any empirical analysis of vulnerability in the face of climate change will show climate's roles. The event occurs, triggering disaster in a social landscape of underlying vulnerabilities. In past studies of climate disasters, social analysts took climate hazards as probabilistic events outside of social agency, requiring no explanation. Risk could only be reduced on the vulnerability side (Blaikie *et al.* 1994, Wisner *et al.* 2004, 49–55). Today that probability is anthropogenic and has a causal structure relevant to disaster reduction – insofar as (1) mitigation is a pathway to hazard control, (2) redressing inequalities that produce effluents may also serve to reduce vulnerabilities and (3) increased hazards shift the threshold of vulnerability and scope for freedom, potentially dampening recursive emancipatory loops. But it is still true that with no vulnerability there is no disaster. Hence, as long as hazards are in the sky, vulnerability and hazard causality remain bifurcated and vulnerability reduction remains a distinct solution. Full bifurcated coverage would include a vulnerability analysis *and* an analysis of the anthropogenic causes of hazard probabilities – right down to the effects of rights and representation or the commodification of risk on hazard generation and control.

---

[34]See Swift 1989, 8 who distinguishes 'between the *proximate or intermediate variables*, which are the direct links to famine, and the *indirect or primary factors*, which are the more general ecological, economic or political processes determining whether communities thrive or decline' – italics in original.

Hazards are now anthropogenic and potentially treatable. The hazards framing, still used widely, cannot provide the full array of options for securing wellbeing. Their linear-causal impact stance sidelines the social causes and solutions of vulnerability. Of course, hazards views, like entitlement-livelihoods approaches, are logical and empirically demonstrable. Each framing trains attention on different observable variables and relations among them. It is not that one is wrong and the other is right. They are frames. Each frames different realities for different objectives with different implications. Researchers need to understand what those objectives are so we can choose framings that lead us to empirics that can illuminate pathways toward individual and collective aspirations – the reduction of pain and suffering and increase in human wellbeing and potential. Ontologies are not objective truth claims. They are chosen framings, chosen objectives. Researchers are not objective. They have objectives – frames they have built to make meaning of the world. Once framed within objectives and assumptions, it is the methods that must maintain the rigor and credibility of analysis.

'No story can be told nor any theory proposed that is not responding to prior (implicit or explicit) questions, and our questions are always the products of our situated selves' (Somers 2008, 10). I choose a vulnerability framing with an ontology that privileges humans and human values and I take a political stance that does so through representation. Humanism privileges human agency in our world, which can be nothing other than social. I choose to see human wellbeing as socially interconnected – through politics and through our anthropogenic sky. Our welfare is ineluctably social, as is our precarious interdependent being (Butler 2011). It is human agency, manifest in acts and in structures, that shapes what we are and what we have, and therefore what we can become and do. This position makes even the choice of a framing that places causality outside of human agency something that is chosen for purposes that are of human value and agency. The hazards framing is not wrong as the chosen frame it represents. Storms do result in damage. But we need to know what it does to choose this framing – what it reveals and occludes and who and what ends are served by that choice. The hazards approach is a snapshot that elides time. It accomplishes the occlusion of history and social cause. The lack of awareness of our ontologies renders them natural, occluding them from ourselves so we can believe they are not frames but the world itself.

Conclusion

... hunger is caused not by a scarcity of food but a scarcity of democracy .... (Francis Moore Lappé 2013, 230)

... democratization must remove the strait-jacket which stifles the peasantry, because any popular movement to transform political life must sever the hold that ruling classes exercise over rural producers. (Michael Watts, 'Entitlements or empowerment?' 1991, 25)

If the ability to maintain minimum nutritional needs is the distinction between rich and poor (Sen 1981, Appadurai 1984, 482), then those whose nutrition does not allow them the surplus energy required to engage in politics should be considered a class – the disenfranchised. Indeed, vulnerability, as Drèze and Sen (1989, 49) argue, is a state that weakens bargaining position – enabling, for example, labor exploitation. That bargaining position is also part of the material basis of citizenship and representation. Emancipation requires sufficient wealth beyond mere subsistence to enable the individual, household, group or community to walk away from daily labor long enough to engage in shaping the political economy that shapes their entitlements. Substantive citizenship is that ability – the ability to influence those who govern. Substantive democracy is when that influence results in response. Many lack the knowledge (often occluded by design) and skills (withered by exclusion), as well as the time (subordinated to survival), needed to exercise influence.

Almost 40 years ago O'Keefe, Westgate and Wisner (1976) wrote 'Taking the natural-ness out of natural disasters' (see Watts 1991, 231, Manyena 2006, 439–40). With the cul-turing of nature, the sociality of wellbeing and crisis should even be more evident today. While one might think that calling our era the Anthropocene would turn attention from nature back to people, it oddly guides gazes back toward hazards. So, in the Anthropocene, the struggle is still to maintain attention on the social and political production and reproduc-tion of risk. The framing outlined in this contribution presents an integrative analysis of the social and political-economic causes of vulnerability – with the hope of generating a socio-centric Anthropocene, so we can perhaps make it to the sociocene or democene. Whatever we call it, in climate analysis as in politics, it is always a struggle to represent the social. As the framing in this paper shows, we need to characterize and know of hazards, but we do not need to explain hazards – even if anthropogenic – to understand the origins of vulnerability, which is produced on the ground. Treating hazards can change outcomes. But mitigation is not vulnerability reduction – although it can help avert damages. Vulnerability resides in the pre-hazard precarity of people.

There are dozens of definitions of vulnerability (Manyena 2006, 440, Miller *et al.* 2010). Like adaptation and resilience, none are complete nor could they be. The definition is contingent on the object at risk and the values that one is hoping to preserve, restore or eliminate (see O'Brien 2011, 545). The choice of theory is itself social – aiming at some outcomes, preserving some values over others (Beymer-Farris *et al.* 2012). But once one is aware of what their objectives are – reduction of hunger and famine or sovereignty and security of agricultural systems – then it is important that they have a coherent analysis of causality. Causal analysis, problematic or not, theoretically singular or fragmented, is necessary. 'The multiplicity of definitions is a reflection of the philosophical and methodo-logical diversities that have emerged from disaster scholarship and research' (Manyena 2006, 440).

Vulnerability reduction measures, of course, do not only derive from understanding causes. Indeed, some causes may be (or appear) immutable, others no longer active, tran-sient or incidental. Redressing direct causes may not always be part of the most effective solutions (Drèze and Sen 1989, 34). The objective of vulnerability analysis is to identify the active processes of vulnerability production and then to identify which are amenable to redress. Other interventions can also be identified that are designed to counter conditions or symptoms of vulnerability without attending to their causes – such as support for risk-pooling strategies or targeted poverty-reduction disaster relief. All forms of available analy-sis should be used to identify the most equitable and effective means of vulnerability eradication.

The contribution of adaptation studies is to complement vulnerability analysis with a fuller picture of how innovations shape our world – and to hammer that knowledge into action. Climate-adaptation scholars are aware that 'adaptations' – an ecological term – are not the natural random work of Darwin's evolution. As Arendt (1960, 460) points out, the miracles of evolution are authored by probability whereas we know the author of the even more frequent miracle of political change through women and men ' … who because they have received the twofold gift of freedom and action can establish a reality of their own'. She places cause (and responsibility) for change and innovation within society. In this sense, 'adaptive capacity' becomes something that must be explained socially. Like any contribution to vulnerability or wellbeing, innovation is socially enabled. It is part of the causal (and reparatory) chain of vulnerability.

The freedoms to act and to innovate follow from rights and representation. As we see vulnerability and adaptation analysts and agencies turning more and more toward 'rights-based' approaches to natural resource management and climate change (Roberts and Parks 2007, McDermott *et al.* 2012, Walsh-Dilley *et al.* 2013), it is important to keep in mind that the fundamental right is the right to influence those who govern and to engage in the making, scrutiny and implementation of rights. Somers (2008) considers the right to have rights the fundamental right that defines citizenship. But the right to shape rights is even more important – this is the right to the means and freedoms to influence those who govern. This is emancipation.

Representation is one means by which individuals, households and groups can shape the political economy that shapes their entitlements. Social movements are another (see Luhmann 2002, 138–41, Peet and Hartwick 2009, 286–7). The criterion Fraser (2008) calls for is participatory parity. This does not mean symbolic forms of participation without real influence over the projects in which people participate (Ribot 1996, Swynge-douw 2005, Mansuri and Rao 2012). The ability to influence authorities and the rules they make and implement produces the very entitlements that spell security and create the flexibility that enables people to buffer themselves against the unpredictable but expected stresses of life. Of course, to be functional, representation requires powers – representatives need discretionary authority, means and resources to respond to people's needs and aspirations; people must have resources and knowledge to act as citizens to influence those who govern (Ribot *et al.* 2008). Poverty is not only a basis of vulnerability but it is also disenfranchising – undermining the ability of the poor to influence those who govern.

To be represented is to be seen and responded to. To demand representation is to see the possibility of response. Making vulnerability legible is part of the process of understanding where those possibilities lie – the job of research and of voice. The legibility that churches and governments produce is matched with occlusions and illusions that divert attention. They do not want citizens to see what they see – they want to externalize causality so that citizens and victims displace their frustrations onto God and nature or turn them on themselves. Citizens must insist that government sees, and they must show that citizens know their rulers know. It is in this context where citizen sanctioning of government could result in response. To insist on security requires knowledge of vulnerability, its causes, and the channels of possible redress. It requires the material resources and time to analyse, organize and exercise the counter-power that translates voice into response. The obvious question, with no obvious answer, is 'how' to create such representation or parity given the asymmetries of power in society and the vesting of authority in science and expertise.[35]

Polanyi (1944) described capitalism's double movement in which capital can destroy its very inputs – labour and land – but people respond to the risks and damages by demanding protections. Fraser (2011) sees a third, emancipatory, movement, demanding that both capitalism and social protections be subject to public scrutiny. Capitalism can be both damaging and emancipatory. The rules that guide it and its effects need to be disciplined and subject to public judgment. Social and environmental protections too provide shelter from the downsides of capitalism – e.g. the systems that generate and shroud risk. These social and environmental protections – social security systems, fortress conservation and

---

[35]Thanks to Tim Forsyth for this pertinent query.

climate policies – also affect redistributions with negative and positive consequences. Rights of recourse and representation must constantly be asserted and re-asserted to make visible and to subject to public scrutiny the links between risk, cause, responsibility and blame as they shape the interdependence that makes sustainability of life possible (Butler 2009, 14, 23).

One of the two fabulous external reviewers for this contribution asked 'how far can a climate process be expected to go in correcting all past wrongs' and 'must all climate researchers also be responsible for analyzing all underlying social issues'. My answer is that any environmental intervention can go very far, and 'yes' this is our responsibility. Without being aware of the past, as in all areas of endeavour, climate researchers are likely to reproduce and deepen past wrongs. Hence, a grasp of the past or serious partnership with vulnerability analysts is not optional. This reviewer continued, 'The kinds of institutions, processes and forums that could enable the fundamental changes you call for do not yet exist', and asked 'What can your paper contribute to helping us imagine them into being?' They do exist in some places at some times for some people. This essay is part of imagining them into broader being. 'Society is positively transformed by showing, through criticism, what most needs changing and in which particular ways' (Peet and Hartwick 1990, 282). If we, as analysts or activists, insist on requiring that all interventions enable democracy, and we insist this demand be enforced, we may help force the hand of practice – by mobilizing liability, sanction or exposure and shame. I do not want to act or be in a world that does not try. Democracy is an ongoing struggle. It is not a state to be arrived at. It will come and go in degrees. Trying is the struggle that produces emancipatory moments – however ephemeral they may be. The fleeting joy and creativity of freedom seem worth it.

References

Adger, W.N. 2006. Vulnerability. *Global Environmental Change*, 16(3), 268–281.

Adger, W.N., *et al.* 2004. New indicators of vulnerability and adaptive capacity. *Tyndall Center for Climate Change Research Technical Report 7.*

Agarwal, B. 1993. Social security and the family: coping with seasonality and calamity in rural India. *Journal of Peasant Studies*, 17(3), 156–165.

Agarwal, A. and S. Narain. 1991. *Global warming in an unequal world.* New Delhi: Centre for Science and Environment.

Agrawal, A. 2005. *Environmentality: technologies of government and the making of subjects.* Durham: Duke University Press.

Agrawal, A. 2010. Local institutions and adaptation to climate change. In: R. Mearns and A. Norton, eds. *Social dimensions of climate change: equity and vulnerability in a warming world.* Washington, DC: The World Bank, pp. 173–198.

Agrawal, A. and J. Ribot. 1999. Accountability in decentralization: a framework with South Asian and African cases. *The Journal of Developing Areas*, 33(4), 473–502.

Agrawal, A. and J. Ribot. 2012. Assessing the effectiveness of democratic accountability mechanisms in local governance. Report commissioned for USAID by Management Systems International (MSI) Project No. 380000.12–500–03–11. Washington, DC: USAID.

Alavi, H. 1965. Peasants and revolution. *Socialist Register*, 2, 241–247.

Anderson, S., J. Morton and C. Toulmin. 2010. Climate change for agrarian societies in drylands: implications and future pathways. In: R. Mearns and A. Norton, eds. *Social dimensions of climate change: equity and vulnerability in a warming world.* Washington, DC: The World Bank, pp. 199–230.

Appadurai, A. 1984. How moral is South Asia's economy? – a review article. *The Journal of Asian Studies*, 43(3), 481–497.

Arendt, B. 2000 [1960]. What is freedom? In: P. Baehr, ed. *The portable Hannah Arendt.* New York: Penguin Books, pp. 438–461.

Arthur, C. 2002. Revealed: how the smoke stacks of America have brought the world's worst drought to Africa. *The Independent*, 12 June. Available from: http://www.freerepublic.com/focus/news/699049/posts [Accessed 2 November 2013].

Bassett, T.J. and C. Fogelman. 2013. Déjà vu or something new? The adaptation concept in the climate change literature. *Geoforum*, 48(0), 42–53.

Bates, R. 1981. *Markets and states in Tropical Africa*. Berkeley: University of California Press.

Batha, E. 2008. Cyclone Sidr would have killed 100,000 not long ago. November 16, 2007. AlertNet [online]. Available from: http://alertnet.org/db/blogs/19216/2007/10/16-165438-1.htm [Accessed 5 December 2008].

Bebbington, A. 1999. Capitals and capabilities: a framework for analysing peasant viability, rural livelihoods and poverty. *World Development*, 27(12), 2021–2044.

Beck, U. 1992. *Risk society: towards a new modernity*. London: Sage.

Beck, U. 2007. *World at risk*. Cambridge, UK: Polity Press.

von Benda-Beckmann, K. 1981. Forum shopping and shopping forums: dispute processing in a Minangkabau village in West Sumatra. *Journal of Legal Pluralism*, 19, 117–159.

Bern, C., *et al.* 1993. Risk factors for mortality in the Bangladesh cyclone of 1991. *Bulletin of World Health Organization*, 71(1), 73–78.

Bernstein, H. 1996. Agrarian questions then and now. *The Journal of Peasant Studies*, 24(1–2), 22–59.

Berry, S. 1993. *No condition is permanent: the social dynamics of agrarian change in sub-Saharan Africa*. Madison: The University of Wisconsin Press.

Beymer-Farris, B.A. and T.J. Bassett. 2012. The REDD menace: resurgent protectionism in Tanzania's mangrove forests. *Global Environmental Change*, 22, 332–341.

Beymer-Farris, B.A., T.J. Bassett and I. Bryceson. 2012. Promises and pitfalls of adaptive management in resilience thinking: the lens of political ecology. In: T. Plieninger and C. Bieling, eds. *Resilience and the cultural landscape*. Cambridge: Cambridge University Press, pp. 283–300.

Birkenholtz, T. 2011. Network political ecology: method and theory in climate change vulnerability and adaptation research. *Progress in Human Geography*, 36(3), 295–315.

Blaikie, P. 1985. *The political economy of soil erosion in developing countries*. London: Longman Press.

Blaikie, P., *et al.* 1994. *At risk: natural hazards, people's vulnerability, and disasters*. London: Routledge.

Bourdieu, P. 1977. *Outline of a theory of practice*. Cambridge: Cambridge University Press.

Brooks, N. 2003. Vulnerability, risk and adaptation: a conceptual framework. *Tyndall Centre for Climate Change and Research Working Paper 38*.

Burga Cahuana, Carol. 2013. Participation and representation: REDD+ in the native communities of Belgica and Infierno in the Peruvian Amazon. Master's Thesis, University of Illinois at Urbana-Champaign.

Burton, I., *et al.* 2002. From impacts assessment to adaptation priorities: the shaping of adaptation policy. *Climate Policy*, 2(2–3), 145–159.

Butler, J. 1997. *Excitable speech: a politics of the performative*. London and New York: Routledge.

Butler, J. 2009. *Frames of war: when is life grievable?* London: Verso Books.

Butler, J. 2011. For and against precarity. *Tidal: Occupy Theory, Occupy Strategy*, Issue 1: The Beginning is Near. Available from: http://tidalmag.org/ [Accessed 3 December 2011].

Cannon, T. 2000. Vulnerability analysis and disasters. In: D.J. Parker, ed. *Floods*. London and New York: Routledge.

Cannon, T., J. Twigg and J. Rowell. 2004. Social vulnerability, sustainable livelihoods and disasters. *Report to DFID, Conflict and Humanitarian Assistance Department, and Sustainable Livelihoods Support Office*. London: DFID.

Cardona, O.D., *et al.* 2012. Determinants of risk: exposure and vulnerability. In: C.B. Field, V. Barros, T.F. Stocker, D. Qin, D.J. Dokken, K.L. Ebi, M.D. Mastrandrea, K.J. Mach, G.-K. Plattner, S.K. Allen, M. Tignor and P.M. Midgley, eds. *A special report of working groups I and II of the Intergovernmental Panel on Climate Change (IPCC)*. Cambridge: Cambridge University Press, pp. 65–108.

Carney, J. 1988. Struggles over land and crops in an irrigated rice scheme: the Gambia. In: J. Davison, ed. *Agriculture, women and land: the African experience*. Boulder: Westview Press, pp. 59–78.

CEDMHA (Center for Excellence in Disaster Management and Humanitarian Assistance). 2007. Cyclone Sidr Update, November 15.

Chakrabarty, D. 2009. The climate of history: four theses. *Critical Inquiry*, 35(2), 197–222.

Chambers, R. 1989. Vulnerability, coping and policy. *IDS Bulletin*, 20(2), 1–7.

Connolly, W.E. 2013. *The fragility of things: self-organizing processes, neoliberal fantasies, and democratic activism*. Durham: Duke University Press.

Deere, C.D. and A. deJanvry. 1984. A conceptual framework for the empirical analysis of peasants. *Giannini Foundation Paper No. 543*. Berkeley: Giannini Foundation, 601–11.

Demeritt, D. 2001. The construction of global warming and the politics of science. *Annals of the Association of American Geographers*, 91(2), 307–337.

Deressa, T., R.M. Hassan and C. Ringler. 2008. Measuring Ethiopian farmers' vulnerability to climate change across regional states. *IFPRI Discussion Paper No. 806*. Washington, DC: IFPRI, Environment and Production Technology Division, 1–22.

Dill, B. 2013. *Fixing the African state: recognition, politics, and community-based development in Tanzania*. New York: Palgrave Macmillan.

Douglas, M. and A. Wildavsky. 1982. *Risk and culture*. Berkeley: University of California Press.

Downing, T.E. 1991. Assessing socioeconomic vulnerability to famine: frameworks, concepts, and applications. Final Report to the U.S. Agency for International Development, Famine Early Warning System Project. 30 January.

Downing, T.E. 1992. Vulnerability and global environmental change in the semi-arid tropics: modeling regional and household agricultural impacts and responses. Paper presented to the International Conference on Sustainable Development in the World's Drylands. 27 Jan.-1 Feb. Fortaleza-Ceará, Brazil.

Downing, T.E. and A. Patwardhan. 2005. Assessing vulnerability for climate adaptation. In: B. Lim, E. Spanger-Siegfried, I. Burton, E. Malone and S. Huq, eds. *Adaptation policy frameworks for climate change*. Cambridge: Cambridge University Press, pp. 67–89.

Drèze, J. and A. Sen. 1989. *Hunger and public action*. Oxford: Clarendon Press.

Duit, A., *et al.* 2010. Governance, complexity, and resilience. *Global Environmental Change*, 20(3), 363–368.

Dworkin, R. 2013. What is equality? Part II: equality of resources. *Philosophy and Public Affairs*, 10(4), 283–345.

European Commission. 2013. Climate change, environmental degradation, and migration. *Commission Staff Working Document 138*. Brussels.

Fassin, D. 2012. *Humanitarian reason: a moral history of the present*. Berkeley: University of California Press.

Figueiredo, P. and P.E. Perkins. 2012. Women and water management in times of climate change: participatory and inclusive processes. *Journal of Cleaner Production*, 60, 188–194.

Folke, C. 2006. Resilience: the emergence of a perspective for social–ecological systems analyses. *Global Environmental Change*, 16(3), 253–267.

Folke, C., *et al.* 2010. Resilience thinking: integrating resilience, adaptability, and transformability. *Ecology and Society*, 15(4), 20. Available from: http://www.ecologyandsociety.org/vol15/iss4/art20/ [Accessed 6 January 2014].

Forsyth, T. 2001. Critical realism and political ecology. In: A. Stainer and G. Lopez, eds. *After post-modernism: critical realism?* London: Athlone Press, pp. 146–154.

Forsyth, T. 2003. *Critical political ecology: the politics of environmental science*. London: Routledge.

Fraser, N. 2000. Rethinking recognition. *New Left Review*, 3(3), 107–120.

Fraser, N. 2008. From redistribution to recognition? Dilemmas of justice in a 'postsocialist' age. In: K. Olsen. ed. *Adding insult to injury: Nancy Fraser debates her critics*. London: Verso Books, pp. 9–41.

Fraser, N. 2011. Marketization, social protection, emancipation: toward a neo-Polanyian conception of capitalist crisis. In: Calhoun, C. and G. Derluguian, eds. *Business as usual: the roots of the global financial meltdown*. New York: New York University Press, pp. 137–158.

Füssel, H.-M. 2007. Vulnerability: a generally applicable conceptual framework for climate change research. *Global Environmental Change*, 17(2), 155–167.

Füssel, H.-M. and R.J.T. Klein. 2006. Climate change vulnerability assessments: an evolution of conceptual thinking. *Climatic Change*, 75(3), 301–329.

Gaillard, J.D. 2010. Vulnerability, capacity and resilience: perspectives for climate and development policy. *Journal of International Development*, 22(2), 218–232.

Gaillard, J.D. and J. Mercer. 2013. From knowledge to action: bridging gaps in disaster risk reduction. *Progress in Human Geography*, 37(1), 93–114.

Gasper, D. 1993. Entitlement analysis: concepts and context. *Development and Change*, 24, 679–718.

Government of Bangladesh. 2008. April. Cyclone Sidr in Bangladesh: damage, loss, and needs assessment for disaster recovery and reconstruction. Prepared by the Government of Bangladesh assisted by the International Development Community with financial support from the European Commission.

Griffiths, J. 1986. What is legal pluralism? *Journal of Legal Pluralism*, 24, 1–55.

Gunderson, L.H. and C.S. Holling. 2002. *Panarchy: understanding transformations in human and natural systems*. Washington, DC: Island Press.

Guyer, J. and P. Peters. 1987. Conceptualizing the household: issues of theory and policy in Africa. *Development and Change*, 18(2), 197–327.

Habermas, J. 1991. *The structural transformation of the public sphere: an inquiry into a category of bourgeois society*. Cambridge: Cambridge University Press.

Hart, G. 1992. Household production reconsidered: gender, labor conflict, and technological change in Malaysia's Muda region. *World Development*, 20(6), 809–823.

Harvey, D. 2008. The right to the city. *New Left Review*, 53, 23–40.

Hayes, A. 2009. Court: army corps of engineers liable for Katrina flooding. CNN [online]. Available from: http://edition.cnn.com/2009/US/11/18/louisiana.katrina.lawsuit/ [Accessed 12 July 2013].

Heltberg, R., P.B. Siegel and S.L. Jorgensen. 2010. Social policies for adaptation to climate change. In: R. Mearns and A. Norton, eds. *Social dimensions of climate change: equity and vulnerability in a warming world*. Washington, DC: The World Bank, pp. 259–275.

Hyvarinen, J. 2012. Loss and damage caused by climate change: legal strategies for vulnerable countries. *October 2012 report of the Foundation for International Environmental Law and Development (FIELD), London*. Available from: http://www.field.org.uk/sites/field.org.uk/files/papers/field_loss__damage_legal_strategies_oct_12.pdf.

IPCC (Intergovernmental Panel on Climate Change). 2012. Managing the risks of extreme events and disasters to advance climate change adaptation. In: C.B. Field, V. Barros, T.F. Stocker, D. Qin, D.J. Dokken, K.L. Ebi, M.D. Mastrandrea, K.J. Mach, G.-K. Plattner, S.K. Allen, M. Tignor and P.M. Midgley, eds. *A special report of working groups I and II of the Intergovernmental Panel on Climate Change*. NY: Cambridge University Press.

IPCC. 2013. Climate change 2013: the physical science basis. *Working Group I Contribution to the IPCC Fifth Assessment Report (Final Draft Underlying Scientific-Technical Assessment)*. Available from: http://www.ipcc.ch/report/ar5/wg1/#.UlAbIGRgY8C [Accessed 5 October 2013].

Isakson, R. 2013. Financialization and the transformation of agro-food supply chains: A political economy. Paper presented at the Yale Program on Agrarian Studies 'Conference on Food Sovereignty: A Critical Dialogue', 15-16 September 2013. Available from: http://www.yale.edu/agrarianstudies/foodsovereignty/papers.html [Accessed 5 October 2013].

Isin, E.F. and B.S. Turner, eds. 2002. *Handbook of citizenship studies*. London: Sage Publications.

deJanvry, A., *et al.* 2001. Access to land and policy reforms. In: J.-P. Patteau, A. deJanvry, G. Gordillo and E. Sadoulet, eds. *Access to land, rural poverty, and public action*. Oxford: Oxford University Press, 1–27.

Jasanoff, S.A. 1999. The songlines of risk. *Environmental Values*, 8(2), 135–152.

Jasanoff, S. 2010. A New Climate for Society. *Theory Culture Society*, 27, 233–253.

Jisheng, Y. 2012. *Tombstone: the great Chinese famine, 1958–1962*. New York: Farrar, Straus and Giroux.

Jones, T. and S. Edwards. 2009. The climate debt crisis: why paying our dues is essential for tackling climate change. *World Development Movement and Jubilee Debt Campaign Report*. Available from: http://wdm.org.uk/sites/default/files/climatedebtcrisis06112009_0.pdf [Accessed 3 August 2013].

Kasperson, R.E., *et al.* 2005. Vulnerable peoples and places. In: R. Hassan, R. Scholes and N. Ash, eds. *Ecosystems and human wellbeing: current state and trends, Vol. 1*. Washington, DC: Island Press, pp. 143–164.

Khan, M.R. and J.T. Roberts. 2013. Adaptation and international climate policy. *WIREs Climate Change*, 4(3), 171–189.

Klein, N. 2012. Geoengineering: testing the waters. *New York Times*. 27 October.

Kymlicka, W. 2002. *Contemporary political philosophy: an introduction*. Oxford: Oxford University Press.

Lappé, F.M. 2013. Beyond the scarcity scare: reframing the discourse of hunger with an eco-mind. *The Journal of Peasant Studies*, 40(1), 219–238.

Leach, M. 2008. Pathways to sustainability in the forest? Misunderstood dynamics and the negotiation of knowledge, power, and policy. *Environment and Planning A*, 40(8), 1783–1795.

Leach, M., R. Mearns and I. Scoones. 1998. Chal- lenges to community based sustainable development: dynamics, entitlements, institutions. *IDS Bulletin*, 28(4), 4–14.

Leach, M., R. Mearns and I. Scoones. 1999. Environmental entitlements: dynamics and institutions in community-based natural resource management. *World Development*, 27(2), 225–247.

Leichenko, R.M. and K. O'Brien. 2008. *Environmental change and globalization: double exposures*. New York, NY: Oxford University Press.

Lock, M. 1997. Displacing suffering: the reconstruction of death in North America and Japan. In: A. Kleinman, V. Das and M. Lock, eds. *Social suffering*. Berkeley: University of California Press, 207–44.

Locke, J. 1823. *Two treatises of government*. London: Thomas Tegg & W. Sharpe and Son.

Luers, A.L., *et al.* 2003. A method for quantifying vulnerability, applied to the agricultural system of the Yaqui Valley, Mexico. *Global Environmental Change*, 13(4), 255–267.

Luhmann, N. 2002. *Risk: a sociological theory*. New Brunswick: Adeline Transaction.

Lund, C. 2008. *Local politics and the dynamics of property in Africa*. Cambridge: Cambridge University Press.

Lund, C. 2013. The past and space: on arguments in African land control. *Africa*, 83(1), 14–35.

Lund, C. and C. Boone. 2013. Land politics in Africa – constituting authority over territory, property and persons. *Africa*, 83(1), 1–13.

Majid, N. and S. McDowell. 2012. Hidden dimensions of the Somalia famine. *Global Food Security*, 1(1), 36–42.

Mamdani, M. 1996. *Citizen and subject: contemporary Africa and the legacy of late colonialism*. Princeton: Princeton University Press.

Manin, B., A. Przeworski, and S. Stokes. 1999. Elections and representation. In: A. Przeworski, S. Stokes and B. Manin, eds. *Democracy, accountability, and representation*. Cambridge: Cambridge University Press, pp. 29–54.

Manor, J. 2005. User committees: a potentially damaging second wave of decentralization? In: J. Ribot and A.M. Larson, eds. *Decentralization of natural resources: experiences in Africa, Asia and Latin America*. London: Frank Cass, pp.192–213.

Mansuri, G. and V. Rao. 2003. Evaluating community driven development: a review of the evidence. *First Draft Report*. February 2003. Washington, DC: The World Bank, Development Research Group.

Mansuri, G. and V. Rao. 2012. Localizing development: does participation work? *World Bank Policy Research Report*. Washington, DC: The World Bank.

Manyena, S.B. 2006. The concept of resilience revisited. *Disasters*, 30(4), 434–450.

Marino, E. and J. Ribot, special issue editors. 2012. Adding insult to injury: climate change, social stratification, and the inequities of intervention (special issue introduction). *Global Environmental Change*, 22(2), 1–7.

Mbeche, R. forthcoming. REDD stakeholder consultation – symbolic or substantive representation in preparing Uganda for REDD? Working Paper for the Responsive Forest Governance Initiative. Council for the Development of Social Science Research in Africa. Mimeo, November 2013.

McDermott, M.H., S. Mahanty and K. Schreckenberg. 2012. Examining equity: a multidimensional framework for assessing equity in payments for ecosystem services. *Environmental Science and Policy*. http://dx.doi.org/10.1016/j.envsci.2012.10.006 [Accessed 2 November 2013].

McElwee, P. 2007. From the moral economy to the world economy: revisiting Vietnamese peasants in a globalizing era. *Journal of Vietnamese Studies*, 2(2), 57–107.

Miller, F., *et al.* 2010. Resilience and vulnerability: complementary or conflicting concepts? *Ecology and Society*, 15(3). http:// www.ecologyandsociety.org/vol15/iss3/art11/

Ministry of Food and Disaster Management of Bangladesh. 2008. Super cyclone Sidr 2007: impacts and strategies for interventions. Bangladesh Secretariat, Dhaka. Available at: http://www.ecologyandsociety.org/vol15/iss3/art11/

Moore, M. 1997. Death without taxes: democracy, state capacity, and aid dependence in the fourth world. Draft. Later published in G. White and M. Robinson, eds. *Towards a democratic developmental state*. 1998. Oxford: Oxford University Press.

Moser, C. 2007. Asset accumulation policy and poverty reduction. In: C. Moser, ed. *Reducing global poverty: the case for asset accumulation*. Washington, DC: Brookings Press, pp. 83–103.

Moser, C. and A. Dani, eds. 2008. *Assets, livelihoods, and social policy*. Washington, DC: The World Bank.

Moser, C. and A. Norton (with T. Conway, C. Ferguson and P. Vizard). 2001. *To claim our rights: livelihood security, human rights and sustainable development*. London: Overseas Development Institute.

Moser, C. and D. Satterthwaite. 2010. Toward pro-poor adaptation to climate change in the urban centers of low-and middle-income countries. In: R. Mearns and A. Norton, eds. *Social dimensions of climate change: equity and vulnerability in a warming world*. Washington, DC: The World Bank, pp. 231–258.

Myers, S.L. and N. Kulish. 2013. Growing clamor about inequalities of climate crisis. *New York Times*. 16 November.

Nozick, R. 1974. *Anarchy, state, and utopia*. New York: Basic Books.

O'Brien, K. 2011. Responding to environmental change: a new age for human geography? *Progress in Human Geography*, 35(4), 542–549.

O'Brien, K. 2012. Global environmental change II: from adaptation to deliberate transformation. *Progress in Human Geography*, 36(5), 667–676.

O'Brien, K., *et al.* 2007. Why different interpretations of vulnerability matter in climate change discourses. *Climate Policy*, 7(1), 73–88.

O'Brien, K. and R.M. Leichenko. 2003. Winners and losers in the context of global change. *Annals of the Association of American Geographers*, 93(1), 89–103.

O'Keefe, P., K. Westgate and B. Wisner. 1976. Taking the naturalness out of natural disasters. *Nature*, 260(5552), 566–567.

Osborne, T. 2011. Carbon forestry and agrarian change: access and land control in a Mexican rainforest. *The Journal of Peasant Studies*, 38(4), 859–883.

Ostrom, E. 2009. A polycentric approach for coping with climate change. *World Bank Policy Research Working Paper No. 5095*. Washington, DC: The World Bank, Development and Economics Research Group.

Peet, R. and E. Hartwick. 2009. *Theories of development: contentions, arguments, alternatives* (2nd ed.). New York: Guilford Press.

Pelling, M. 2003. *The vulnerability of cities: natural disasters and social resilience*. London: Earthscan.

Pelling, M. 2011. *Adaptation to climate change: from resilience to transformation*. Abingdon: Routledge.

Polanyi, K. 1944. *The great transformation*. New York: Rinehart.

Poteete, A. and J. Ribot. 2011. Repertoires of domination: decentralization as process in Botswana and Senegal. *World Development*, 39(3), 439–449.

Prowse, M. 2003. Toward a clearer understanding of 'vulnerability' in relation to chronic poverty. *CPRC Working Paper No. 24*. Manchester: University of Manchester, Chronic Poverty Research Centre.

Prowse, M. and L. Scott. 2008. Assets and adaptation: an emerging debate. *IDS Bulletin*, 39(4), 42–52.

Rebotier, J. 2012. Vulnerability conditions and risk representations in Latin-America: framing the territorializing of urban risk. *Global Environmental Change*, 22(2), 391–398.

Ribot, J. 1995. The causal structure of vulnerability: its application to climate impact analysis. *GeoJournal*, 35(2), 119–122.

Ribot, J. 1996. Participation without representation: chiefs, councils and forestry law in the West African Sahel. *Cultural Survival Quarterly*, 20(1), 40–44.

Ribot, J. 1998. Theorizing access: forest profits along Senegal's charcoal commodity chain. *Development and Change*, 29(2), 307–341.

Ribot, J. 2001. Integral local development: 'accommodating multiple interests' through entrustment and accountable representation. *International Journal of Agricultural Resources, Governance and Ecology*, 1(34), 327–350.

Ribot, J. 2004. *Waiting for democracy: the politics of choice in natural resource decentralization*. Washington, DC: World Resources Institute.

Ribot, J. 2007. Representation, citizenship and the public domain in democratic decentralization. *Development*, 50(1), 43–49.

Ribot, J. 2011. Vulnerability before adaptation: toward transformative climate action. *Global Environmental Change*, 21(4), 1160–1162.

Ribot, J. 2013. Vulnerability does not just fall from the sky: toward multi-scale pro-poor climate policy. In: M.R. Redclift & M. Grasso, eds. *Handbook on climate change and human security*. Cheltenham: Edward Elgar, pp. 164–199.

Ribot, J. and R. Mearns. 2008. Steering community driven development? A desk study of NRM choices. *WRI Environmental Governance in Africa Working Paper*. Available from: http://www.wri.org/publication/steering-community-driven-development

Ribot, J. and N.L. Peluso. 2003. A theory of access: putting property and tenure in place. *Rural Sociology*, 68(2), 153–181.

Ribot, J., A.R. Magalhães and S. Panagides, eds. 1996. *Climate change, climate variability and social vulnerability in the semi-arid tropics*. Cambridge: Cambridge University Press.

Ribot, J., A. Chhatre and T. Lankina. 2008. Institutional choice and recognition in the formation and consolidation of local democracy. *Conservation and Society*, 6(1), 1–11.

Roberts, J.T. and B.C. Parks. 2007. *A climate of injustice: global inequality, North-South politics, and climate policy*. Cambridge, MA: MIT Press.

Rodney, W. 1973. *How Europe underdeveloped Africa*. London and Dar es Salaam: Bogle-L'Ouverture Publications.

Rose, N. 1996. The death of the social? Re-figuring the territory of government. *Economy and Society*, 25(3), 327–356.

Rose, N. 1999. *Powers of freedom: reframing political thought*. Cambridge: Cambridge University Press.

Sachs, W. 2008. Climate change and human rights. *Development*, 51, 332–337.

Sato, J. 2013. Resource politics and state-society Relations: why are certain states more inclusive than others? Mimeo, May 2013.

Schipper, E.L.F. 2010. Religion as an integral part of determining and reducing climate change and disaster risk: an agenda for research. In: M. Voss, ed. *Climate change: the social science perspective*. Wiesbaden, Germany: VS-Verlag, pp. 377–393.

Schroeder, R. 1999. Community forestry and conditionality in the Gambia. *Africa*, 69(1), 1–22.

Scott, J. 1976. *The moral economy of the peasant*. New Haven: Yale University Press.

Sen, A. 1980. Famines. *World Development*, 8(9), 613–621.

Sen, A. 1981. *Poverty and famines: an essay on entitlement and deprivation*. Oxford: Oxford University Press.

Sen, A. 1984. Rights and capabilities. In: A. Sen, ed. *Resources, values and development*. Oxford: Basil Blackwell, pp. 307–204.

Sen, A. 1997. Editorial: Human capital and human capability. *World Development*, 25(12), 1959–1961.

Sen, A. 1999. *Development as freedom*. New York: Knopf.

Shackelton, S.E. and C.M. Shackelton. 2012. Linking poverty, HIV/AIDS and climate change to human and ecosystem vulnerability in southern Africa: consequences for livelihoods and sustainable ecosystem management. *International Journal of Sustainable Development and World Ecology*, 19(3), 275–286.

Shanin, T. 1971. *Peasants and peasant societies: selected readings*. Harmondsworth: Penguin.

Smith, N. 2006. There's no such thing as a natural disaster. Blog posting. Available from: http://libcom.org/library/there%E2%80%99s-no-such-thing-natural-disaster-neil-smith [Accessed 2 January 2014].

Smucker, T.A. and B. Wisner. 2008. Changing household responses to drought in Tharaka, Kenya: vulnerability persistence and challenge. *Journal Compilation, Overseas Development Institute*. Oxford: Blackwell.

Somers, M. 2008. *Genealogies of citizenship: markets, statelessness, and the right to have rights*. Cambridge: Cambridge University Press.

Stiglitz, J.E. 2001. Foreword. In: K. Polanyi. *The great transformation*. Boston: Beacon Press, pp. 7–17.

Swift, J. 1989. Why are rural people vulnerable to famine? *IDS Bulletin*, 20(2), 8–15.

Swyngedouw, E. 2005. Governance innovation and the citizen: the Janus face of governance-beyond-the-state. *Urban Studies*, 42(11), 1991–2006.

Taylor, C. 1994. The politics of recognition. In: A. Guttman, ed. *Multiculturalism*. Princeton: Princeton University Press.

Turner, M.D. 2000. Drought, domestic budgeting and wealth distribution in Sahelian households. *Development and Change*, 31(4), 1009–1035.

Turner II, B.L., *et al.* 2003a. Illustrating the coupled human-environment system for vulnerability analysis: three case studies. *Proceedings of the National Academy of Sciences of the United States of America*, 100(14), 8080–8085.

Turner II, B.L., *et al.* 2003b. A framework for vulnerability analysis in sustainability science. *Proceedings of the National Academy of Sciences of the United States of America*, 100(14), 8074–8079.

Vaughan, M. 1987. *The story of an African famine: gender and famine in twentieth century Malawi*. Cambridge: Cambridge University Press.

Vayda, A.P. 1983. Progressive contextualization: methods for research in human ecology. *Human Ecology*, 11(3), 265–281.

Wallerstein, I. 1974. *The modern world system I: capitalist agriculture and the origins of the European world-economy in the sixteenth century*. New York: Academic Press.

Walsh-Dilley, M., W. Wolford and J. McCarthy. 2013. Under Review. Rights for resilience: bringing power, rights and agency into the resilience framework. Paper prepared for Oxfam America. Mimeo.

Watts, M.J. 1983a. *Silent violence*. Berkeley: University of California Press.

Watts, M.J. 1983b. On the poverty of theory: natural hazards research in context. In: K. Hewitt, ed. *Interpretations of calamity*. London: Allen Unwin, pp. 231–261.

Watts, M.J. 1991. Entitlements or empowerment? Famine and starvation in Africa. *Review of African Political Economy*, 51, 9–26.

Watts, M.J. and H. Bohle. 1993. The space of vulnerability: the causal structure of hunger and famine. *Progress in Human Geography*, 17(1), 43–68.

White House. 2006. The federal response to Hurricane Katrina. February 2006. Available from: http://www.whitehouse.gov/reports/katrina-lessons-learned.pdf [Accessed 5 December 2008].

Wilkinson, I. 2010. *Vulnerability in everyday life*. London: Routledge.

Wisner, B. 1976. Man-made famine in eastern Kenya: the interrelationship of environment and development. *Discussion Paper No. 96*. Institute of Development Studies at the University of Sussex, Brighton, England.

Wisner, B., *et al.* 2004. *At risk: natural hazards, people's vulnerability and disasters* (2nd ed.). New York: Routledge.

Wolf, E. 1981. *Europe and the people without history*. Los Angeles: University of California Press.

Wolford, W. 2007. Land reform in the time of neoliberalism: a many-splendored thing. *Antipode*, 39 (3), 550–570.

Yohe, G. and R.S.J. Tol. 2002. Indicators for social and economic coping capacity — moving toward a working definition of adaptive capacity. *Global Environmental Change*, 12(1), 25–40.

Jesse Ribot is a Professor of Geography, Natural Resources and Environmental Science, and faculty of Beckman Institute for Advanced Science and Technology at University of Illinois since 2008. Before 2008 he worked at World Resources Institute and taught at MIT. He has been a fellow at The New School for Social Research, Yale Program in Agrarian Studies, Rutgers Center for the Critical Analysis of Contemporary Culture, Max Planck Institute for Social Anthropology, Woodrow Wilson Center for International Scholars and Harvard Center for Population and Development Studies. He is an Africanist studying local democracy, resource access and social vulnerability.

The problem of property in industrial fisheries

Liam Campling and Elizabeth Havice

Fisheries systems are widely considered to be 'in crisis' in both economic and ecological terms, a considerable concern given their global significance to food security, international trade and employment. The most common explanation for the crisis suggests that it is caused by weak and illiberal property regimes. It follows that correcting the crisis involves the creation of private property rights that will restore equilibrium between the profitable, productive function of fishing firms and fish stocks in order to maximize 'rent'. In this approach, coastal states are seen as passive, weak, failed and/or corrupted observers and facilitators of the fisheries crisis, unless they institute private property relations. This paper offers an alternative analysis by using the perspective of historical materialism to re-examine longstanding debates over the problem of property and its relation to ground-rent in industrial fisheries. It identifies coastal states as modern landed property, enabling an exploration of the existence of and struggles over surplus value, and drawing attention to the role of the state and the significance of the environmental conditions of production in understanding political-ecological conditions in fisheries. As on land, property in the sea is a site of social struggle and will always remain so under capitalism, no matter which juridical interest holds the property rights.

1. Introduction

It is well established that many marine fish populations and the fisheries production systems based upon them face dire and deteriorating ecological conditions. Some three-fourths of the world's fisheries are at or beyond 'full exploitation' (FAO 2010), indicating the likelihood that many fish populations, and the ecosystems of which they are a part, will decline (or continue to do so) with current and expanded levels of competitive extraction. The combination of growing demand for fisheries products and environmental change at scales from local fisheries to global climate change, will continue, if not intensify, such patterns.

Authorship of this paper is fully collaborative. We thank the anonymous reviewers, Henry Bernstein, Chris Carr, Gavin Capps, Ben Fine, Louise Fortmann, Phil Steinberg and, especially, Dick Walker for comments on prior drafts. We thank Amanda Henley for carefully making the maps. We also thank participants at the following events for feedback and discussion: 'Symposium on economic (in)security and the global economy' at Queen Mary University of London, 'Grabbing green' at the University of Toronto, 'Food sovereignty: a critical dialogue' at Yale University, 'International relations, capitalism and the sea' at Birkbeck, University of London, and Agrarian Change Seminar at SOAS, University of London. Remaining errors are ours.

The significance of these trends extends beyond the realm of the ecologic. In 2009, fish accounted for 16.6 percent of the global population's intake of animal protein; for about 3 billion people, fish provides 20 percent of their animal protein intake. The per capita supply of fish for food has skyrocketed from less than 3 kilograms in 1950 to 17.2 kilograms in 2009. Capture fisheries accounted for around 58 percent of total fish production in 2011, while aquaculture systems supplied the remaining 42 percent (FAO 2012a). Fish are used as fertilizer and fish meal fattens chickens and pigs, which in turn are for human consumption. In international exchange, fish had a first-sale value of USD 109 billion in 2010 (FAO 2010) and an estimated 120 million people depend directly on commercial capture fisheries for their livelihoods (World Bank 2010).

These patterns present a 'crisis'; a narrative frequently deployed by institutions, non-government organizations and scientists to typify the combination of ecological decline and the economic significance of fisheries systems, and to relate the urgency of rectifying the fishing and management practices underlying them. Most recently, the World Bank has identified 'weak' fisheries management and the lack of private property rights as key causes of the fisheries crisis; it has spearheaded projects and regulatory guidelines to rectify such 'poor governance' (World Bank 2009). Likewise, the United Nations (UN) Special Rapporteur on the Right to Food has made a case for reforming 'ocean grabbing': aggressive industrial fishing tactics that divert fisheries resources away from local and national populations (De Schutter 2012). The Special Rapporteur, alongside the Food and Agriculture Organization (FAO), call on *the state* to strengthen tenure to ensure the right to food and right to fish (FAO 2012b).

This nexus of crisis, property and the state is grounded in familiar (and contested) debates over 'good governance'. In fact, the logic that 'poor governance' is linked to economic inefficiencies and drives biological decline in fisheries systems played an important role in founding the oft-cited concept of the 'tragedy of the commons' (Gordon 1954, Scott 1955, Hardin 1968), a narrative as prevalent today as when first debated.

But what are property forms in fisheries systems, and how is property related to the socio-economic and ecological outcomes of production? This paper explores this question in industrial fishing activities that take place within states' 200-nautical-mile exclusive economic zones (EEZs) and outside of the 12-nautical-mile 'territorial seas' (see Figure 1a).[1] Fishing inside EEZs yields around 90 percent of global marine catch – the remainder is caught in the high seas (OECD 2010). Industrial fishing is frequently vilified for vacuum cleaning the oceans of fisheries resources. Meanwhile, states are chastised for failing to efficiently regulate industrial fishing. Property relations in coastal and small-scale fisheries are equally contentious but not addressed here, not least because the often ingenious solutions to the spatial challenges of managing coastal environments have been well documented (e.g. McCay and Acheson 1987, Cordell 1989, Ostrom 1990).

To better understand property in industrial fisheries, we situate our analysis in a rich intellectual tradition around the political economy of landed property and rent, which has generally been developed in relation to agriculture, mining and real estate (e.g. Patnaik 1983, Haila 1990, Fine 1994, Harvey 2006). By contrast, mainstream approaches to property and rent have played the dominant role in shaping fisheries policy (e.g. Gordon 1954, Homans and Wilen 1997). In Section 2 we explore the orthodox position on property in industrial fisheries, demonstrating that EEZ fisheries are not 'commons' or 'open access',

---

[1]The 'territorial seas' have a higher level of *political* sovereignty than the rest of the EEZ and are often reserved for smaller-scale fishing activities.

but state property. Section 3 works through the historical emergence of property relations in the ocean, using Marx's category of modern landed property as a conceptual tool. Section 4 explores more fully the relationship between property and rent in EEZ fishing. In Sections 3 and 4, we draw on examples from industrial tuna fishing to illustrate how actually-existing property relations shape struggles over surplus value in a production system built off of a resource that is mobile, renewable and exhaustible. The analysis supports a central claim of this paper that property is a site of social struggle over surplus value, and will always remain so under capitalism, no matter which juridical player holds those property rights. It contributes to studies of property and rent by drawing attention to the significance of the state as the 'owner' of natural resources in exclusive economic zones and identifying how the features of the resource are reflected in the institutions and relations associated with property. Importantly, this paper is not a 'complete explanation of property dynamics at sea, but an exploratory initiative and an invitation for further grounded research on fisheries dynamics through the lens of materialist approaches to property and rent.

2.   The tragedy of the orthodoxy: conceptual confusion around the commons

Today's dominant approach to fisheries management centers on codifying and deploying 'rights' to create long-range incentives for economically 'efficient' production practices. These origins emerged after World War II: US government scientists aimed at managing catch to achieve the biological objective of maximum sustainable yield, while economists promoted limited entry into a fishery by creating and allocating durable fishing rights (Scheiber and Carr 1998). Both focused on developing ways to limit catch. Over time, attention has trained on the development of quotas (and most recently, markets for quotas) because a quota can act as a property right.

This management orthodoxy is an attack on open access conditions – the lack of recognizable, enforceable, and often private, property rights – that predates Hardin's (1968) tragedy of the commons thesis. Gordon (1954) argued that problems in fisheries stem from the fact that open access conditions yield no economic rent. Echoing Ricardo's approach, Gordon (1954) saw variation in rent as a reflection of the productivity of a fishing ground relative to others and asserted that without property rights, fish left uncaught have no value to the fisher because they can be caught by another. Under such conditions, a competitive race for fish ensues: fishers over-invest in various modes of fishing effort (e.g. bigger boats and nets), which when used erode natural resource rent. The logic follows that withholding fishing today is an investment that can only pay off if the fisher has some form of property right that excludes others from extraction and is related to future fishing effort. These ideas provide the conceptual justification for designating property rights based on a proportion of fish (quotas) or spatial areas (territorial use-rights); they are foundational to a recent emphasis in mainstream fisheries economics and management circles for incremental movement towards more 'complete' rights, culminating in private property rights, where possible. In theory, maximizing natural resource rent is critical because lost rents drive overexploitation. Social goals beyond economic efficiency are not explicit; economic efficiency is assumed to generate socially desirable conditions by stimulating productivity and limiting waste.

The World Bank is chief among the institutional players steeped in this orthodox approach. Rent calculations from a World Bank (2009) modeling exercise quantified potential and actual economic rents in marine fisheries sectors; the resulting figures are the cornerstone and empirical justification for a swathe of new Bank-led fisheries reform projects.[2] The analysis is guided by what the Bank calls 'economic rents in the traditional (Smith-Ricardian) sense' (59, fn 4). This is problematic in two ways. First, Adam Smith and David Ricardo developed entirely different theories of rent. There is not space here to elaborate these distinctions in detail, but a simplified overview is instructive. Smith referred to 'neat rent' " ... as the produce of those powers of nature, the use of which the landlord lends to the farmer", ... and is what "remains free to the landlord" ... after deduction of [production costs], estate maintenance costs "and other necessary charges"' (Smith as cited by Gee 1981, 5). It is significant that he recognized that the potential for rent emerges from property ownership (the concept of absolute ground-rent, see below), which he delineated from that rent which arises from the greater application of capital – which he termed 'interest' (Patnaik 1983). By contrast, Ricardo defined rent as 'that compensation which is paid to the owner of land for the use of its original and indestructible powers' (Ricardo 1996, 46). While Ricardo recognized the juridical role of land ownership, his theoretical emphasis is on the powers of the soil – rent as 'a technical-economic phenomenon' (Haila 1990, 277). Ricardo ignored that land ownership has an economic effect. In his vision, rent is only a reflection of variable natural qualities of resources (land) because in his theorization, prices of production and values are identical. In short, his conceptualization *cannot* accommodate the economic impact of property ownership (Patnaik 1983).[3]

Second, and related, the Bank's definition diverges from Smith and Ricardo – and the significance of their distinct treatment (or lack thereof) of property relations – in favour of the neoclassical understanding of rent that dominates the modern field of economics. In the report, the Bank initially asserts that economic rents (used interchangeably with resource rent) are a measure of net economic benefits (profitability after accounting for costs) from, in this case, the harvest of wild fish stocks (World Bank 2009, 30). Later, in the appendix, the Bank elaborates the concept of economic rent in fisheries, defining rent as the *payment* to a factor in fixed supply, and illuminating that economic rent represents rental income to the owners of the factor in fixed supply, who rents it out to users. In fisheries, 'fixed supply' is mediated by either open access conditions or restrictive management regimes and rent is reduced to a 'normal' or 'natural' payment for the differential productivity of resources (World Bank 2009, 57–9), ignoring the role of ownership in shaping the distribution of economic surplus (i.e. absolute rent, a critical component for Smith).

The Bank's project is not to think about the distribution of rents, but to use its definition of rent to measure 'economic performance' in marine fisheries: the difference between the *maximum* rents obtainable from fisheries and the *actual* rents currently obtained. Rent is a technical market payment which the Bank quantifies using globally aggregated data on maximum sustainable yields in fisheries and biomass carrying capacity, as well as the global industry's production costs, revenues and assumed profit rates. The Bank undertakes this analysis to reveal the economic and biological health of the global fishery, the first step towards the 'economic objective of maximizing the net economic benefits (sustainable rents) flowing from the fishery' (World Bank 2009, 29). Accordingly, the analysis offers an 'economic justification for fisheries reform' by

---

[2]For example, in 2012 the Bank launched the 'Global Partnership for the Oceans', an alliance of hundreds of governments, international organizations, civil society groups and private sector interests to address threats to the health, productivity and resilience of the oceans.
[3]For more on the differences between the two see Gee (1981) and Patnaik (1983). For further debate on rent theory see Fine (2013).

revealing that lost rent in fisheries is around USD 50 billion annually and that current practices in fisheries have drained trillions of dollars from the global economy over the last 30 years. The World Bank attributes these lost rents to two interrelated factors: depleted fish stocks and fleet over-capitalization in which too many fishers chase too few fish (World Bank 2009, xix).

How, then, to combat these two rent-depleting factors? Though the size, scale and structure of fisheries systems are highly diverse, the World Bank uniformly ties its problem definition and solution to the state. On the former, fish stock depletion constitutes a loss of the *nation*'s 'natural capital', and thus a loss of *national* wealth. On the latter, recovering and capturing lost rents is also a project of the state. According to the report:

> Most marine wild fisheries are considered to be property of nations. Governments are generally entrusted with the stewardship of these national assets, and their accepted role is to ensure that these assets are used as productively as possible, for both current and future generations. ... The scale of these losses – the sunken billions – justifies increased efforts by national economic policy makers to reverse this perennial haemorrhage of national and global economic benefits. (World Bank 2009, 50)

The Bank identifies the state as responsible for strengthening property rights in order to restore an economic logic to the fishing sector (World Bank 2009, xxi). This is because 'the "tragedy of the commons" suggests that where forms of open access persist (which is the case in many of the world's fisheries), profits will be dissipated' (World Bank 2009, 38). Accordingly, the Bank estimates that billions in lost 'economic rent' will be recovered globally if states define, strengthen and manage property rights so that 'biomass (the fish stock) and the capital stock (fleet) are in equilibrium' (World Bank 2009, 40).

Imprecision around property is rife in this narrative, as in the assertion of a 'Smith-Ricardian' theory of rent. The World Bank (2009) at once defines fisheries as: property of nations (50), as operating under open access conditions (i.e. no one's property) (xxi), and as vulnerable to the tragedy of the commons in which forms of open access persist (suggesting that common property and open access are the same thing) (38). This inexactitude over property in fisheries management is common and underwrites the urgency to create and strengthen property systems, especially *private* property rights (Bromley 2008a),[4] but it is notable that the Bank reconciles its own inconsistency over property relations in fisheries through a particular lens of 'governance': states have failed to act on their control over fisheries as property, and have thus contributed to the fisheries crisis by creating *conditions* of open access. In this vision, before property is allocated to industry, 'property' – particularly state property – is not a powerful institution; certainly not powerful enough to achieve the goal of maximizing rents from fisheries.

3. The emergence of state-landed property in the oceans

3.1. *Modern landed property and extractive industries*

The orthodoxy is not the only way to approach the emergence and characteristics of property – and rent – relations in the ocean. Insights are plentiful from classic and ongoing

---

[4]On conceptual confusion on property regimes and natural resources, see Schlager and Ostrom (1992); for a broad overview of debates over access and property in fisheries, see Campling et al. (2012).

debates in political economy over property relations in agriculture and extractive industries. Our thinking is guided in particular by Capps' (2010, 2012a, 2012b) development
of the category of modern landed property in his path-breaking work on mining in
South Africa.

For Marx, the category of modern landed property is an essential relation of capitalism.[5] This relation forms the basis for a 'third class' of landowners that is separate from,
but can only exist in concert with, capitalists and workers. Modern landed property's
role in capitalist production is as lessor of land, issuer of land and extractor of ground-
rent (Neocosmos 1986), whether the land is used as a *means* of production (e.g. agriculture,
mining) or as a *condition* of production (e.g. as a site for factories, retail, server hubs,
housing). The rise to dominance of capitalism in Britain created the conditions for the
class modern (rather than feudal) landed property as 'a specific historical form' (Marx
1976, 1981). This process saw the simultaneous 'freeing' of the immediate producers
from their feudal obligations to landowners as producers of surplus on the land and the dis-
embedding of *feudal* landed property's dual function of control of the land *and* agricultural
production.[6] The separation of these functions established the 'class basis for a new collectivity of landlords defined by their possession of (bourgeois) property rights *alone*' (Capps
2012a, 317, emphasis added).

Once a landowning class has established juridical property rights (normally bestowed
by a sovereign capitalist state), it mediates capital's access to landed resources.
However, unlike capital and wage labor, modern landed property exists *outside* of the
process of production (Marx 1981, 776). At the same time, landed property's legal claim
to 'particular portions of the globe as exclusive spheres of their private will to the exclusion
of all others' allows it to extract a portion of the surplus value created in the production
process (Marx 1981, 752, 908). In the abstract, this portion takes the form of *ground-
rent*, i.e. 'the form in which landed property is … valorized' (Marx 1981, 756). In
Section 4 we detail the distinction between this conceptualization of ground-rent and the
Bank's neoclassical vision of natural resource rent.

Building on Ben Fine's (1994) theoretical and comparative empirical work on mining,
Capps develops the category of landed property to explain particular configurations of
control and contestation over mineral access rights in South Africa during and after Apartheid. Drawing on Banaji's (e.g. 2010) insight that multiple forms of exploitation can exist
under a dominant mode of production (e.g. the persistence of forced labor under capitalism), Capps (2010) shows logically and historically that though landed property is most
commonly a private landlord, the phenomenal form that it takes can be any juridical
actor or organization, including the state (Fine 1994, Harvey 2006, Capps 2010), as we
suggest is the case in EEZ fisheries.

Capps' work on the platinum mining industry, the chieftaincy and the state in South
Africa exemplifies sensitivity to the 'historical conditions of existence of landed property'
(Fine 1979, 248). He develops the category of 'tribal-landed property' to denote the elite
(chieftaincy's) capture of property rights from occupants ('the tribe') residing on land of

---

[5]Along with capital and wage labor. It is beyond the scope of this paper to discuss fisheries production
sensu stricto or other dynamics of the capital-labor relation.
[6]In contrast to approaches to political economy that are not historicised, the 'modern' in landed property denotes that the category exists only under conditions of generalised commodity production.
Landed property has existed through history, but its characteristics under capitalism are specific, at
least in 'its purely economic form' (Marx 1981, 751, 755), i.e. the struggle for surplus value in the
form of ground-rent.

interest to mining capital.[7] In addition to the similarity of a non-private individual perform-ing the class role of landed property, his category of tribal-landed property also lends lessons to property relations in EEZ fisheries because the juridical right is mediated by ongoing and multi-scalar social and political struggles between classes (discussed below).

Approaching property from a class-analytic lens stands in stark contrast to mainstream economics-driven accounts of EEZ fisheries. This point has not gone unnoticed: while not using Marx's categories, legal scholar Bromley (2008b) laments that models and policy pre-scriptions in fisheries economics fail to account for relations between capitalists and land-owners. He notes that, 'Just as the owner of agricultural land is paid rent by a tenant, the owner of the wealth of ocean fish must be paid for surrendering those fish to the private sector' (2008b, 43). Further,

> the standard narrative fails to tell us whether or not fishing firms are actually paying the owner of the fish for the benefits received by firms harvesting our fish and then selling them on the market. Of course, fisheries economists feel no need to raise this little detail, because they apparently believe – after approximately three decades of state property under EEZs – that no one owns the fish until those critters have been captured. (41)

Likewise, Rieser (1997) points to class-based complications in private property models for fisheries. She notes that while such models (e.g. transferable quotas) include exclusivity and alienation rights, management rights remain vested in the government. Writing from a legal perspective, she defines quotas as usufruct: the right to use and enjoy the profits and advan-tages of something belonging to another. Because quotas are usufruct, she argues, they reduce the right holder's incentive to invest in the resource for the long term or to ameliorate rent-seeking behavior.

Though these accounts highlight that a 'state-ownership' relation underlies the for-mation of private property systems in fisheries, each lacks systematic treatment of the role of the state in fisheries management. This gap, along with the World Bank's assertion that state property is only 'efficient' when it is used to develop private property, ignores that states *are* economic actors in EEZ fisheries. Most directly put,

> To argue that fisheries have collapsed because individual fishermen failed to control their be-havior is to whitewash the role of governments and scientists in establishing policies that encouraged the building of the ultimate industrial capitalist system, the global fishing industry. (Finley 2011, 8)

In short, states and firms mutually created 'the tragedy' driven by the competitive dynamics of capital accumulation. Despite the terminology of 'tragedy of the commons', in the absence of a legal conception of property instituted by the state, fisheries are an open access resource: no one's property. But how then did fish, which live in the sea outside of the terrestrial-territorial boundaries of the state, come to be state property?

3.2. *The historical emergence of property in the sea*

Efforts to define property in the sea must be understood as projects associated with territory making and unmaking. Examples from processes of early state formation in Europe are

---

[7]Tribal-landed property could be deployed, with modifications to reflect particular historical and social conditions, to analyze some coastal fisheries managed as common property and overseen by customary authorities (often social and political elites).

illustrative. The ascendency of sea tenure between the thirteenth and seventeenth centuries demonstrates that rights in the sea reflect the power that certain social classes have to influence others' ability to access resources. In parts of Europe in the Middle Ages, feudal law determined marine tenure practices and feudal social relations bestowed ownership and use-rights for whales, bluefin tuna and other socially valuable species (Cordell 1989, Longo and Clark 2012).

Despite this, Grotius' work in the early seventeenth century is most commonly marked as the beginning of the struggle over property relations in the sea. His proposal in *The free sea* to ensure open access conditions across the oceans was an attempt to support the Dutch capitalist trading regime and prevent rival European states from gaining control of shipping lanes and increasing English appropriation of herring fishing grounds. As part of this effort, Grotius maintained that fisheries were inexhaustible and should be open to all peoples (Grotius 1916). Throughout the seventeenth century, major fishing nations – England, France, Spain, Portugal and the Netherlands – were propelled into marine boundary contests that have left lasting geopolitical legacies (Cordell 1989). In short, by the time Grotius was developing his treatise on property in the sea, pressures to define ocean use-rights, territory and ownership were deeply entangled with the development of emerging production practices, trade patterns, shipping lanes and geopolitics in the world market of early capitalism. For example, in the context of the transition to capitalism in Sicily in the 1600s, merchant bankers took private control of bluefin tuna fisheries that were historically organized around feudal ownership and property structures (Longo and Clark 2012). Along the Scottish coast in the 1700s, entire fishing towns were built by landlords (modern landed property) and peopled by displaced crofters (Howard 2012).

The long process of creating property relations in the sea is most recently marked by the largest single enclosure in history: state-sovereignty over exclusive economic zones (EEZs) (see Figure 1a). In the 1970s, multiple individual states declared their EEZs. Later this customary law was institutionalized when most states ratified the 1982 United Nations Convention on the Law of the Sea (UNCLOS) which instills state sovereignty over national EEZs and the resources in them by recognizing a series of rights that individual states have over fishing activities including, *inter alia*, the right to: charge access and fishing fees (ground-rent) to fishing firms, define the conditions of production (i.e. resource management) and prohibit or exclude fishers. Since property is a bundle of rights implemented and conferred by social relations and reflected in juridical practices (MacPherson 1983), these sovereign rights mean that EEZ fish are state property.

Overall, then, marine fisheries have been moving towards the enclosure of open access regimes for hundreds of years. This has been far more significant in ocean management than the polemic that 'open access' is rampant and the driver of problems in fisheries systems, though several have noted that historical property relations in fisheries have been designed as strong, weak, private or open access according to the interests of the group doing the defining (McCay 1981, Cordell 1989). However, policy makers, though aware that fisheries are state property, often do not treat them as a part of any property relation until *private* property regimes are applied in the sector (Mansfield 2008). State property is not ignored, but is at best seen as a step toward 'real' (private) property rights – and the related individual incentive for profit, stewardship and improved management – and at worst as a cover for open access (Mansfield 2004).

This clarification is not to belie the technical complexity of creating, allocating and enforcing property relations in fisheries systems, a complexity that emerges in part from the material characteristics of fisheries. The lack of physical boundaries in the oceans and the inability to fence off individual fish makes it difficult for states, firms and fishers to translate sovereign control over fisheries into its terrestrial equivalent. Instead, it clarifies that EEZ fisheries are a part of state property regimes and that overfishing, which intensified with new industrial technologies and techniques from the 1950s onwards, has occurred in the presence of property relations. As duly noted but insufficiently conceptualized by the World Bank, the state is the starting point for understanding property in industrial fisheries. Yet the complex roles and multiple logics of states have largely been elided unsystematically into the story of fisheries crisis. To contribute to addressing this gap, we use the category of modern landed property to think through the role(s) of the state in EEZ property relations.

3.3. *State-landed property and industrial tuna fisheries*

Whether retaining control over fisheries property relations or using them to create and deploy private property regimes, states do not act solely as functional rent maximizing agents, and are unlikely ever to do so. Instead, they are active players in struggles over the creation and distribution of surplus value from the production of fisheries commodities, and are involved in mediating domestic and foreign interests and the relations among them. As state-landed property, coastal states sit at the nexus of rent appropriation and other distributional struggles around surplus value, (perceived) 'national interest',[8] geopolitics, resource management and industry regulation in EEZs. Our discussion suggests the historical and political naiveté of imploring states to promote good governance by instating and enforcing idealized private property relations, and the limitations of proposals for 'equilibrium' conditions and 'rent maximization' in EEZ fisheries. Further, drawing attention to property as a site of social struggle informs the potential malleability of modern landed property as an analytical category. We also highlight how the environmental conditions of production specific to particular fisheries inform the institutional form of property relations.

We explore these dynamics by investigating the state as landed property in industrial tuna fisheries at the national scale (within EEZs) and international scale (following the fish as they move between and outside of EEZs). Our examples cannot be generalized to explain the actions that states take in their position as landed property in all marine fisheries. In fact, methodologically, the category 'landed property' must be worked through the 'historical conditions of existence' to illuminate relations at play. We see the state's role in shaping the conditions of production – such as fisheries management – not as driven solely by functional logics such as: rent appropriation, guaranteeing the existence and reproduction of the capital-relation, and/or legitimating itself in the eyes of civil society. Instead we see the state's role as mediated by socio-economic and political struggles, ideologies and bureaucratic realities (on these dynamics in relation to questions of nature, see O'Connor 1998). Conflict and collaboration *within* the state (e.g. among different state agencies), as well as among the state, capital and other social forces, can lead to contradictory policy outcomes and unintended consequences, making a singular theorization of the actions and logics of 'the state' in the management of fisheries implausible.

---

[8]We use scare quotes here because the national interest is an ideological construct: what is purported to be of interest to 'the nation' is very often of interest to the ruling elite or elements thereof.

The industrial tuna sector in the global South offers only one of many potential ways that a state can act upon its role as landed property in fisheries systems. In this example, coastal states endeavor to maximize their appropriation of rent in the name of perceived 'national interest' and to assert strength in international politics, though – as we show here and elsewhere (e.g. Havice and Campling 2013) – doing so involves a range of compromises and contradictions that both serve and limit these broad objectives as well as those of (differentiated) capitalist enterprises. By contrast, in other fisheries, the state may choose to actively facilitate capital accumulation through a 'Ricardian reform' (Capps 2012a), in essence eliminating the antagonistic capital-landed property relationship by subsidizing national fishing enterprises and limiting rent appropriation (e.g. Alaskan Pollock and other US fisheries; see MRAG 2013). We emphasize that our approach requires attention to specific historical conditions, which in turn can help to inform a broader understanding of the multiple logics of state interventions in fisheries, and how they are linked to those conditions.

Under open access conditions, no actor assumes the role of landed property. Thus, the potential for capture of ground-rent is eliminated and this portion of surplus value is captured as surplus profits by fishing enterprises. The development of property relations through the EEZ – an 'alien force' that disrupts the movement of capital in the sea – marked the possibility of states capturing ground-rent, primarily in the form of an access payment, which firms pay to fish in a state's EEZ (see below). Following UNCLOS, the struggle over surplus profits was no longer exclusively between capitalists.

EEZ declarations can be situated in the world-historic context of 'Third Worldist' demands to correct asymmetries in the world economy, e.g. the 1970s' call for a New International Economic Order and assertions of 'resource nationalism' (Schurman 1998). UNCLOS was seen as an institutional mechanism to redistribute marine resources away from historically dominant distant water fishing fleets (primarily from the global North) to developing country coastal and island states (Pontecorvo 1988). It had two major implications for tuna fisheries. First, countries declared sovereignty over the vast majority of the planet's tuna stocks in the largest and most rapid state-led enclosures in human history. Second, newly established sovereignty over EEZs gave coastal states a juridical basis as landed property: suddenly, they were able to extract a portion of surplus value (ground-rent) from industrial fishing capital previously fishing without regulation or rent payment.

Optimism over the potential for coastal developing states to use their new sovereignty to appropriate ground-rent was quickly complicated by economic and political power, 'variables that critically mediate the relationship between property rights and income distribution' in the sector (Schurman 1998, 133). The appropriation of nature in capture fisheries in EEZs is mediated through rent relations marked by struggles between capital and landed property evident in (1) the difficulties of determining and realizing property relations over a mobile resource in the sea, and (2) the geo-economic and political power of states representing (and in some cases subsidizing[9]) the profit-seeking interests of industrial fishing enterprises. The current ecological status and production dynamics of tuna fisheries are products of four decades of struggle over control of property and appropriation of rents that play out simultaneously in international, national, public and private forums.

---

[9] A form of redistributive rent (Walker 1974).

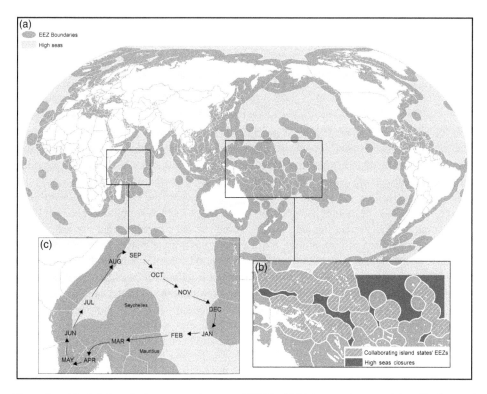

Figure 1. (a) global exclusive economic zones (EEZs); (b) high seas closures; (c) stylized annual tuna migration.

3.3.1. *Example 1: Environmental conditions of production and negotiated sovereignty*

The ecological characteristics of tuna – that they are a highly migratory species that transcend any single state's jurisdiction – mean that multiple states can claim an interest in resources inside of individual states' EEZs. This shapes, and offers the potential to both erode and augment, the powers that individual states confer through sovereignty and, in turn, the ways that landed property affects capitalist accumulation in the oceans. We refer to these dynamics as 'negotiated' national sovereignty:[10] an ongoing, inter-state, political process of determining the bundle of rights associated with sovereignty. Three historically sequential examples illustrate the relationship between the environmental characteristics of the resource and negotiated sovereignty.

First, conflicting approaches to sovereignty over, and management of, highly migratory species – especially tuna – were central in UNCLOS negotiations (Carr 2004). On one hand, distant water fishing nations argued that because highly migratory species move from one country's EEZ to another (as well as into the high seas) no single state has sovereignty over them and, as such, access should be open to all. On the other hand, developing coastal states argued that they had sovereign control over any fish while in its EEZ (Joseph 1977). Perhaps the most dramatic example of how these tensions played out comes from the US, whose fleet pioneered industrial tuna fishing in the Eastern and Western and Central

---

[10]This speaks to efforts to theorize state sovereignty (including over transboundary resources) as formulated with extra-territorial influences (see e.g. Ong 2006, Lunstrum 2013).

Pacific regions. The US opted out, and to date is still not a signatory, of UNCLOS; part of its initial reluctance was refusal to accept coastal state sovereignty over highly migratory tuna stocks. But how then could its large fleet retain access to tuna fishing grounds?

At first, the US state bluntly rejected UNCLOS principles and used geo-economic and political power to contest coastal states' efforts to regulate US boats' fishing activities (Kronmiller 1983). In response, coastal states apprehended US vessels fishing without payments; but this did little to deter the fleet because the US government reimbursed all of the expenses incurred by what it defined as 'illegal' seizures of the US vessels. Further, the value of this penalty was subtracted from foreign aid that would have otherwise gone to the coastal state (Van Dyke and Nicol 1987), and the 'Pelly Amendment' prohibited the US from providing defense assistance to any country that seized or fined a US vessel for fishing beyond 12 miles from its coast (Hollick 1981). The US only began to respect the EEZ when Kiribati threatened to lure Soviet fleets into its waters at the height of Cold War politics (Teiwaki 1987).

Second, UNCLOS requires that state sovereignty over highly migratory species like tuna is realized through *cooperative* management among 'resource owning' states, and 'resource using' distant water fishing states (Article 64, 65). To comply, the UN adopted the Fish Stocks Agreement in 1995, which outlines guidelines for creating regional fisheries management organizations (RFMOs). In these organizations, coastal and distant water fishing states *jointly* determine management provisions for highly migratory stocks; provisions that often apply both to the open access high seas and *inside* EEZs.

In this context, in the Western and Central Pacific, Pacific island countries safeguarded sovereignty with a provision proposed to and passed in the RFMO Convention: all agreements are to be undertaken 'without prejudice to the sovereign rights of coastal states for the purpose of exploring and exploiting, conserving and managing highly migratory fish stocks within areas under national jurisdiction' [Convention on the Conservation and Management of Highly Migratory Fish Stocks in the Western and Central Pacific Ocean 2004, Article 10 (1)]. But Pacific island state sovereignty remains up for negotiation. For example, catch *allocation* decisions, as well as conservation mandates, are made by consensus among RFMO members, suggesting that Pacific island countries can be subject to RFMO decisions that will impact fishing practices in their own waters (Havice and Campling 2010).

Inter-state relations do not only constrain sovereignty, they can also be deployed to strengthen the power of landed property vis-à-vis capital. For example, in the 2000s, a group of eight neighboring Pacific island countries used their role as landed property to reconfigure property relations *outside* of their EEZs. These countries collectively prohibited any vessel fishing inside *any* of the countries' EEZs from fishing in pockets of the high seas nestled between their interlocking EEZs (Figure 1b). By blocking fishing in the high seas pockets, they challenged international law that designates the pockets as nobody's property and channeled fishing effort into their EEZs where they can charge for and regulate fishing (Havice 2013). Fishing enterprises must 'follow the fish' to survive commercially, so they comply. These three examples illustrate how state property and the struggles over distribution of surplus value mediate, and are mediated by, the natural features of tuna resources and unequal geo-economic and political power in the system of states.

3.3.2. *Example 2: Struggles over surplus value in tuna access arrangements*

Identifying, much less measuring, the exact sources of coastal state revenue captured from industrial fishing is riddled with problems. In practice, access revenue can be the result of redistribution through government subsidies and/or increased or eroded by global competition (reflected in fish or fish product prices), rather than being wholly generated through 'pure' ground-rent. However, individual coastal states attempt to deploy sovereignty to shift the struggle over surplus value in their favor. In return for direct and indirect revenue, individual coastal states sell (primarily) foreign firms access to their resource; the terms and conditions of the sale are determined through fishing access negotiations. Foreign states often intervene on behalf of their 'national' capital to negotiate (i.e. limit) the terms and extent of coastal state rent appropriation.

Immediately following the recognition of EEZs, coastal states' ability to capture ground-rent in access negotiations was largely overwhelmed by the economic and political power of industrial fishing capital (Schurman 1998). Foreign firms and states far more experienced in international negotiations than developing coastal states pursued a range of strategies to secure access under the new property regime. Distant water fleets kept the terms and conditions of access negotiations highly secretive and played individual coastal states off of each other to secure the most favorable terms of access. To support access negotiations for 'their' distant water fleets, foreign states often paid a portion of the access fee on behalf of industrial fishing capitals, a practice that continues today. For example, the EU engages in direct bilateral government-to-government tuna access negotiations with African states on behalf of (largely) French and Spanish fleets and heavily subsidizes these agreements. The US state has a multilateral access agreement with 14 Pacific island countries; the government pays a portion of the access fee – which comes in the form of development assistance – to secure fishing licenses for the US fleet. Japanese industry associations enter into agreements with coastal states, but Japanese government agencies, present and active in negotiations, formally and informally couple aid to access negotiation outcomes (Campling 2012a, Havice 2010, Havice and Campling 2010).

Coastal states have drawn from their position as 'landlord' to confront the hierarchy of states and corporate power in access arrangements. In the first instance, coastal states allocate an annual vessel license or a number of fishing days, rather than a durable private property right, to a fishing interest. This is not because they are illiberal actors, but because they seek to retain and increase control over fishing and associated rent appropriation, rather than permanently pass rights, and accumulation, off to foreign industry.

Over time, coastal states have deepened their individual and collective bargaining power as landed property to strengthen the terms and conditions of access agreements and their capture of surplus value. Seychelles, for example, individually ratcheted up demands on distant water fleets as the significance of the fishery in its EEZ became clearer through the 1980s and early 1990s (Campling 2012a). Others, as noted above, engaged in cooperative strategies – essentially creating a collectivity of landlords – to improve control over the tuna in their waters. In 1979, the independent Pacific island countries formed the Pacific Islands Forum Fisheries Agency (FFA), an intergovernmental organization devoted to strengthening regional cooperation so members can control and develop tuna fisheries to their benefit, especially in relation to foreign fishing capital.

In each example, coastal states deployed their position as landed property vis-à-vis industrial capital to capture an enhanced share of surplus value. One rough proxy for rent appropriation is the rate of return on the landed value of catch, a metric commonly used in negotiating arrangements. In industrial tuna fisheries in the Western and Central Pacific region, this increased from around 3 percent in the 1980s to over 10 percent in the 2010s as a direct result of coastal states deploying their sovereignty in the context of changing world-market and environmental conditions (Havice 2013). Coastal states' (in) ability to capture or increase their portion of surplus value over time indicates that resource access relations are a site of political and social struggle among states and firms, not a technical category determined by the market.

4. Back to rents

4.1. *What is ground-rent?*

We have so far conceptualized the state as modern landed property able to extract rent as a re-distributive portion of surplus value through this property relation alone. Thinking about property relations in this way tells a different story of the origins, dynamics and solutions to the fisheries crisis than that offered in the orthodox, liberal conception. Further illuminating the challenges to the World Bank's orthodox approach to the fisheries crisis requires delving into the limitations of its core, but conceptually underdeveloped, economic logic: a theory of rent as 'net economic benefits' (which parallels the neoclassical theory of natural resource rent summarized elsewhere, see Khan 2000).[11] In this vision, rent is a given and is determined by the market – it is a 'normal' or 'natural' payment for the differential productivity of land and other natural resources. As such, property rights play the technical role of producing a market for access to rents. In contradistinction, our discussion of state landed-property reveals 'rent' to be a distribution of surplus value that is political and contested.

In the abstract, ground-rent is the portion of surplus value taken by modern landed property. Rather than being ahistorical and 'normal', it 'is a product of society and not of the soil' (Marx, cited by Perelman 1975, 703). In other words, 'natural forces'[12] – whether especially fertile soil, river power or an abundant fishery – enter Marx's theoretical work to explain how one source of 'surplus profits' (those above the system-wide average) are produced and 'how the laws that apply to industrial capital in general are modified by the existence of landed property' (Fine 1979, 242).[13] It is precisely the dependency of rent 'upon historically and socially specific relations between capitalists and landlords' (Milonakis and Fine 2009, 67) that means that rent

> is immediately linked to the historical conditions of existence of landed property. Just as these conditions differ so the effects of landed property differ. *There is therefore no general theory of rent, nor can the conclusions reached for one instance in which a rent relation exists be automatically applied to others.* (Fine 1979, 248 emphasis added; see also Ball 1980)

In other words, examining rent relations requires empirical analysis of the specific forms that modern landed property takes in particular places and times.

Marxist political economy elaborates four types of ground-rent – all of which come from surplus profit produced in the labor process and appropriated from nature (Marx 1981, Moore 2011). We separate them to think systematically about the relationships at

---

[11]Neoclassical natural resource rent – unlike neoclassical monopoly rent, for example – is normally seen as having an efficiency-enhancing role, but only where property rights (not necessarily private ones) stop the dissipation of rent by encouraging producers (e.g. fishers) to focus on the marginal rather than the average cost of their activities, with the latter producing the so-called 'tragedy of the commons' (Khan 2000).

[12]Just as capital appropriates a 'free gift' from labor in the form of surplus labor, it appropriates free gifts of nature. While it costs money to fish and these expenses are the focus of mainstream economics' calculations of rent, the fish reproduce without capitalist investment. The gift of natural reproduction can be described as an 'ecological surplus' – 'the gap between appropriated and capitalized natures' (see Moore 2011, 129; for a partial extension to tuna fisheries, Campling 2012b). Debates on the 'free gifts' of nature, fish as a means of production and implications for ground-rent are beyond the scope of this contribution.

[13]For a theoretical discussion of how surplus profit is produced and then transformed into ground-rent, see Marx (1981, 780–7).

play, though in practice maintaining boundaries between them is an empirical impossibility. Capitalist ground-rent originates in legal ownership: 'For a thing to be sold, it simply has to be capable of being monopolized and alienated' (Marx 1981, 772, 785). Landed property is able to use legal claim to 'particular portions of the globe' to systematically block capital from equalizing the rate of profit (Harvey 2006). This property relation (upheld by the capitalist state) gives landlords the *active* possibility 'to create an artificial scarcity of land by their behavior as a class in demanding a positive return on *all* land in use' (Walker 1974, 53 emphasis added), regardless of whether it is of the lowest quality, as Ricardo's theory would have it. This is what Marx calls absolute rent (Harvey (2006), Patnaik (1983). This is distinct from differential rents (see below) and monopoly rent – where the latter 'results from conditions of selling and competition' – and each, unlike absolute rent, is *passive*, accruing 'to the landowner independent of his actions' (Walker 1974, 54).

Marx's category of differential rent I is a development of Ricardo's theory of rent.[14] It focuses on the natural conditions of land (and other 'natural forces') and how they form a basis for 'the exceptionally increased productivity of labor' (Marx 1981, 786). Marx argues that differential rent is *created* not by a natural force alone because natural force 'has no value, since it represents no objectified labor and hence no price' (Marx 1981, 787); the 'only precondition' for the form of surplus profit that is the basis for differential rent I 'is the inequality of types of soil' (or fishery), and 'the result of the varying productivity of equal capitals when invested on the land' (Marx 1981, 798, 812).

Marx also thought that surplus profit could be enhanced relatively through investment. The category differential rent II arises through different levels of investment in productivity where 'the variation in fertility is supplemented by difference in the distribution of capital (and creditworthiness) among the farmers' (Marx 1981, 815). After such an investment is made, it stays with the land, rather than in the hands of the capitalist.[15] If there are two plots of land of the same quality (differential rent I) and a capitalist farmer invests in one (for example, drainage ditches in one of the plots), that plot is more productive. But all of the benefit does not accrue to the farmer. The landlord can capture the surplus profit in the form of ground-rent, for example the next time the land lease is up. As such, differential rent II can be a major axis of struggle between landed property and capital.[16]

This simplified overview of Marx's categories indicates the complexity of theorizing 'rent' not as a given that is set by the market, but as a redistributive portion of surplus value (extracted from the process of capitalist production); a qualitative relation that is an always already existing component of the totality of capitalist development.[17] Since

---

[14]For Ricardo (1996, 46), rent is 'that compensation which is paid to the owner of land for the use of its original and indestructible powers'. While Ricardo assumed the qualities of soil are 'indestructible', Marx's study of agronomy and soil science revealed natural qualities as changeable and noted the 'brutal exhaustion of the soil' being driven down by 'the entire spirit of capitalist production … orientated towards the most immediate monetary profit' (Marx 1981, 756, 754 fn 27).

[15]'[T]he land bears the rent not because capital has been invested in it but rather because the capital investment has made the land a more productive field of investment than before' (Marx 1981, 880).

[16]One of Marx's extensions of Ricardo's theory of rent is to demonstrate that it is impossible to delineate between the two types of differential rent in practice because they interact rather than simply being additive (e.g. a capital investment in the form of differential rent II might enhance differential rent I) (Harvey 2006, 354–7).

[17]While in theory types of rent can be delineated, in practice, actual payments by capital to landed property may include more than surplus profits. For example, actual payment of ground-rent can contain 'foreign component[s]' such as a landlord's capture of a portion of the average profit and/ or of normal wages. Marx uses the category of *lease price* to indicate these phenomenal possibilities

rent is the material basis of landed property, it is a part of the contested struggle between and within classes over surplus value and can influence 'the pace and pattern of capital accumulation' (Fine 1994, 280).

4.2. *Ground-rent in the EEZ*

The EEZ is an example of the historical and institutional specificity of landed property, and industrial fisheries within them can help to illustrate how different kinds of rents are entangled in actually-existing property relations. We established that the institutionalization of EEZs created conditions for absolute rent extraction by state land-property *regardless* of the quality of the fishery. We have also seen that distinct parcels have different conditions that make them relatively more or less valuable, including natural conditions that impact relative labor productivity compared to an otherwise equal parcel (differential rent I). The capitalist fishing enterprise pays the coastal state ground-rent for the right to access a parcel of the ocean and extract the resource.

However, unlike agricultural land, minerals and trees, fish move between 'landowners', making calculations of different rent I particularly complex. Tuna, a highly migratory species, are an extreme example. Figure 1c, a highly simplified annual migratory pattern of tuna populations in the Indian Ocean, shows that tuna pass through the EEZs of Madagascar, Mauritius, Seychelles and several other coastal and island territories on their annual migratory routes. As a result, differential rent I available in a given state's EEZ varies month by month, and often year by year. Some states negotiate consistently higher rents because of the sheer volume of species' movement through national waters through the year (e.g. Seychelles). Rates of rent appropriation are dependent upon a wide range of human, oceanographic and climatic variables which can be environmental and cyclical (e.g. El Niño) or anthropogenic (e.g. prior catch intensity). But, given the institution of the EEZ, capitalists must enter into access relations with a number of different national 'landlords' in the Indian Ocean in order to be productive throughout the year (i.e. to maximize profitable investment in and rate of exploitation from constant and variable capital).

Differential rent II denotes that surplus profits can be *enhanced* through investment, rather than only being a function of especially high labor productivity generated in concert with some 'natural force'. But what kinds of investments are available in EEZ fisheries, and do these constitute permanent investments in ocean 'property'? Which actors capture value from such investments, and do they augment or diminish surplus value, and by association rent possibilities, in the short and long term?

Industrial fishing capitalists make investments in the landlord's property.[18] One example is the deployment of fish aggregating devices (FADs): a man-made floating object that is a form of (fixed) constant capital that depreciates over time. Fish are naturally attracted to floating objects in the water, so industry have deployed thousands of FADs that are tagged with global positioning system (GPS) devices and sonar to detect volumes of fish aggregating under them. Firms that have paid to access EEZs use FADs to enhance the productivity of tuna extraction, which is measured as catch per unit of fishing effort. In the

---

(Marx 1981, 763). We recognise the importance of the category lease price when discussing actual payments from fishing firms to coastal states, but use the term ground-rent for parsimony.
[18]Fishing boats and gear are investments in the means of production rather than investments that enhance the productivity of the fishery. In other words, they are investments in extraction, rather than in enhancing the conditions for extraction.

Western and Central Pacific, the world's largest tuna fishery, between 60 and 70 percent of all catch in the sector is now taken on FADs (Miyake et al. 2010). With FADs, the landlord (the state) can capture more rent from the EEZ because catch per unit of fishing effort is (temporarily) increased.

Meanwhile, a government's investment in its EEZ reveals the multifaceted role that state-landed property can play as resource manager, industry regulator and landlord. For example, states invest in fisheries management to protect and stabilize existing stocks and their value in the face of extractive pressure. Available management tools include fishing regulations (e.g. limits on FADs and fishing effort) and the creation of marine protected areas. States also invest in enforcement: monitoring, control and surveillance mechanisms such as patrol vessels and satellite vessel monitoring systems. It is worth noting that restrictions on fishing can limit ground-rent that states appropriate in the short run and act as a drain on state coffers because investment in management and enforcement can be costly. In the long run, these investments may allow for higher surplus profits (and thus greater rent appropriation) when the stock is stabilized, and may also reflect the political interest of some states in demonstrating a response to pressure from environmental lobbies.

But state investment in the property that is home to a resource that moves and can be depleted generates tensions. If one state invests in protecting spawning grounds for tuna, boats active in other waters (including the high seas, where no landed property exists) can benefit from the resulting stock stablization and/or enhancement (i.e. more fish) without paying ground-rent to the state who made the initial investment. Likewise, fishing and licensing practices in one EEZ can affect catch rates in another. Conversely, capitalist investment to enhance profits, such as FADs, potentially harms tuna reproduction in the long term, ultimately diminishing the natural basis for surplus value production and capture available to capital *and* landed property: fish caught on FADs are often smaller than fish caught without.[19] To counteract potential long-term ecological damage from FADs, eight Pacific island countries formed a collectivity of landlords and implemented an annual, region-wide, 3-month ban on their use (PNA 2010). This indicates, once more, that state-landed property simultaneously plays multiple, highly politicized roles in the oceans.

5. Conclusion

Questions of property and rent have long been at the heart of debates over the growing fisheries crisis, a debate that is gaining attention because of the import of fisheries in ecological systems, food security and economic development. In the mainstream analyses of fisheries that dominate policy debates, when states do not implement private property systems, they are responsible for creating conditions of open access, which, in turn, skews rent relations and drives the crisis. The tragedy of this orthodoxy is multi-faceted. First, it insists that an ahistorical, idealized, uniform and technical policy proposal will address highly complex and differentiated socio-ecological problems. Second, despite insisting on the development of the category of private property, it fails to carefully delineate actually existing property relations. In doing so, it contributes to, rather than corrects, the conceptual confusion around property that has long plagued fisheries systems. Finally, it is technically imprecise: it conflates Smith and Ricardo's theories of rent and posits that rent is a 'normal' market payment, despite that it is politically constructed, and continuously struggled over.

---

[19]On the socio-ecological implications of FAD use in the Indian Ocean, see Campling (2012b).

In contrast, we have used examples from industrial fisheries to illustrate the historical development of property relations in the oceans, the role of (changing) powers in shaping them, and a historically, politically and ecologically contingent theory of rent. Our examples reveal that property relations have emerged out of multilateral negotiations, geo-political contestation and struggles over the influence of global production dynamics and the distribution of socio-economic benefits. In practice, states' ability to appropriate ground-rent through sovereign rights over EEZs can be both threatened and augmented. Further, state-landed property is not always strictly territorially bounded: coastal states can at times have influence beyond the geographic boundaries of legal lines in the sea.

Our analysis contends with the conceptual confusion around property in fisheries, clarifying that it matters that most industrial fisheries are state property: the complex roles and multiple logics of states deserve careful conceptual and empirical attention in the narrative of the fisheries crisis. The paper also extends in three ways Capps' (2010) work on modern landed property. First, it illuminates state-landed property and in particular the peculiar, negotiated institutional form of 'sovereign rights' over marine resources in the 200-mile EEZ. Second, it identifies how environmental conditions of production in fisheries (a renewable but exhaustible resource that moves) come to bear on property and rent relations. We further specify this insight at a species level by analyzing the impact of highly migratory movement on inter- and transnational dynamics. It is the combination of the peculiarity of these environmental conditions of production and the 'historical conditions of existence of landed property' that makes a general theory of rent impossible (e.g. Fine 1979, 248). Finally, it offers an original, albeit partial, attempt at working through Marx's categories of ground-rent in relation to fisheries.

In sum, our analysis reveals that rather than being passive, weak, failed and corrupted observers or facilitators of the fisheries crisis, coastal states are engaged in a struggle to appropriate ground-rent in industrial fisheries. The landed property-capital relation is embedded in a combination of several complex roles and multiple (sometimes contradictory) logics of the state in fisheries systems. As on land, property in the sea is a site of social struggle and will always remain so under capitalism, no matter which juridical interest holds those property rights.

References

Ball, M. 1980. On Marx's theory of agricultural rent: A reply to Ben Fine. *Economy and Society*, 9(3), 304–26.

Bromley, D.W. 2008a. The crisis in ocean governance: Conceptual confusion, spurious economics, political indifference. *Maritime Studies*, 6(2), 7–22.

Bromley, D.W. 2008b. Rejoinder: The crisis in ocean governance: Clarifications and elaborations. *Journal of Maritime Studies MAST*, 6(2), 39–51.

Campling, L. 2012a. The EU-Centered commodity chain in canned tuna and upgrading in Seychelles. Thesis (PhD), University of London.

Campling, L. 2012b. The tuna 'commodity frontier': Business strategies and environment in the industrial tuna fisheries of the Western Indian Ocean. *Journal of Agrarian Change*, 12(2–3), 252–278.

Campling, L., E. Havice and P. Howard. 2012. The political economy and ecology of capture fisheries: Market dynamics, resource access and relations of exploitation and resistance. *Journal of Agrarian Change*, 12(2–3), 177–203.

Capps, G. 2010. Tribal-landed property: The political economy of the Bafokeng chieftaincy, South Africa, 1837–1994. Thesis (PhD), London School of Economics.

Capps, G. 2012a. A bourgeois reform with social justice? The contradictions of the minerals development bill and black economic empowerment in the South African platinum mining industry. *Review of African Political Economy*, 39(132), 315–333.

Capps, G. 2012b. Victim of its own success? The platinum mining industry and the apartheid mineral property system in South Africa's political transition. *Review of African Political Economy*, 39(131), 63–84.

Carr, C.J. 2004. Transformations in the law governing highly migratory species: 1970 to the present. *In*: D. Caron and H.N. Scheiber, eds. *Bringing new law to ocean waters*. Berkeley: Law of the Sea Institute, pp. 55–94.

Cordell, J., ed. 1989. *A sea of small boats*. Cambridge: Cultural Survival.

De Schutter, O. 2012. *'Ocean grabbing' as serious a threat as 'land grabbing'*. New York: United Nations.

FAO. 2010. *The state of world fisheries and aquaculture*. Rome: United Nations Food and Agriculture Organization.

FAO. 2012a. *The state of world fisheries and aquaculture*. Rome: United Nations Food and Agriculture Organization.

FAO. 2012b. *Voluntary guidelines on the responsible governance of tenure of land, fisheries and forests in the context of national food security*. Rome: United Nations Food and Agriculture Organization.

Fine, B. 1979. On Marx's theory of agricultural rent. *Economy and Society*, 8(3), 241–278.

Fine, B. 1994. Coal, diamonds and oil: Toward a comparative theory of mining? *Review of Political Economy*, 6(3), 279–302.

Fine, B., ed. 2013. *The value dimension: Marx versus Ricardo and Sraffa*. New York: Routledge.

Finley, C. 2011. *All the fish in the sea: Maximum sustainable yield and the failure of fisheries management*. Chicago: University of Chicago Press.

Gee, J.M.A. 1981. The origin of rent in Adam Smith's *wealth of nations*: An anti-neoclassical view. *History of Political Economy*, 13(1), 1–18.

Gordon, H.S. 1954. The economic theory of a common property resource: The fishery. *Journal of Political Economy*, 62(2), 124–142.

Grotius, H. 1916. *Mare liberum (the free sea)*. New York: Oxford University Press.

Haila, A. 1990. The theory of land rent at the crossroads. *Environment and Planning D: Society and Space*, 8(3), 275–296.

Hardin, G. 1968. The tragedy of the commons. *Science*, 162(3859), 1243–1248.

Harvey, D. 2006. *The limits to capital*. London: Verso.

Havice, E. 2010. The structure of tuna access agreements in the Western and Central Pacific Ocean: Lessons for vessel day scheme planning. *Marine Policy*, 34(5), 979–987.

Havice, E. 2013. Rights-based management in the Western and Central Pacific Ocean tuna fishery: Economic and environmental change under the vessel day scheme. *Marine Policy*, 42(November), 259–267.

Havice, E. and L. Campling. 2010. Shifting tides in the Western Central Pacific Ocean tuna fishery: The political economy of regulation and industry responses. *Global Environmental Politics*, 10(1), 89–114.

Havice, E. and L. Campling. 2013. Articulating upgrading: Island developing states and canned tuna production. *Environment and Planning A*, 45(11), 2610–2627.

Hollick, A.L. 1981. *US foreign policy and the law of the sea*. Princeton: Princeton University Press.

Homans, F.R. and J.E. Wilen. 1997. A model of regulated open access resource use. *Journal of Environmental Economics and Management*, 32(1), 1–21.

Howard, P.M. 2012. Sharing appropriation? Share systems, class and commodity relations in Scottish fisheries. *Journal of Agrarian Change*, 12(2–3), 316–343.

Joseph, J. 1977. The management of highly migratory species: Some important concepts. *Marine Policy*, 14(3), 275–288.

Khan, M.H. 2000. Rents, efficiency and growth. *In*: M.H. Khan and K. Sundaram-Jomo, eds. *Rents, rent-seeking and economic development: Theory and evidence in Asia*. Cambridge: Cambridge University Press, pp. 21–69.

Kronmiller, T. 1983. Exclusive economic zone of the United States: The question of sovereign rights relative to tuna. Statement before the US Senate Committee on Foreign Relations. *Ocean News*, 27(23).

Longo, S.B. and B. Clark. 2012. The commodification of bluefin tuna: The historical transformation of the Mediterranean fishery. *Journal of Agrarian Change*, 12(2–3), 204–226.

Lunstrum, E. 2013. Articulated sovereignty: Extending Mozambican state power through the Great Limpopo Transfrontier Park. *Political Geography*, 36(September), 1–11.

MacPherson, C.B. 1983. *Property: Mainstream and critical positions*. Toronto: University of Toronto Press.

Mansfield, B. 2004. Neoliberalism in the oceans: 'Rationalization', property rights, and the commons question. *Geoforum*, 35(3), 313–326.

Mansfield, B. 2008. Comments on Daniel Bromley's paper. *Maritime Studies*, 6(2), 23–26.

Marx, K. 1976. *Capital: A critique of political economy volume I*. London: Penguin Books.

Marx, K. 1981. *Capital: A critique of political economy volume III*. London: Penguin Books.

McCay, B.J. 1981. Development issues in fisheries as agrarian systems. *Culture and Agriculture*, 11, 1–8.

MacCay, B.J. and J.M. Acheson. 1987. *The question of the commons: The culture and ecology of communal resources*. Tuscon: University of Arizona Press.

Milonakis, D. and B. Fine. 2009. *From political economy to economics: Methods, the social and historical in the evolution of economic theory*. London: Routledge.

Miyake, P.M., et al. 2010. Recent developments in the tuna industry: Stocks, fisheries, management, processing, trade and markets. *Fisheries and Aquaculture Technical Paper Number 543*, Rome: United Nations Food and Agriculture Organization.

Moore, J.W. 2011. Ecology, capital and the nature of our times: Accumulation and crisis in the capitalist world-ecology. *Journal of World Systems Research*, 17(1), 108–147.

MRAG. 2013. *The scramble for fish – draft final report* (unpublished). Washington, DC: The World Bank.

Neocosmos, M. 1986. Marx's third class: Capitalist landed property and capitalist development. *The Journal of Peasant Studies*, 13(3), 5–44.

O'Connor, J.R. 1998. *Natural causes: Essays in ecological Marxism*. New York: Guilford Press.

OECD. 2010. *Globalisation in fisheries and aquaculture: Opportunities and challenges*. Paris: OECD.

Ong, A. 2006. *Neoliberalism as exception: Mutations in citizenship and sovereignty*. Durham: Duke University Press.

Ostrom, E. 1990. *Governing the commons: The evolution of institutions for collective action*. Cambridge: Cambridge University Press.

Patnaik, U. 1983. Classical theory of rent and its application to India: Some preliminary propositions, with some thoughts on sharecropping. *Journal of Peasant Studies*, 10(2), 71–87.

Perelman, M. 1975. Natural resources and agriculture under capitalism: Marx's economic model. *American Journal of Agricultural Economics*, 57(4), 701–704.

PNA. 2010. *A third arrangement implementing the Nauru Agreement setting forth additional terms and conditions of access to the fisheries zones of the Parties*.

Pontecorvo, G. 1988. The enclosure of the marine commons: Adjustment and redistribution in world fisheries. *Marine Policy*, 12(4), 361–372.

Ricardo, D. 1996. *Principles of Political Economy and Taxation*. New York: Prometheus.

Rieser, A. 1997. Property rights and ecosystem management in US fisheries: Contracting for the commons. *Ecology Law Quarterly*, 24(1), 813–832.

Scheiber, H.N. and C.J. Carr. 1998. From extended jurisdiction to privatization: International law, biology and economics in the marine fisheries debates, 1937–1976. *Berkeley Journal of International Law*, 16(1), 10–54.

Schlager, E. and E. Ostrom. 1992. Property-rights regimes and natural resources: A conceptual analysis. *Land Economics*, 68(3), 249–262.

Schurman, R.A. 1998. Tuna dreams: Resource nationalism and the Pacific Islands' tuna industry. *Development and Change*, 29(1), 107–136.

Scott, A. 1955. The fishery: The objectives of sole ownership. *Journal of Political Economy*, 63(2), 116–154.

Teiwaki, R. 1987. Access agreements in the South Pacific: Kiribati and the distant water fishing nations 1979–1986. *Marine Policy*, 11(4), 273–284.

Van Dyke, J.M. and C. Nicol. 1987. US tuna policy: A reluctant acceptance of the international norm. *In*: D. Doulman, ed. *Tuna issues and perspectives in the Pacific Islands region*. Honolulu: East-West Center, pp. 105–132.

Walker, R.A. 1974. Urban ground rent: Building a new conceptual framework. *Antipode*, 6(1), 51–58.

World Bank. 2009. *The sunken billions: The economic justification for fisheries reform*. Washington, DC: The World Bank.

World Bank. 2010. *The hidden harvests: The global contribution of capture fisheries*. Washington, DC: The World Bank.

Liam Campling teaches at the School of Business and Management, Queen Mary University of London. His main research interests are in international political economy and uneven development, commodity chain analysis (with an empirical emphasis on tuna), the history and theory of capitalism and the political economy of development in island states in the Western Indian and Pacific oceans. He is a Book Reviews Section Co-editor of the *Journal of Agrarian Change*, and a Corresponding Editor of *Historical Materialism*.

Elizabeth Havice is an assistant professor of international development and globalization in the Geography Department at the University of North Carolina – Chapel Hill. Her work is on the political economy of resource regulation, production and consumption in natural resource systems. Much of her empirical research is on the tuna industry that spans the Pacific Rim. Campling and Havice have co-authored articles that have appeared in the *Journal of Agrarian Change*, *Global Environmental Politics* and *Environment and Planning A*. They also recently co-edited a 12-article special issue of the *Journal of Agrarian Change* on the political economy and ecology of capture fisheries.

The role of rural indebtedness in the evolution of capitalism

Julien-François Gerber

Few studies have attempted to systematize the broader consequences of ordinary indebtedness – the inevitable other side of credit. My purpose here is to suggest four preliminary theses on the role of indebtedness in the evolution of capitalism, with special reference to the rural sphere. I argue that across time and space, credit/debt relations have not only been a key factor behind social differentiation through the control of land, labour and capital (Thesis I). They have also fostered market discipline by forcing the borrower – whether a poor peasant or a company manager – to calculate, pay, trade, work, intensify (Thesis II). Interest-bearing and guarantee-based loans have thus generated pressures for economic growth, short-termism and innovations, but have also undermined traditional community bonds and environmental conditions (Thesis III). Through its remarkable reward-or-punish nature, the credit/debt couple represents a powerful mechanism of social selection that has, in the long run, crucially shaped the evolution of capitalism (Thesis IV).

Introduction

Much has been written on rural indebtedness, usually in relation to particular case studies. The land dispossession and bonded labour effects of debt have been documented extensively by agrarian analysts. However, the broader economic, behavioural and social-environmental consequences of debt remain surprisingly under-theorized. Graeber (2009, 107) noted that anthropologists 'have had especially little to say about the phenomenon of debt' and Carruthers (2005, 355) wrote that credit is 'a topic that sociology has mostly ignored'.[1] This lack of interest may be explained by the ambivalence and complexity of the credit/debt couple – at times a survival requirement, a source of formidable potential or the cause of great burden.

In the present contribution, instead of adding new empirical material to an already long list of available case studies,[2] my purpose is to step back and provide a tentative overview

---

I thank Rolf Steppacher and Stephen Marglin for our numerous discussions, as well as Amit Bhaduri, Tom Brass, David Graeber, Shivani Kaul and Joan Martínez-Alier. All remaining mistakes are mine. Four anonymous reviewers and the support of the Swiss National Science Foundation are also acknowledged.

[1]However, it is interesting to note that two of the most influential sociologists and anthropologists – Pierre Bourdieu (*et al.* 1963) and Clifford Geertz (1962) – started their career by working on credit issues.

[2]I have done so elsewhere (Gerber 2013).

of the role of indebtedness in the evolution of capitalism, with special emphasis on the rural world. In order to do so, my discussion takes the form of four theses that seek to integrate a large part of the available data. My focus is not on the national level but on the community one – households and firms – within early modern Europe and today's developing countries. There, I take into account not only peasant sectors but also rural entrepreneurs and the élite, all of them being typically involved in credit/debt relations. Because the topic is so vast,[3] my theses should be considered working hypotheses open for further discussions and research.

Rural indebtedness is certainly not a new phenomenon. It is in fact concomitant to the appearance of money and perhaps even paved the way for its creation. The phenomenon was widespread in Ancient Near Eastern civilizations, Greece and the Roman Empire, as well as China and India (Harris 2008, Graeber 2011). It was also present in medieval Europe (Duby 1998, Schofield and Mayhew 2002). But by the second half of the sixteenth century, after much politico-ethical controversy, a new factor entered the post-medieval picture of most of Western Europe: interest-bearing loans started to be legalized and enforced. The morality of charging an interest was relegated to individual conscience and the state largely ceased to hamper 'usury' (Jones 1989). As a result, interest-bearing loans expanded dramatically in a favourable context characterized by new emerging property relations, enlargements of urbanizations and an increase in population size. The credit engine was launched – largely unintentionally – and it never stopped.

Since the 1990s, the number of studies focusing on credit/debt in early modern Europe has grown rapidly (they are still very rare outside Europe). Virtually all of them emphasize the new extent and importance of the credit economy that, depending on the region, arose between the fourteenth and the seventeenth centuries. Nearly invisible in the work of previous historians, credit is now regarded as dominating England's commercial life already by the mid-sixteenth century (Hoppit 1990, Muldrew 1998, Finn 2003). In the same vein, a reinterpretation of French economic history has been suggested based on the discovery of large amounts of credit throughout the country as early as the seventeenth century, sometimes earlier (Hoffman 1996, Postel-Vinay 1998, Fontaine 2008). Much like in early modern England, generalized indebtedness appears to be no less than 'a structural feature of early modern France' (Fontaine 2008, 50). The same situation has been observed in early modern Italy, Spain, Low Countries, Germany and Switzerland (see references in Berthe 1998, Fontaine 2008).

Although much variation existed between the different regions, the use of credit became routine throughout society, partly because the money supply expanded much more slowly than the need for it, and partly because lending became openly profitable. Over time, interest-bearing and guarantee-based loans increased, while interest-free debt – an indicator of neighbourly non-commercial exchange – declined with the expansion of credit networks. Thus, generalized indebtedness seems to correlate with the birth and spread of capitalism, an observation that has received too little attention. Does it reveal causality? What is its significance? What are the implications? My contribution is intended to tackle these questions, not only historically but also within contemporary developing economies. Indeed, much like early modern Europe, ordinary indebtedness also appears to be a structural feature of many rural areas of the global South.

---

[3]Due to space constraints, I had to limit the bibliography to some of the most relevant works.

After a brief review of the agrarian literature on rural indebtedness, I discuss four theses on the role of credit/debt relations in the evolution of capitalism, before concluding with some remarks.

The classical theses on rural indebtedness

Studies addressing the implications of rural indebtedness can broadly be divided into four camps. The 'stagnationists' emphasize the poverty-generating and stagnation effects of rural indebtedness (Braudel and Labrousse 1977, Kriedte *et al.* 1981, Goubert 1986), while the 'formalists' stress the credit-based improvement of production and the associated entrepreneurial spirit (Hoffman 1996, Postel-Vinay 1998, Schofield and Lambrecht 2009). The 'culturalists' have highlighted the gradual shift from an embedded 'moral economy' of local lending practices to an impersonal large-scale system of formal credit (Schrader 1997, Muldrew 1998, Finn 2003, Fontaine 2008). And fourth, while acknowledging a shift, the Marxist camp has emphasized exploitation and the social differentiation effect of credit/debt (Kautsky 1900, Lenin 1967, Brass 2010). This tradition has produced some of the earliest and most elaborated insights on rural indebtedness, including in the *Journal of Peasant Studies*. Accordingly, this brief review section will largely focus on the fourth camp.

Having said this, it appears nonetheless that Marxian scholars have mostly focused on the proprietor/employees relationship – the classical ingredient in most definitions of capitalism – rather than on credit/debt relations. On one hand, this is understandable as Marx himself largely concentrated on the proprietor/employees relationship and admitted in the third volume of *Capital* that 'It lies outside the scope of our plan to give a detailed analysis of the credit system' (Marx 1992, 525). On the other hand, the neglect of credit/debt relations is surprising since Marx clearly pointed out that the surplus value is appropriated by *both* the owners of capital who receive profit of enterprise and those who collect interest (Marx 1992, ch. 23). For him, circulating capital (including credit) and production ('industrial') capital represent the two constitutive pillars of capitalism. As circulating capital is necessary – but not sufficient – for the formation of production capital, it always historically precedes the latter. However, in his analysis of the credit system, Marx mainly focused on inter-capitalist relations within a (relatively) modern banking system and did not study in much detail the role of credit/debt in rural settings. Very briefly, he saw 'usury' as one factor – among others – 'assisting in establishment of the new mode of production by ruining the feudal lord and small-scale producer, on the one hand, and centralizing the conditions of labour into capital, on the other' (732). In this concentration phenomenon, the producer pays the capitalist his surplus labour in the form of interest and this exploitative process prevents the producer, Marx thought, from increasing productivity and improving technological conditions. As we will see, his influential idea associating indebtedness with stagnation needs to be qualified.

Subsequently, some Marxist analysts of the rural world refined and expanded on Marx's considerations. Kautsky (1900) put more emphasis on the role of debt. He identified peasant indebtedness as *the* key mechanism alienating peasants from their means of production, stressing, like Marx, the counter-productive effect of debt on the development of productive forces. For Lenin (1967), one of the main consequences of modern credit is the social differentiation induced within pre-capitalist communities. On this, both Chayanov and Lenin agreed. Chayanov (1966, 262) wrote that 'widely developed mortgage credit, the financial arm of circulating capital, [counts among the key] new ways in which capitalism penetrates agriculture […] convert[ing] the farmers into a labour force working with other people's means of production'. Importantly, Lenin clearly stressed both potentials of credit, not only the stagnation effect of rural indebtedness but also its implications in terms of economic growth. Although poor peasants consume relatively less than rich ones, they spend relatively more on basic goods and debt typically ensues if they lack cash to meet these needs. Wealthy peasants are simultaneously less dependent on the market for basic goods and more dependent on it to sell their production. The bulk of their cash expenditures go towards sustaining or increasing their production. As they routinely use investment credit, they hold a much larger mass of total debt. The emergence of agrarian capitalism thus gave rise to two different types of debt: one was a sign of precariousness, the other of increasing consolidation and capitalization.

Among more recent authors, Banaji (1977) and Roseberry (1978) mainly updated these ideas with additional fieldwork data. They illustrate the idea that usury capital represents *the* strategic form of circulating capital in many rural areas because peasants are increasingly dependent on credit to reproduce their households. The interest rate becomes the primary mechanism for the extraction of surplus value and, accordingly, credit/debt relations can be regarded as 'concealed wage labour' (Banaji) or as a distinctive form of 'production relation' (Roseberry).[4] Bernstein's (1979) notion of 'simple reproduction squeeze' elaborates on similar ideas. Extremely vulnerable to external demands and ecological conditions, poor peasants become ever more dependent on credit, frequently mortgaging their land. They may eventually become full proletarians if they fail to generate sufficient income by supplying (cheap) labour and/or commodities. In this context often characterized by 'contractual interlocking', Bhaduri (1983) showed that peasant indebtedness gives rise to an exploitative system of 'forced commerce'. Brass (1999, 2010), for his part, developed a theory of unfree labour based on a 'deproletarianization' of the workforce (i.e. its shift from the status of free to unfree labour). He argues that unfree labour – typically made possible through debt – is not only compatible with capitalism but that it is capitalism's production relation of choice when the class struggle allows it. Indeed, unfree labour is today present in a multitude of rural regions, including in highly commercialized agricultural sectors.

These Marxian insights form an important substratum on which it is possible to elaborate further. In this paper, I seek to integrate them with specific aspects of the other three camps (i.e. stagnation versus entrepreneurial behaviour, and economic evolution). Drawing on the work of German and Swiss heterodox economists (Heinsohn and Steiger 1996, 2003, Altvater 1997, Steppacher 2008, Steppacher and Gerber 2012), my objective is to include but also to go beyond the well-known consequence of rural indebtedness as a mechanism of capital, land and labour control. I also hope to contribute something to the debate on the transition to capitalism which strikingly lacks any substantial discussion of credit and debt (a point also made by McIntosh 1988).

Thesis I: indebtedness restructures ownership relations

Background

One well-known effect of rural indebtedness has been the reconfiguration of land ownership, especially its concentration. The phenomenon has a long history, starting with the spread of rural credit around the Mesopotamian city-states 5000 years

---

[4]In a similar albeit more general vein, Roemer (1982) argued that credit markets function on the basis of surplus value extraction no less than labour markets.

ago.[5] Reviewing the different mechanisms of ownership concentration during the transition to capitalism in early modern England, Dobb (1947, 181) is particularly clear:

> When we examine the actual changes that were occurring in fifteenth- and sixteenth-century England, it is evident that economic distress at various periods both of large feudal landowners and of certain sections of smaller ones, placing them in the position of distress-sellers and involving them in mortgage and debt, must have played a major role in facilitating easy purchase of land by the *parvenu* bourgeoisie.

The decline of the smallholder and the rise of the landless labourer has been one of the most important themes in the historiography of early modern England as well as of other parts of Western Europe. While a great majority of studies acknowledge the role of indebtedness in land concentration, few actually address the issue in detail. Those that do focus on this question tend to reveal the vast extent of the phenomenon. In fifteenth-century Italy, for instance, a massive process of debt-driven land concentration is documented (Gaulin and Menant 1998). The same applies to rural Sweden between 1750 and 1850 (Ågren 1994).

The effects of these debt-driven processes seem comparable – and even more powerful in many regions – to those of the different 'enclosure movements', whose real scope and significance remain a topic of controversy (Wordie 1983, Allen 1992). When the credit system was favourable to the dominant classes, its effects tended to be reinforced (main scenario), but when it became too dangerous for the same classes, legal 'subterfuges' were found to protect themselves against the danger of dispossession. The aristocracy introduced land laws preventing its estates from being seized by creditors – these were the 'strict settlement' laws in England, the *fideicommis* in Italy and France, and the *major-azgo* in Spain (Fontaine 2008). Overall, however, the phenomenon is difficult to evaluate because it often leaves the written trace of a standard 'voluntary' land sale, glossing over the fact that the owner was forced to sell in order to pay creditors (Laffont 1999). This is still one of the most common motives behind land sales in the global South today.

However, creditors have sometimes no incentive for acquiring land or assets or for being reimbursed. They prefer to remain interest collectors (*crédirentiers*) in money terms or in kind. This was for instance true in early modern France, where long-term debts formed the major source of credit in much of the rural world (Postel-Vinay 1998). These long-term debts consisted of perpetual and life annuities (*rentes*) that lenders with extra funds 'bought' from individuals needing money. The debtors then paid interest on the capital, either 'in perpetuity' or until the death of the creditor. To a large extent, this is still true of modern banks. During the first decades of the twentieth century, the vice-president of a major Californian bank recognized that 'No bank which wants to maintain its solvency will loan for even [a year] on farm mortgage security. [...] What the bank wants is to keep the borrower working' and paying interest year after year (Chenowith 1923, quoted in Henderson 1998, 110). In other words, the capitalist uses credit to keep

---

[5]It is interesting to note that Sumerian and later Babylonian kings, faced with the uncontrolled consequences of rural indebtedness, periodically announced general amnesties cancelling all outstanding consumer debt (commercial debts were not affected), giving back all lands to their original owners, and allowing all debt-peons to return to their families (Hudson 2002). Similar policies were also promoted by the early Judaeo-Christian tradition and are currently witnessing a revival among anti-debt activists.

on extracting value from the debtor's work. This applies to various loans arrangements and has been observed virtually everywhere.

One classical form of labour control is through advances on sales. The creditor lends a certain amount of money that is repaid through the peasant's production at a price that includes interest. This mechanism has historically pushed peasants into monoculture cash crop production as it represented the best way to secure monetary repayment. Advances on wages follow the same pattern. During the early transition to industrial capitalism, this kind of control device played a central role (Dobb 1947). After having been dispossessed through debt mechanisms, former peasants were now once again subdued by way of chronic indebtedness, as in the putting-out system:

> The role played by wage advances deserves more attention than it has received, for at least in some trades it appears to have been an important means by which the capitalist maintained his hegemony. Wage advances were to the capitalist what free samples of heroin are to the pusher: a means of creating dependence. It is of little moment that one was a legal and the other a physiological dependence. Both represent an addiction from which only the exceptionally strong-willed and fortunate escape. (Marglin 1974, 80)

Many variants of debt-bondage have been described, including numerous mechanisms for keeping workers in perpetual debt to their creditors, sometimes over generations.[6]

Theoretical propositions

Within the transition to capitalism, debt mechanisms have thus played a central role in shaping the emergent property structure. In early modern Europe as in the global South today, indebtedness represents, together with (legal or extra-legal) expropriation and purchase, the third main mechanism of 'primitive accumulation'. However, the effects of credit/debt relations on land ownership should not simply be understood as a process whereby the poor debtor gets expelled from his land by the rich creditor:

- First, the modern laws of mortgage, starting in the early seventeenth century, were a key institutional innovation that contributed to the elimination of yeoman farming (Allen 1992). This type of credit allowed landlords to simply buy up freeholds and heritable copyholds – but it also drastically increased the level of indebtedness of the landed classes.
- Second, the indebtedness of the élite drove some of its members to modify ownership relations in order to facilitate the repayment of their debts, notably though land expansion (including imperialism) and by selecting the most cost-efficient tenants. Debt pushed many landlords to enclose with the aim of generating extra income to reimburse their creditors (Habakkuk 1994).

---

[6]Drawing on the Latin-American context, Knight (1986, 46) identified three subcategories of debt-based labour recruitment. First, there is free wage labour linked to cash advances. This is the classic example of 'debt-peonage' and serves to attract labour (voluntarily) from the subsistence to the commercial sector. Second, there is 'traditional peonage' characterized by workers linked to a single hacienda and residing on it for most of their lives. They do so for lack of a better alternative rather than because of extra-economic coercion, and debt functions more 'as a perk rather than a bond' (46). And, third, there is 'debt servitude', which reproduces aspects of chattel slavery despite the formal abolition thereof. For obvious reasons, debt servitude is the least studied form of labour trafficking today, yet it is the most widely used method of enslaving people (Kara 2009).

- Third, debt not only shaped property relations during the land dispossession stage, it also continued afterwards through different mechanisms of labour control, including 'deproletarianization' whenever possible (Brass 1999).
- And fourth, if debt was always so harmful, why are there many documented cases of peasants voluntarily taking out investment loans (Hoffman 1996, Postel-Vinay 1998, Fontaine 2008)? Credit allows, *without previous savings*, to project one's imagination into future improvements, a powerful potential that explains why credit-based economic activities were – and still are – so attractive to many people. 'Success stories' must have contributed to make credit tempting and worth considering. Of course, the creditors – usually belonging to the élite – had an advantage in this attractiveness of credit, but the debtor's activities could also work out well. In other words, credit/debt relations generated social differentiation in every rural class.

Although credit/debt relations predate capitalism, the latter offered them a particularly flourishing ground, amplifying their impacts both qualitatively and quantitatively. Credit/debt relations became in fact a central element of the 'capitalist social-property relations' defined by Brenner (1985a, 1985b, also Wood 2002) as the particular class relations generating capitalism's relentless drive to accumulate through the market dependence of economic actors. As Heinsohn and Steiger (1996, 2003) pointed out, capitalist property entails the unique and simultaneous potential of (a) alienating, leasing and accumulating assets such as land,[7] (b) encumbering property as collateral for obtaining money as capital, (c) collecting interest,[8] (d) burdening property in issuing money[9] and (e) enforcing agreements using state power (judicial system and police forces). Taken together, these institutional possibilities – enabling a wide variety of credit instruments – generate specific pressures and features that are typical of capitalism (more on that later).[10] Non-capitalist societies also have webs of mutual obligations binding their members together, but their forms of credit are behaviourally totally different (Testart 2005).

Credit/debt relations may thus provide a basis for some rapprochement between world-system theory and Brenner's agrarian thesis. For the record, Wallerstein (1974) argued that capitalism developed in Europe as a result of the growth of world trade and markets, while for Brenner (1985a), the transition to capitalism should be understood through internal pressures within feudalism itself, and not via some external force. Articulating the two positions, it appears that merchants as well as landlords and tenants got trapped in the growing credit/debt relations and sometimes benefited from them. Credit/debt mechanisms were a key lever behind the emergent classes of

---

[7]The possibility to alienate land is an historical and cultural oddity that spread with the advance of capitalism.

[8]The practice of charging interest for debts appears to have been invented in third-millennium-BC Mesopotamia, and to have spread quite slowly (Hudson 2002). It does not appear to have ever been practiced in Pharaonic Egypt, for instance, and Tacitus claims the Germans of his day were still unaware of the institution (first century AD). It is hardly universal.

[9]Modern banknotes are anonymous property titles (see Heinsohn and Steiger 1996, 2003, for further discussions). Before banknotes were invented, debts frequently circulated as an early form of paper money. Today, most of the money in use is debt-bearing money created by commercial banks (Lietaer *et al.* 2012).

[10]It is striking that many researchers seem unaware of these specificities of capitalist property. They indiscriminately speak of 'property' whether they refer to a tribal community or a capitalist firm. It is in order to clarify this aspect that Heinsohn and Steiger (1996, 2003) have called any systems that do not present these characteristics *possession*-based regimes – as opposed to such *property*-based regimes.

commercial landlords and tenants, notably by fostering their market dependence and therefore the expansion and/or profitability of their holdings. Forced to survive the constraints of debt, landlords and tenants but also merchants and entrepreneurs had no choice but to cut costs, grow and innovate (more on that later). Credit/debt relations, in this sense, are clearly more than a mere 'assistant' to the birth of capitalism (Marx). They were an essential lever in the transition from market as opportunity to market as compulsion, as we will see next.

Thesis II: indebtedness shapes capitalist rationality and culture

Background

The expansion of credit in early modern England correlates remarkably with the rise of a more commercial mentality. Habakkuk (1994, 315) wrote that 'there is a strong case for supposing that debt was a stimulus to development' ('development' meaning here productivity increase, commercial activities and industrialization). In order to illustrate the phenomenon, let us take the example of the Scottish rural élite as studied by Watt (2006). In the inflationary economic conditions of the late sixteenth and early seventeenth centuries, the Scottish aristocracy borrowed extensively because of high levels of expenditure (rather than deficiency of income). A century later, the same aristocracy was in severe financial distress due to the debts accumulated, the collapse of incomes during the mid-century civil war and a deflationary economic environment. As a result, a fundamental shift from traditionalism to commercialism occurred. A direct consequence of debt was indeed

> a more commercial attitude to the running of estates [...]. To meet existing interest payments or to reduce the overall debt burden, productivity of estates had to rise and chiefly expenditure had to fall. [...] The droving trade in highland cattle was growing from the early seventeenth century but major expansion was associated with the rapid growth of London after the Restoration and is likely to have been stimulated by the financial problems of the chiefs. The ninth earl of Argyll introduced short leases from five to nineteen years to ensure accountability and developed businesses in coal, salt, fishing, shipping and lime and slate quarrying. His very substantial debts were surely the principal reason for this focus on commerce. The marquis of Atholl introduced more commercial estate management and John Campbell of Glenorchy, first earl of Breadalbane, invested in steelbow and the exploitation of timber. [...] both men were responding to the realities of financial crisis. (Watt 2006, 47–8)

Other responses to debt by the Scottish élites included the development of commercial networks, notably investments in colonial ventures. Indebtedness also increased pressure to convert rentals into cash payment, to raise rents and therefore to shift the debt burden onto tenants. 'Debt might be viewed as an acid dissolving the obligations of the past. The financial crisis precipitated by the combination of indebtedness and deflation was therefore a pivotal aspect in the process by which highland chiefs adopted the values of [capitalist] landlords' (Watt 2006, 51).

At the farmer level, rural indebtedness also plays a key role in Post's (2011) account of the transition to capitalism in the northeastern United States. According to him, the transition in the countryside proceeded unevenly over the first four decades of the nineteenth century, but in all regions

> the burdens of mortgages, taxes and debts ensured that Northern farmers marketed both their surplus and portions of their subsistence-output. Put simply, northern-US farmers became

dependent upon successful market-production for their economic survival – they became agrarian petty-commodity producers who had to specialise output, accumulate land and capital, and introduce new tools and methods in order to obtain, maintain, and expand landed property. (Post 2011, 191)[11]

Going back to sixteenth-century England and focusing on the changes in mentalities, Muldrew (1998, 95) wrote that 'It was credit, above all, which dominated the way in which the market was structured and interpreted'. According to him, credit imposed upon everyone in the society new and similar constraints. Initially, in the absence of a fully functioning legal enforcement system, people had to trust that those with whom they dealt would honour their word and possessed the means to repay their loans. Since credit was an imperative, people lived always with an eye to reputation, monitoring conduct so as to avoid a bad name. There is much evidence revealing how concerns with indebtedness weighed very heavily on the minds of people who filled their letters and diaries with such worries (Muldrew 1998, Finn 2003, Watt 2006).

During this early phase of capitalism, a new meaning for 'honesty' appeared: it became synonymous with creditworthiness. Every member of a community was encouraged to become creditworthy, namely to live soberly and orderly and to work hard (Hoppit 1990, Muldrew 1998, Newton 2003). There was a torrent of moral literature warning against the perils of the 'prodigal' indebted life and fostering people to discipline in order ensure payments. In the early eighteenth century, Daniel Defoe wrote that a government, a businessman, a farmer or a labourer could properly handle credit only if they endorsed a specific pattern of 'good behaviours' notably characterized by a calculating and industrious attitude (Hoppit 1990, 319). Marglin (2008) described the dominant capitalist system of knowledge as being essentially 'algorithmic', emphasizing its calculating, rational and maximizing character, as opposed to the culturally and ecologically embedded 'experiential knowledge' of non-capitalist sectors and societies. When using credit, one is simply forced to master 'algorithmic knowledge' in order to survive socioeconomically, if not physically.

Such personal discipline applies in different intensities to *all* credit/debt relations involving significant costs and guarantee. Defaulting is no minor matter for most borrowers. Until the nineteenth century, imprisonment was not rare, including the incarceration of members of the debtor's family. Slavery and prostitution remain classic outcomes of indebtedness (Kara 2009, Graeber 2011). Today, property selling or foreclosure are the most frequent means to recover an unpaid debt, and the resulting loss of house or land has led to countless suicides worldwide – not only in India. There is little doubt that such pressures make debtors work hard and do everything they can to repay loans on time.[12] The first thing indebted farmers do is to implement a 'structural adjustment programme' in their exploitation through intensifying production, engaging in commercial activities and enrolling in temporary wage labour. In this sense, indebtedness is far from being only a factor of stagnation.

---

[11]In the same vein, Atack and Bateman (1987, 271, quoted in Post 2011, 90) wrote that 'Before the Civil War, […] our evidence indicates [that family-farmers] deliberately sought to produce for the market and to move away from the generalists' life of self-sufficiency toward specialisation. Debts incurred to establish and maintain a farm often forced that choice upon them'.

[12]As Brass (2010, 78) showed, the claim that indebted peasants-workers are 'inefficient' misses the point: 'a labourer who is unfree works as hard as (or harder than) one who is free' for fear of the possible punishments. This theme is forcefully present in many cultural texts situated in rural India, for instance in the classic movie *Mother India* (1957), or the novel *Godaan* (1936) by Premchand.

Theoretical propositions

In sum, credit/debt relations have crucially promoted the specific rationality and values that prevail under capitalism – and continue to do so. Once an economic actor – whether rich or poor – has entered an interest-bearing and guarantee-based credit contract, he/she is compelled (to different degrees) to think and to behave in a particular way in order to secure timely repayments. The typical debtor must fundamentally focus on the potential demand of moneyholders. He/she is forced to produce commodities that, from the very beginning, are not designed for personal consumption but for the purpose of obtaining money.[13] Moreover,

> The demand for a rate of interest forces upon [the debtor] a value of production, expressed in terms of quantity, time, money or price, which must be greater than the money proper advanced as capital. This demand thus necessitates a value surplus in the production of commodities, the rate of *profit*. [Emphasis in the original]. (Heinsohn and Steiger 2003, 511)

As a result, because debtors must reimburse the principal *plus* pay interest, indebtedness brings about the accumulation so characteristic of capitalism. It requires economic growth to occur (more on that later). In fact, the constraints generated by indebtedness – that is, to remain solvent, to respect time limits and to make a profit – define the entire hierarchy of economic decision-making and the valuation process associated with it (Steppacher 2008, Steppacher and Gerber 2012). Above all, the debtor must focus on a monetary cost/benefit evaluation of all economic transactions and resources, based on the current market prices. He/she must think in individual terms and prioritize short-term benefits. Accordingly, land, harvests, labour time and natural resources are monetarily evaluated while surrounding sociocultural and ecological considerations remain secondary. Summarizing the argument, Altvater (1997, 60) noted that

> debts force a profitability appropriate to interest obligations and thus the corresponding acquisitive economic rational structuring of the production process, i.e. the take-over and completion of 'capital accounting', the adequate choice of equipment, and a distribution between wages and profit which enables the interest to be derived. Debt service in the economy thus forces *economic rationalisation*. Monetary indebtedness forces capitalist economising, and space and time are subordinated to monetary rationality. [Emphasis in the original].

A note on the different categories of rural credit and their consequences

Credit/debt relations are a central economic feature of the contemporary countryside of developing countries. Although there is a wide variety of rural loans, it is nonetheless possible to identify a limited number of broad categories (Gerber 2014). Due to space restriction, it is not possible here to describe in detail each of the categories listed in Table 1. The latter schematically links these different types of credit with some of their main consequences. The table highlights a qualitative *gradation* in the consequences resulting from the different forms of credit/debt relations: broadly speaking, as we move down the table towards the modern forms of credit, there is an increased potential for fostering capitalist dynamics characterized by a relentless drive to concentrate, invest, grow and expand market dependence. In contrast, the reciprocity of lending practices tends to diminish.

---

[13]In these conditions, a market becomes necessary. Heinsohn (2008, 251) suggested that the (Mesopotamian) origin of the market can be traced back to the phenomenon of debt: 'It is indebted proprietors that *constitute* the [primordial] market'. [Italics in the original].

Table 1. Schematic representation of the characteristics and consequences of the main categories of rural loans in today's developing world (Y: yes; P: possibly).

		Origins (M = modern; A = ancient)	Informal (I) or formal (F) credit	Interest (I) + security (S) involved	Loan in money (M) or in kind (K)	Main material advantages		Some consequences from the debtor's perspective						
						For the creditor	For the debtor	Consumption (food, seeds ...)	Investment (accumulation)	Market discipline (attitude)	Forced commercialization	Land loss	Unfree labour	Debtor-creditor reciprocity
Personal	between equals	A	I	Ø	M/K	cash, goods (if mutual)		Y						Y
loans	in patron-clients rel.	A	I	Ø/I	M/K	Ø/profit/control	cash, goods	Y	P				P	P
ROSCA loans*		A	I/F	Ø	M/K	cash, goods (if mutual)		Y	P	P				Y
Delayed	purchases	A	I/F	Ø/I	M/K	profit on interest	goods	Y	P	P				
payments for	wages**	A	I/F	Ø	M/K	income	labour control	P			P		P	
Pawnbroker loans		A	I/F	I+S	M/K	profit on interest	cash	Y	P					
Advances on sales or wages		A	I/F	I+S	M/K	labour control	cash, goods	P	P		Y		P	
Moneylender loans		A	I/F	I+(S)	M/K	profit on interest	cash or capital	P	P	P	P	P	P	
Subcontracting credit		M	F	I+S	M/K	labour control	income		Y	Y	Y	P	P	
Small-scale credit union loans		M	F	I+S	M	cash or capital (if mutual)		P	P	Y	P			P
Bank credit		M	F	I+S	M	profit on interest	land or capital	P	Y	Y	P	P		
Microcredit		M	F	I+S	M	profit on interest	cash	P	P	Y	P	P		

* ROCSA: rotating savings and credit association.
** Here, the employer-debtor owes the employee-creditor some wage money (the consequences are here from the employee-creditor's perspective).

The disciplining effects of indebtedness are evidently weak with low levels of debt (as compared to the borrower's income). They then increase in intensity with higher debt levels, which strengthens the creditor's power position over the debtor. Then, with very high levels of debt, these effects may decrease again, illustrating Keynes' (1945, 258) famous remark: 'Owe your banker £1000 and you are at his mercy; owe him £1 million and the position is reversed'. In the same vein, some large companies or governments live with huge debts that are never paid off but instead are rolled over into new debts as they mature.

Thesis III: indebtedness undermines community and the environment

Credit and debt have played a leading role in eroding community-based ways of life. Very directly, debt undermines community life by dispossessing defaulting debtors and fostering social differentiation (see Thesis I). It is also an important factor behind temporary, permanent and circular migrations. In addition, credit destabilizes customary commons because it frequently requires that alienable portions of the community's land are provided as collateral. In early modern Europe, communal indebtedness dispossessed villagers from their commons

(Cooper 1978, Badosa Coll 1990, Pichard 2001). And today, commons continue to be dissolved as peasants turn to cash crop production and therefore to collateral-based credit.

More indirectly, credit/debt relations also erode traditional bonds between community members by individualizing economic responsibilities within dyadic relations. 'The culture of credit was generated through a process whereby the nature of the community was redefined as a conglomeration of competing but interdependent households' (Muldrew 1998, 4). The atomistic character of credit/debt relations is visible, for instance, in the English debate on usury, where there has been a gradual shift in the discussions away from the primacy of community-based custom. By the mid-sixteenth century, it was increasingly recognized that usury was a matter of internal conscience and that paying usurious rates was not always sinful. This new appreciation provided a religious 'rationale' for the emerging capitalist credit/debt relations. In this way,

> by removing the centre of moral judgment from the community to the individual conscience [Protestant theologians] admitted that what each person intended by one's actions could only be judged by intention. This had the practical effect of freeing individual action. (Jones 1989, 174)

However, some authors (e.g. Hoffman *et al.* 1999, Finn 2003) argued that the early modern spread of credit has, in fact, enforced interpersonal relationships, notably through the development of 'trust'. While nobody would deny that new relations are formed, this inevitably takes place within a context of discipline, suspicion and power imbalances. 'Trust' in this sense is a strategy, *faute de mieux*, and certainly not a solidarity value.

With the initial sixteenth-century credit boom, credit/debt relations clearly exacerbated tensions between community households. These tensions resulted partly from the fact that the new credit-based economy increased social inequalities and the number of desperately poor (Muldrew 1998). In this context, it is not surprising that there was an explosion of debt litigations in local courts seeking to enforce agreements. In some regions of rural England, the average was more than one suit per household per decade between about 1550 and 1650. Regarding the consequences on community, Graeber (2011, 334–5) noted that

> One can only imagine the tensions and temptations that must have existed in a community – and communities, much though they are based on love, in fact, *because* they are based on love, will always also be full of hatred, rivalry and passion – when it became clear that with sufficiently clever scheming, manipulation, and perhaps a bit of strategic bribery, [people] could arrange to have almost anyone they hated imprisoned or even hanged. [...] The effects on communal solidarity must have been devastating. The sudden accessibility of violence really did threaten to transform what had been the essence of sociality into a war of all against all. [Emphasis in the original].

Besides the impact on community solidarity, indebtedness also creates adverse pressures on the environment as a result of the accumulation and cost cutting it requires. Ecologically damaging growth may represent a response to prior levels of indebtedness. This has been reported in early modern France where commercial agriculture expanded because of mounting communal debts (Pichard 2001) as well as in early modern Scotland where the aristocracy's indebtedness stimulated extractive and agricultural activities and caused deforestation (Watt 2006). More recently, Canada's National Farmers Union (NFU 2010, 19–20) declared that

> Debt repayment deadlines push farmers to make choices based on short-term cash flow, rather than on the needs of the soil or of the next generation. Farms have traditionally been places

where long-term thinking and holistic decision-making prevailed. Debt forces farmers to adopt the short-term thinking common to corporate boardrooms, with predictable results for the environment, fertility, and the future.

In addition, ecological destruction may also occur as a consequence of access to rural credit and microcredit which are repeatedly used to purchase unsustainable technologies. Many farmers from developing countries get into debt in order to buy 'green revolution' technologies that drastically accelerate pressures on the environment. From the 1980s onwards, for instance, Paraguayan Mennonite farmers experienced a 'growth miracle' much celebrated by the political class. But the other side of the coin was a high level of indebtedness resulting from the agro-industrialization process. Regehr (1994, 141) wrote that 'This indebtedness pressures the Mennonite economy to decrease costs, use more aggressive economic methods, and lower its sensitivity to the social needs and the environmental concerns'.

Thesis IV: indebtedness is a motor behind the evolution of capitalism

Background

By generating or enhancing some of the key economic pressures that prevail in capitalism, the dynamics of credit/debt relations shape some of the typical features of capitalist evolution. As Steppacher (2008) put it, interest-bearing and collateral-based loans entail (1) the pressure for increasing production in order to reimburse the principal and pay interest, that is, the imperative of *economic growth*, (2) the *time pressure* imposed by the period for which the credit is granted and (3) the pressure to improve monetary cost/benefit conditions – through *technological or institutional innovations* – in order to guarantee solvency. Let us briefly review each of these three points.

The idea that financial development – with credit as its main instrument – plays a leading role in fostering economic growth traces its origins at least to Schumpeter (1934) and it has subsequently been confirmed by a variety of studies. Historians have highlighted the crucial role of 'financial revolutions' in the growth of seventeenth-century Holland (de Vries and van der Woude 1997), eighteenth-century England (North and Weingast 1989) and the nineteenth-century United States (Rousseau and Sylla 2004). For the twentieth century, higher levels of banking system development have been found to be positively associated with faster rates of accumulation and economic activity, suggesting that finance does not follow economic growth, but *leads it* (King and Levine 1993, Levine 2004).

However, contrary to the present paper and to other accounts (e.g. Rowbotham's 'forced economic growth', 1998), these studies do not emphasize debt-driven disciplining mechanisms as a major explanation for this link. They rather stress that adequate credit conditions encourage the 'natural' entrepreneurial spirit to flourish. In fact, economic growth does not simply result from 'unlimited wants' as expressed in an innate 'drive for profits' made possible via investment credit. It results – perhaps above all – from the obligation to take out loans and from the subsequent constant threat of defaulting in a competitive context. The threat of bankruptcy is the key complement to the familiar 'whip of market competition' explaining capitalist dynamics.

In this context, productive activities have not only to expand; they also have to do so *quickly*. Regarding the second point, as early as the sixteenth century, 'there was a great deal of importance placed on keeping time in meeting obligations' (Muldrew 1998, 174). Unsurprisingly, 'the proverb that time is money seems to have already been common by the late sixteenth century, thus even time had become a commodity of measurement which could be used to judge the worth of competing individual householders' (380).

On the third point, there is evidence that one of the main effects of indebtedness was the drive to increase land productivity and cost-cutting techniques (Habakkuk 1994, Watt 2006). Accordingly, many productive innovations appeared in different spheres of early modern rural economies (Hoffman 1996, Post 2011). Exemplifying this for the United States, Post (2011, 152) describes how, after 1840, indebtedness and taxes forced family-farmers to compete in the market, a process that 'unleashed a dynamic of productive specialisation, technical innovation and accumulation'.

Institutionally, things evolved in an equally specific way. As we have seen for early modern England, the initial credit expansion correlates with a dramatic boom of debt litigations in local courts. Formal bankruptcy proceedings came to be applied widely. Later on, however, rates of litigation declined as the individuals' skills, knowledge and way of thinking seem to have become better adapted to the management of credit (Muldrew 1998, Finn 2003). This highlights how credit may shape 'correct' behaviours and mentalities over time.[14] Simultaneously, the credit boom also shaped the legal system itself. Creditors continuously pushed for legal reforms to strengthen their position and facilitate law enforcement: 'From 1540 to the early eighteenth century, the authority of the state grew because of the demand on the part of private civil disputants who required a reservoir of authority to mediate and resolve their many [debt-related] disputes' (Muldrew 1998, 331). Accordingly, legal institutions have been gradually transformed to the benefit of more secured credit/debt relations – an evolution fostered by increasing longer-distant and therefore more impersonal relationships. The mutuality of exchange relations seems to have diminished over time and there was a general shift, in the late seventeenth century, from informal to formal credit (Finn 2003).

Credit and debt thus constituted a major mechanism by which capitalism reshaped the political and legal institutions in the early modern and modern period. It played a central role in the shift from the feudal conventions of social obligations to the capitalist conceptions of contractual enforcement – that is, from 'status' to 'contract'.

Theoretical propositions

Through the pressures they create and their remarkable reward-or-punish nature, interest-bearing and guarantee-based loans (in all their variety) have a deep impact on the trajectory of capitalism. Starting with the idea that the institutions regulating 'owning' (i.e. full control) and 'borrowing' (i.e. temporary control, or uncompleted exchange) are perhaps the most fundamental economic institutions, I suggest that in market economies, and especially in capitalism, credit/debt relations represent an economic equivalent of natural selection in biology.[15] Where the latter is defined as the process by which heritable traits are passed from one generation to the other as a result of differential reproductive success, credit/debt selection is the process by which ownership (and thus power) is acquired and passed from one generation to the other as a result of differential success in dealing with credit and debt. This selection mechanism seems a more powerful factor of economic evolution than the development of trade per se (i.e. completed exchanges): whereas trade is a finite operation between 'equals', credit/debt relations generate durable constraints and have the potential to involve a broader spectrum of actors – including very unequal ones – as they require in principle no previous savings or production on

---

[14]Today, rural creditors use training methods to build a 'credit culture'. One typical technique (widely used in microfinance) is to increase loan sizes according to repayment performances.

[15]This analogy should not cause alarm. I do not intend to suggest an overarching mono-causal 'law of change': even in modern biology, natural selection is far from explaining everything.

the borrower's side. Credit/debt mechanisms seem also more fundamental than innovations, the favourite factor of evolutionary economists (who have been surprisingly silent on credit and debt). As we have seen, the pressure for innovations may be explained, to some extent, by the imperatives of credit themselves.

Like natural selection, credit/debt selection is 'everywhere' in the market sphere. By using credit in order to subsist, subdue or grow, some actors gain, some lose, some maintain themselves and some are eliminated. The interest of the dominant classes is to monopolize their power position – routinely using credit to accumulate or to subordinate others – while lower classes are often forced to borrow but may also try to climb the social ladder using productive credit. On an historical scale, credit/debt selection represents what Marx (1982, 778) called an 'enormous social mechanism' that not only shaped the ownership structure and laws but also behaviours and ways of thinking. Peasants, landlords or entrepreneurs who entered a credit/debt relation had to start to evaluate more things in money terms (land, labour, nature, etc.) and to keep track of their budget. If they failed to do so, the possibility of sanction was never far behind. Accordingly, as we have seen, debtors had to engage in commercial activities, specialize, sell their labour power, innovate and/or increase productivity. After a while, some debtors went bankrupt and others were lucky or disciplined enough to come out unharmed. A handful even became rich – hence the emergence of social differentiation, among other consequences.

Credit/debt selection transforms or reinforces patterns of capital ownership and therefore the power structure. It concerns (to different degrees) everyone taking out a loan – from a short-term, informal, consumptive microloan to a long-term, bank-funded, productive investment – as long as significant costs and guarantees are involved. It operates at different levels, from the individual to the government, including the household and the firm. It takes many forms, from the most benign to the most selective ones – sometimes leading to suicide or murder. It is both 'passive' – impersonal market forces eliminate defaulting debtors – as well as 'active' – creditors carefully select who will benefit from a loan. Indeed, formal and informal lenders typically select people with specific skills and values compatible with profit generation, namely the ability to calculate, to work hard, to fit into market institutions and, if needed, to dis-embed their economic activities from the community life or from the local ecosystem's long-term sustainability (Gerber 2013). Creditors then monitor their clients, pressure them through guarantees and apply sanctions when needed. This process sorts those already presenting the required abilities and encourages others to adopt and develop them. Echoing the present argument, Marx (1992, 735–6, my emphasis) wrote that

> when a man without fortune receives credit in his capacity of industrialist or merchant, it occurs with the expectation that he will function as capitalist […]. *He receives credit in his capacity of potential capitalist.* The circumstance that a man without fortune but possessing energy, solidity, ability and business acumen may become a capitalist in this manner […] is greatly admired by apologists of the capitalist system. [The mechanism] reinforces the supremacy of capital itself, expands its base and enables it to recruit ever new forces for itself out of the substratum of society. In a similar way, the circumstance that the Catholic Church in the Middle Ages formed its hierarchy out of the best brains in the land, regardless of their estate, birth or fortune, was one of the principal means of consolidating ecclesiastical rule and suppressing the laity. The more a ruling class is able to assimilate the foremost minds of a ruled class, the more stable and dangerous becomes its rule.

The idea of a selection based on credit has been noted in passing by some neoclassical economists. Stiglitz (1994, 25), for example, pointed out that 'one of the most important functions of financial institutions is to select among alternative projects and to monitor the use of the funds'. However, credit/debt selection certainly does not lead to any 'optimum' in the neoclassical sense: its outcomes can be as much a result of luck as of skills, and they can be as constructive as destructive. The mechanism proposed here is at the same time Marxian (credit/debt relations can be a class struggle over the property structure), Veblenian (credit/debt institutions *create* related interests, values and behaviours), Lamarckian (acquired property and sociocultural capital are often inherited) and Darwinian (principles of variation, selection and retention of 'credit-friendly' abilities and innovations).

Concluding remarks

Edmund Leach famously said of the Kachin of highland Burma that 'when a Kachin talks about the "debts" [...], he is talking about what an anthropologist means by "social structure"' (1954, 141). This statement seems to apply as well to the economic sphere of capitalist societies. Credit/debt relations 'glue' together the entire economic structure from international financial organizations and central banks down to smallholders and agricultural workers, via commercial banks, industries and the various dealers and sub-dealers. Under capitalist property, the potentials of credit could be revealed fully, implying an intimate connection between the two and a particularly strong influence of credit/debt selection in shaping the economy.

Together with the proprietor/employees relations, the credit/debt couple is a fundamental component of the capitalist social-property relations. Credit/debt relations have compelled borrowers to behave similarly to *Homo economicus* and to generate some of the typical features of capitalism, namely its constant pressures for solvency, concentration, growth, short-termism and innovations, but also two of its most emblematic impacts: the undermining of community and the degradation of the environment. The development of credit has equally led to increasingly serious financial crises through an excrescence of the 'virtual' level of the economy. As Douglas (1920) and Soddy (1926) pointed out long ago, it is easy for the financial system to increase private and public debts in a multiplicity of ways, and to mistake this expansion for the creation of real wealth.

With the worsening of financial crises and household indebtedness, debt-centred organizing offers the potential to reinvigorate radical struggle in the twenty-first century (Kleiner 2012) – just like environmental disruptions. Indeed, today's class consciousness often falls along creditor/debtor lines rather than capitalist/worker lines, especially in the rural world. Instead of letting credit/debt selection rule, democratic control should take back the economy's trajectory through the definition of socially useful objectives and the corresponding training and financing of the relevant actors. Interest-free credit and monetary systems are not as radical as it may seem and have already started under different forms of mutual credit and alternative currencies (Lietaer *et al.* 2012, Gerber 2014). Keynes (1936, 376) himself described a post-capitalist 'quasi stationary community' which would be characterized by a stable population, the absence of wars, full employment and, one may add, ecological sustainability. In such a society, he argued, the marginal efficiency of capital would fall to zero, leading to a near zero interest rate, and, consequently, to the 'euthanasia of the rentier'.

References

Ågren, M. 1994. Land and Debt: On the Process of Social Differentiation in Rural Sweden, circa 1750–1850. *Rural History*, 5(1), 23–40.

Allen, R.C. 1992. *Enclosure and the Yeoman*. Oxford: Clarendon Press.

Altvater, E. 1997. Financial Crisis on the Threshold of the 21st Century. In: *Socialist Register 1997*, ed. L. Panitch. London: The Merlin Press, pp. 48–74.

Atack, J. and F. Bateman. 1987. *To Their Own Soil: Agriculture in the Antebellum North*. Aimes: University of Iowa Press.

Badosa Coll, E. 1990. Endeutament Col·lectiu i Desaparició de Bens Comunals a Catalunya a la Secona Meitat del Segle XVIII. *Pedralbes*, 10, 51–66.

Banaji, J. 1977. Modes of Production in a Materialist Conception of History. *Capital and Class*, 6, 1–44.

Bernstein, H. 1979. African Peasantries: A Theoretical Framework. *Journal of Peasant Studies*, 6(4), 421–43.

Berthe, M., ed. 1998. *Endettement Paysan et Crédit Rural dans l'Europe Médiévale et Moderne*. Toulouse: Presses Universitaires du Mirail.

Bhaduri, A. 1983. *The Economic Structure of Backward Agriculture*. London: Academic Press.

Bourdieu, P., L. Boltanski and J.-C. Chamboredon. 1963. *La Banque et sa Clientèle. Introduction à une Sociologie du Crédit*. Paris: Centre de Sociologie Européenne.

Brass, T. 1999. *Towards a Comparative Political Economy of Unfree Labour*. London: Frank Cass.

Brass, T. 2010. Capitalism, Primitive Accumulation and Unfree Labour. In: *Imperialism, Crisis and Class Struggle*, ed. H. Veltmeyer. The Hague: Brill, pp. 67–131.

Braudel, F. and E. Labrousse, eds. 1977. *Histoire Economique et Sociale de la France, Tome 1: 1450–1660*. Paris: Presses Universitaires de France.

Brenner, R. 1985a [1982]. The Agrarian Roots of European Capitalism. In: *The Brenner Debate*, eds. T. Aston and C. H. E. Philpin. Cambridge: Cambridge University Press, pp. 213–327.

Brenner, R. 1985b. The Social Basis of Economic Development. In: *Analytical Marxism*, ed. J. E. Roemer. Cambridge: Cambridge University Press, pp. 25–53.

Carruthers, B.G. 2005. The Sociology of Money and Credit. In: *The Handbook of Economic Sociology*, eds. N.J. Smelser and R. Swedberg. Princeton: Princeton University Press, pp. 355–78.

Chayanov, A.V. 1966 [1925]. *The Theory of Peasant Economy*. Homewood: Richard Irwin.

Chenowith, C.G. 1923. *Long-Term Agricultural Credit in California*. Master's Thesis. Berkeley: University of California.

Cooper, J.P. 1978. In Search of Agrarian Capitalism. *Past and Present*, 80, 20–65.

Dobb, M. 1947. *Studies in the Development of Capitalism*. New York: International Publishers.

Douglas, C.H. 1920. *Economic Democracy*. London: Palmer.

Duby, G. 1998 [1962]. *Rural Economy and Country Life in the Medieval West*. Philadelphia: University of Pennsylvania Press.

Finn, M.C. 2003. *The Character of Credit: Personal Debt in English Culture, 1740–1914*. Cambridge: Cambridge University Press.

Fontaine, L. 2008. *L'Economie Morale: Pauvreté, Crédit et Confiance dans l'Europe Préindustrielle*. Paris: Gallimard.

Gaulin, J.-L. and F. Menant. 1998. Crédit Rural et Endettement Paysan dans l'Italie Communale. In: *Endettement Paysan et Crédit Rural dans l'Europe Médiévale et Moderne*, ed. M. Berthe. Toulouse: Presses Universitaires du Mirail, pp. 35–67.

Geertz, C. 1962 [1956]. The Rotating Credit Association: A 'Middle Rung' in Development. *Economic Development and Cultural Change*, 10(3), 241–63.

Gerber, J.-F. 2013. The Hidden Consequences of Credit: An Illustration from Rural Indonesia. *Development and Change*, 44(4), 839–60.

Gerber, J.-F. 2014. An Overview of Local Credit Systems and Their Significance. Submitted.

Goubert, P. 1986 [1962]. *The French Peasantry in the Seventeenth Century*. Cambridge: Cambridge University Press.

Graeber, D. 2009. Debt, Violence and Impersonal Markets: Polanyian Meditations. In: *Market and Society: The Great Transformation Today*, eds. C. Hann and K. Hart. Cambridge: Cambridge University Press, pp. 106–32.

Graeber, D. 2011. *Debt: The First 5,000 Years*. New York: Melville House.

Habakkuk, J. 1994. *Marriage, Debt, and the Estates System: English Landownership, 1650–1950*. Oxford: Oxford University Press.

Harris, W.V., ed. 2008. *The Monetary Systems of the Greeks and Romans*. Oxford: Oxford University Press.

Heinsohn, G. 2008. Where Does the Market Come From? In: *Property Economics: Property Rights, Creditor's Money and the Foundations of the Economy*, ed. O. Steiger. Marburg: Metropolis-Verlag, pp. 243–61.

Heinsohn, G. and O. Steiger. 1996. *Eigentum, Zins und Geld: Ungelöste Rätsel der Wirtschaftswissenschaft*. Reinbek bei Hamburg: Rowohlt.

Heinsohn, G. and O. Steiger. 2003. The Property Theory of Interest and Money. In: *Recent Developments in Institutional Economics*, ed. G.M. Hodgson. Cheltenham: Edward Elgar, pp. 484–517.

Henderson, G.L. 1998. Nature and Fictitious Capital: The Historical Geography of an Agrarian Question. *Antipode*, 30(2), 73–118.

Hoffman, P. 1996. *Growth in a Traditional Society: The French Countryside, 1450–1815*. Princeton: Princeton University Press.

Hoffman, P., G. Postel-Vinay and J.-L. Rosenthal. 1999. Information and Economic History: How the Credit Market in Old Regime Paris Forces us to Rethink the Transition to Capitalism. *The American Historical Review*, 104(1), 69–94.

Hoppit, J. 1990. Attitudes to Credit in Britain, 1680–1790. *Historical Journal*, 33(2), 305–22.

Hudson, M. 2002. Reconstructing the Origins of Interest-Bearing Debt and the Logic of Clean Slates. In: *Debt and Economic Renewal in the Ancient Near East*, eds. M. Hudson and M. Van De Mieroop. Baltimore: CDL Press, pp. 7–58.

Jones, N.L. 1989. *God and the Moneylenders: Usury and Law in Early Modern England*. Cambridge, MA: Basil Blackwell.

Kara, S. 2009. *Sex Trafficking: Inside the Business of Modern Slavery*. New York: Columbia University Press.

Kautsky, K. 1900 [1899]. *La Question Agraire*. Paris: Giard et Brière.

Keynes, J.M. 1936. *The General Theory of Employment, Interest and Money*. London: Macmillan.

Keynes, J.M. 1945. Overseas Financial Policy in Stage III. In: *The Collected Writings of John Maynard Keynes*. Vol. 24. London: Macmillan.

King, R.G. and R. Levine. 1993. Finance and Growth: Schumpeter Might Be Right. *Quarterly Journal of Economics*, 108(3), 717–37.

Kleiner, D. 2012. Debt Revolt. In: *Beautiful Trouble*, eds. A. Boyd and D.O. Mitchell. New York: OR Books, pp. 226–27.

Knight, A. 1986. Mexican Peonage: What Was It and Why Was It? *Journal of Latin American Studies*, 18(1), 41–74.

Kriedte, P., H. Medick and J. Schlumbohm. 1981. *Industrialization before Industrialization: Rural Industry in the Genesis of Capitalism*. Cambridge: Cambridge University Press.

Laffont, J.-L., ed. 1999. *Le Notaire, le Paysan et la Terre dans la France Méridionale à l'Epoque Moderne*. Toulouse: Presses Universitaires du Mirail.

Leach, E.R. 1954. *Political Systems of Highland Burma: A Study of Kachin Social Structure*. London: G. Bell.

Lenin, V.I. 1967 [1899]. *The Development of Capitalism in Russia*. Collected Works, Vol. 3. Moscow: Progress Publishers.

Levine, R. 2004. Finance and Growth: Theory and Evidence. In: *Handbook of Economic Growth*, Vol. 1a, eds. P. Aghion and S. Durlauf. Amsterdam: Elsevier Science, pp. 865–934.

Lietaer, B., C. Arnsperger, S. Goerner and S. Brunnhuber. 2012. *Money and Sustainability: The Missing Link*. Club of Rome Report. Axminster: Triarchy Press.

Marglin, S.A. 1974. What do Bosses do? The Origins and Functions of Hierarchy in Capitalist Production. *The Review of Radical Political Economics*, 6(2), 60–112.

Marglin, S.A. 2008. *The Dismal Science: How Thinking Like an Economist Undermines Community*. Cambridge, MA: Harvard University Press.

Marx, K. 1982 [1867]. *Capital*. Vol. 1. New York: Penguin Classics.

Marx, K. 1992 [1894]. *Capital*. Vol. 3. New York: Penguin Classics.

McIntosh, M.K. 1988. Money Lending on the Periphery of London, 1300–1600. *Albion*, 20(4), 557–71.

Muldrew, C. 1998. *The Economy of Obligation. The Culture of Credit and Social Relations in Early Modern England*. London: Macmillan.

National Farmers Union (NFU). 2010. Losing Our Grip. Online report by Canada's NFU.

Newton, T. 2003. Credit and Civilization. *British Journal of Sociology*, 54(3), 347–71.

North, D. and B. Weingast. 1989. Constitutions and Commitment: The Evolution of Institutions Governing Public Choice in Seventeenth-Century England. *Journal of Economic History*, 44(4), 803–32.

Pichard, G. 2001. L'Espace Absorbé par l'Economique? Endettement Communautaire et Pression sur l'Environnement en Provence (1640–1730). *Histoire et Sociétés Rurales*, 16, 81–115.

Post, C. 2011. *The American Road to Capitalism: Studies in Class-Structure, Economic Development and Political Conflict, 1620–1877*. Leiden and Boston: Brill.

Postel-Vinay, G. 1998. *La Terre et l'Argent: L'Agriculture et le Crédit en France du XVIIIe au Début du XXe Siècle*. Paris: Albin Michel.

Regehr, W. 1994. Mennonite Economic Life and the Paraguayan Experience. In: *Anabaptist/Mennonite Faith and Economics*, eds. C. Redekop, V.A. Krahn and S.J. Steiner. Lanham: University Press of America, pp. 127–52.

Roemer, J.E. 1982. *A General Theory of Exploitation and Class*. Cambridge, MA: Harvard University Press.

Roseberry, W. 1978. Peasants as Proletarians. *Critique of Anthropology*, 3(11), 3–18.

Rousseau, P. and R. Sylla. 2004. Emerging Financial Markets and Early U.S. Growth. *Explorations in Economic History*, 42(1), 1–26.

Rowbotham, M. 1998. *The Grip of Death: A Study of Modern Money, Debt Slavery and Destructive Economics*. Charlbury: Jon Carpenter.

Schofield, P.R. and T. Lambrecht, eds. 2009. *Credit and the Rural Economy in North-western Europe, c.1200–c.1850*. Turnhout: Brepols.

Schofield, P.R. and N.J. Mayhew, eds. 2002. *Credit and Debt in Medieval England, c.1180–c.1350*. Oxford: Oxbow.

Schrader, H. 1997. *Changing Financial Landscapes in India and Indonesia: Sociological Aspects of Monetization and Market Integration*. Hamburg: LIT Verlag.

Schumpeter, J.A. 1934 [1911]. *The Theory of Economic Development*. Cambridge, MA: Harvard University Press.

Soddy, F. 1926. *Wealth, Virtual Wealth and Debt*. London: George Allen & Unwin.

Steppacher, R. 2008. Property, Mineral Resources and 'Sustainable Development'. In: *Property Economics: Property Rights, Creditor's Money and the Foundations of the Economy*, ed. O. Steiger. Marburg: Metropolis-Verlag, pp. 217–41.

Steppacher, R. and J.-F. Gerber. 2012. Meanings and Significance of Property with Reference to Today's Three Major Eco-Institutional Crises. In: *Towards an Integrated Paradigm in Heterodox Economics*, eds. J.-F. Gerber and R. Steppacher. Basingstoke: Palgrave-Macmillan, pp. 111–26.

Stiglitz, J.E. 1994. The Role of the State in Financial Markets. *Proceedings of the World Bank Annual Conference on Development Economics*, 1993, 19–52.

Testart, A. 2005. *Éléments de Classification des Sociétés*. Paris: Errance.

de Vries, J. and A. van der Woude. 1997. *The First Modern Economy: Success, Failure, and Perseverance of the Dutch Economy, 1500–1815*. Cambridge: Cambridge University Press.

Wallerstein, I. 1974. *The Modern World-System: Capitalist Agriculture and the Origins of the European World-Economy in the Sixteenth Century*. Vol. 1. New York: Academic Press.

Watt, D. 2006. 'The Laberinth of Thir Difficulties': The Influence of Debt on the Highland Élite c.1550–1700. *The Scottish Historical Review*, 85(219), 28–51.

Wood, E.M. 2002. *The Origin of Capitalism*. Second Edition. New York: Verso.

Wordie, J.R. 1983. The Chronology of English Enclosure, 1500–1915. *Economic History Review*, 36(4), 483–505.

Julien-François Gerber is a Fellow at the Jawaharlal Nehru Institute of Advanced Study, Jawaharlal Nehru University, New Delhi, India. Prior to that, he was visiting scholar at Harvard University's Department of Economics for two years. He is interested in the integration of the anthropology of development, heterodox economics and human ecology, as well as in the relationship between science and activism. He has done fieldwork in Cameroon, Ecuador and Indonesia.

Food and finance: the financial transformation of agro-food supply chains

S. Ryan Isakson

This article draws upon existing literature to document and describe the rise of finance in food provisioning. It queries the role of financialization in the contemporary food crisis and analyzes its impacts upon the distribution of power and wealth within and along the generalized agro-food supply chain. A systematic treatment of key nodes in the supply chain reveals four key insights: (1) the line between finance and food provisioning has become increasingly blurred in recent decades, with financial actors taking a growing interest in food and agriculture and agro-food enterprises becoming increasingly involved in financial activities; (2) financialization has reinforced the position of food retailers as the dominant actors within the agro-food system, though they are largely subject to the dictates of finance capital and face renewed competition from financialized commodity traders; (3) financialization has intensified the exploitation of food workers, increasing their workload while pushing down their real wages and heightening the precarity of their positions, and (4) small-scale farmers have been especially hard hit by financialization, as their livelihoods have become even more uncertain due to increasing volatility in agricultural markets, they have become weaker vis-à-vis other actors in the agro-food supply chain, and they face growing competition for their farmland. The paper concludes by identifying themes for future research and asking readers to reimagine the role of finance in food provisioning.

Introduction: the rise of finance

The contemporary food crisis, which rose to prominence with the dramatic increase in food prices in 2007–2008, continues to ravage the world's poor. Nearly one billion people are chronically malnourished while another billion suffer from the constant uncertainty of whether there will be a next meal and from where it will come. Many analysts point to the 'financialization of food' as a key culprit for the ongoing crisis. When doing so, they often focus upon the increasing participation of financial actors in agricultural derivatives markets and the resulting impact upon food prices. Yet the rise of finance capital is not novel. Many political economists maintain that it is a recurring feature of capitalist development (Lenin 1974, Arrighi 1994, Hobson 2010). Moreover, the current phase of financialization, which has been unfolding since the late 1970s (Palley 2007, Krippner 2011), has

I would like to thank Jun Borras and Jacqueline Morse for their encouragement and insights in conceptualizing this project. I am also grateful for the helpful comments and suggestions from two anonymous reviewers.

permeated nearly every aspect of food provisioning. Financial actors are playing an increasingly active role in food retailing (Burch and Lawrence 2009, 2013), food processing (Rossman 2010), commodity trading (Murphy *et al.* 2012), the determination of food prices and the distribution of agricultural risk (Breger Bush 2012, Ghosh *et al.* 2012, Spratt 2013, Clapp 2014), the provisioning of agricultural inputs (Ross 2008, The Economist 2009), and the ownership and control of farmland (HighQuest 2010, Cotula 2012, Fairbairn 2014). At the same time, enterprises that operate in each of these activities are increasingly active in financial markets and earning a growing share of their revenues from financial activities. How, if at all, has this financialization of food and agriculture contributed to the contemporary food crisis? How is it reshaping social relations and the distribution of wealth and power within and along the generalized agro-food supply chain? Reviewing the existing literature, this paper documents the finanancialization of the major nodes of the value chain and explores these questions. In doing so, I take a political economy approach that questions both how and why financialization is redistributing costs and benefits along the agro-food value chain and gives particular attention to the impacts of financialization upon small-scale agricultural producers and laborers within the agro-food sector. Reviewing the literature through a political economy lens reveals four key insights: (1) the line between finance and food provisioning has become increasingly blurred in recent decades, with financial actors taking a growing interest in food and agriculture and agro-food enterprises becoming increasingly involved in financial activities; (2) financialization has reinforced the position of food retailers as the dominant actors within the agro-food system, though they are largely subject to the dictates of finance capital and face renewed competition from grain traders who are well-positioned to capitalize upon the financial transformation of food; (3) financialization has intensified the exploitation of food workers and heightened the precariousness of their employment, and (4) small-scale farmers have been especially hard hit by financialization, as their livelihoods have become even more uncertain and their market power vis-à-vis other actors in the agro-food supply chain has eroded.

Paraphrasing Epstein (2001, 2005), financialization refers to the increasing importance of financial motives, financial actors, financial markets and financial institutions in the operation of economies and their governing institutions, at both the domestic and international level. For Krippner (2011, 4), this process can be understood as 'the tendency for profit making in the economy to occur increasingly through financial channels rather than through productive activities' (cf Arrighi 1994). She differentiates financial activities – where liquid capital is provisioned (or transferred) in expectation of future interest, dividends or capital gains – from the real economy where the actual production and trade of commodities occurs.

In line with Krippner's understanding of financialization, several studies have documented the growing share of financial profits in most of the world's largest economies since the late 1970s (Epstein and Jayadev 2005, Krippner 2005, 2011, Palley 2007, Orhangazi 2008). Four trends have defined the rise of finance. First, the share of total domestic profits earned by traditional financial firms has increased, indicating the growing importance of private institutional investors like banks, mutual funds, hedge funds, pension funds and private equity funds (Palley 2007, Krippner 2011).[1] Second, traditionally non-financial firms are also engaging in financial activities and earning a noticeably larger

---

[1]In the United States, the ratio of financial to non-financial profits rose from 20 percent in 1983 to 50 percent in 2001, a 150 percent increase (Krippner 2011, 41).

share of their revenues from consumer credit and other financial activities (Orhangazi 2008, Krippner 2011).[2] Third, since the so-called 'shareholder revolution' of the 1990s, managers for firms of all types have reoriented the direction of their enterprises such that their top priority is to satisfy their shareholders' demands for dividends (Crotty 2009, Baud and Durand 2012). Finally, as financialization has transferred income from the real sector to the financial sector, the process has contributed to stagnating wages for workers and increased income inequality, resulting in slower economic growth and rising volatility (Stockhammer 2004, Palley 2007, Baud and Durand 2012, Wolff 2013).

Following Krippner (2011), it is possible to identify three types of explanations for the rise of finance, each roughly adhering to one of the primary schools of economic thought. Stemming from the tradition of neo-classical cum neo-institutional economics, the orthodox explanation for financialization celebrates it as a solution to the so-called 'agency problem' of corporations, whereby the interests of a firm's manager may not necessarily be in line with its owners (i.e. shareholders). The rapidly expanding practice of compensating executives in stock options combined with the emerging threat of private equity takeovers since the 1980s have helped to align managers' and shareholders' interests, providing an efficient solution to the agency problem (Palley 2007, Krippner 2011). A second type of explanation emerges directly from (post) Keynesian thinking about speculative bubbles. Accordingly, adherents argue that insufficiently regulated financial markets are inherently prone to speculative excess and that investors' ignorance about the functioning of an economy will drive them to engage in herd behavior (Palley 2007, Crotty 2009, Krippner 2011). Finally, Marxist scholars portray financialization as a cyclical feature of capitalism. They attribute the contemporary rise of finance to an overconsumption crisis in the 1970s: faced with insufficient demand for their products and declining profits, US and European firms redirected their surplus capital from productive activities to financial markets (Arrighi 1994, Harvey 2010, Krippner 2011).

Regardless of one's theoretical inclination, the rise of finance has occurred in the context of stagnant wages, a rising volume of debt and financial deregulation. In the United States and other financial hubs, the real wage received by workers has remained flat since the 1970s; this has occurred despite rising productivity (Palley 2007, Wolff 2013). In part, this scenario can be attributed to the aforementioned shareholder revolution, wherein managers have prioritized profits and dividends over the welfare of their workers. However, in keeping with the Marxian thesis, it also created a situation where insufficient demand from cash-strapped households discouraged investment in the real sector and spurred investors to deposit their surplus funds in the financial sector. To be sure, some of these funds were loaned back to consumers, allowing them to improve their standard of living despite stagnant wages, though also contributing to a dramatic increase in household and mortgage sector debt (Palley 2007). But working-class households also adapted by sending additional household members into the paid workforce which, for firms, helped to increase demand for goods and services that were previously provisioned in the household, including a demand for prepared and processed foods. As will be discussed later, this transition contributed to the rise of the food manufacturing sector as one of the most powerful links within the agro-food supply chain during the 1970s and 1980s. Beginning in the 1990s, supermarkets were able to usurp some of that power as consumers began to

---

[2]In the US, for instance, financial assets as a share of non-financial firms' tangible assets remained fairly steady at around 35 percent from 1952 to 1983, but have sharply increased since then and, as of 2003, were around 100 percent (Orhangazi 2008).

switch to fresher and healthier prepared foods (Busch and Bain 2004, Burch and Lawrence 2009).

The neoliberal restructuring of economies in both the North and South also enabled the rise of finance. Faced with economic turbulence in the 1970s and early 1980s, governments in the US, Britain and other financial hubs dismantled Keynesian-influenced regulations that had governed the financial sector for nearly half a century. Influenced by the rise of neo-classical economics, particularly Milton Friedman's 'efficient markets hypothesis', states erected a 'New Financial Architecture' (NFA) that reflected the era's prevailing belief that minimal government regulation enables markets to generate efficient and socially optimal outcomes. 'The NFA', economist James Crotty (2009, 564) writes, 'is based on light regulation of commercial banks, even lighter regulation of investment banks, and little if any regulation of the "shadow banking system" – hedge and private equity funds and bank-centered Special Investment Vehicles'. As financial institutions that were previously prohibited from doing so consolidated and amassed massive pools of investment funds, their technicians devised new financial products that would purportedly allow financial markets to regulate themselves and identified new arenas for investment, including foreign direct investment (FDI) in liberalized southern markets and, importantly for this paper, agricultural derivatives, farmland and agro-food enterprises. As the following sections discuss, this financialization of the agro-food sector has dramatically transformed food provisioning. It has consolidated financial elites' wealth and power at the expense of agricultural producers and workers and exacerbated the fragility of the global industrial food system.

The financialization of food retailing

Since the onset of the 'supermarket revolution' in the early 1990s, food retailers have emerged as the most powerful actors *within* the agro-food system (Reardon and Berdegué 2002, Busch and Bain 2004, Burch and Lawrence 2005). Faced with saturated markets and stagnant sales at home, Northern food retailers found opportunities to expand their sales by investing in Southern markets that had recently been required to liberalize FDI under their structural adjustment programs (Reardon *et al.* 2009). US and European supermarkets merged with and acquired their Southern counterparts while setting up new stores to expand their customer base in areas where food trade had previously occurred in open-air markets and other settings (Reardon and Berdegué 2002). The result has been an increasingly globalized and concentrated food retail sector dominated by a handful of Northern-based supermarkets. Supermarkets' share of food retail sales in Latin America, for example, exploded from 10–20 percent in 1990 to 60 percent by 2001; the five largest supermarket chains now account for nearly two-thirds of food sales on the continent (Reardon and Berdegué 2002, Clapp 2012). A similar process is unfolding in Africa and Asia – especially in India, China, and Vietnam – as food retailers tap into markets where rapid economic growth is facilitating the adoption of Western food practices (Reardon *et al.* 2009).[3]

---

[3]Domestic and regional supermarkets first spurred the consolidation of supermarkets in Latin America. Relatively large domestic retailers merged with and acquired their smaller counterparts. When even larger transnational grocers entered regional markets they acquired these large domestic chains, a process that Reardon and Berdegué (2002) refer to as 'big eats small then bigger eats big'. Apparently a similar process is unfolding in China, where domestic chains are growing in tandem with transnational retailers.

The globalization and concentration of food retailing has not only boosted sales for a small number of supermarket chains, it has also enhanced their economic power vis-à-vis other actors in the agro-food supply chain. Indeed, food retailers can now be characterized as the 'masters of the food system' (Burch and Lawrence 2013). Their *oligopolistic* hold over downstream food consumers (and food service providers) augments the *oligopsonistic* power that they hold over upstream food actors,[4] thereby giving them tremendous power over the types and quality of food produced, the manner in which food is produced, the location of food production, the terms of food exchange and, ultimately, the distribution of surplus within the food system (Busch and Bain 2004, Burch and Lawrence 2005). Yet, as Burch and Lawrence (2013) ask, how do these power relations shift as the 'masters of the food system' confront the financial titans who are the contemporary 'masters of the universe'? Are they, as Burch and Lawrence (2009) contend in an earlier piece, the best-positioned actors within the agro-food supply chain to exploit the benefits of financialization? This section addresses these questions.

Like the financialization process in general, the financialization of food retailing has blurred the boundaries between the financial and real sectors. Even as financial actors have come to play a more prominent role in food sales, food retailers are earning a greater share of their revenues from financial activities. Investigating financialization in the more conventional sense (i.e. financial actors entering the retail sector), Burch and Lawrence (2009, 2013) analyze how private equity takeovers of supermarkets transform the food retail sector. Using a private equity consortium's takeover of the UK-based Somerfield Supermarkets as an example, Burch and Lawrence identify four strategies that financial actors are employing as a means of realizing shareholder value: (1) narrowing the retailer's product line and reducing the number of suppliers, thereby streamlining the sourcing process (including the closure of some distribution centers) and reducing costs; (2) reducing the number of employees (many of whom previously worked in the distribution centers) and increasing the workload (i.e. rate of exploitation) of the remaining workers; (3) disregarding previous commitments to environmental quality and the well-being of food producers in the global South, including Somerfield's withdrawal from the Ethical Trading Initiative (ETI), which sets labor standards for developing country suppliers, and (4) de-bundling and repackaging assets, including the introduction of an operating company/property company (opco/propco) arrangement, whereby Somerfield sold its real estate properties to a newly created subsidiary of itself that, in turn, leased the property back to Somerfield. The opco/propco model separates the use value of land from its exchange value (Fairbairn 2014), thereby allowing food retailers to (a) receive special tax considerations and (b) repackage their properties into real estate investment trusts (REITs) where investors can deposit their funds.[5]

In addition to the obvious and detrimental impacts upon food workers, the reconfiguration of food retailing to the objectives of finance capital is likely to negatively affect small-

---

[4]An oligopoly is a market form in which there are a small number of sellers, which means, among other things, that sellers are able to exercise market power over buyers (in this instance, food retailers are able to exercise power over food consumers). An oligopsony, in contrast, refers to a market where there are a limited number of buyers. Once again, the small number of buyers gives them economic power over sellers (in this case food retailers have power over food processors and traders). In other words, food retailers hold economic power over both their upstream and downstream counterparts in the agro-food supply chain.

[5]As described in Fairbairn (2014) and below, REITs are one mechanism through which farmland is financialized.

scale agricultural producers. Somerfield's withdrawal from the ETI demonstrates the triumph of shareholder values, where returns on financial investment trump the welfare of food producers and, in fact, push the bar lower as other supermarkets may consider following suit. Given the sway of food retailers over the agro-food system, their further disregard for labor and environmental quality could prove particularly harmful to agricultural producers and overall human health. Similarly, the streamlining of supply chains may reduce supermarkets' costs, but it also limits the number of buyers for agricultural produce. More concentrated markets, in turn, reduce the relative market power of farmers and may eliminate the market altogether for small-scale producers who do not harvest a sufficient quantity of output for buyers who prefer to purchase in volume.

Even as private equity consortiums and other financial actors are reorganizing food retailing, food retailers are diversifying into new financial activities. Typical for non-financial firms in the contemporary era, some have diverted their cash flows from investment in fixed capital into financial investments. For instance, the world's largest food retailer, Wal-Mart, launched a $25-million-dollar private fund to provide equity for the growth of its women- and minority-owned suppliers (Wal-Mart 2005), while its closest competitor, Carrefour, has branched out into a variety of financial investments (Baud and Durand 2012). More prevalent, however, have been retailers' development of their own financial activities. Since the financial deregulation of the 1980s and 1990s, supermarkets have offered an increasingly wide array of financial products to their customers, including credit and pre-paid debit cards, savings and checking accounts, insurance programs and even home mortgages. Retailers such as Wal-Mart and Tesco champion their initiatives as offering financial services to underserved/unbanked populations (Juhn 2007, Werdigier 2009). In doing so, they downplay the fact that they are often profiting from their customers' debt.

Along with the foreign expansion described above, financial activities have provided an important boost to food retailers' profits, enabling them to satisfy their impatient shareholders in the face of sluggish sales. Baud and Durand (2012) analyzed the interdependencies between globalization and financialization in the retail sector. Focusing upon the world's 10 leading retailers (eight of whom earn a significant share of their revenues from food sales)[6] they observed that, even though the corporations have experienced slowing revenue growth in their domestic markets since the 1990s, the growth rate of their shareholders' return on equity has increased, with particularly rapid growth since the 2001 dot-com crisis. Indeed, despite falling domestic sales growth, shareholders in retail corporations have fared very well in recent years. Their share of the leading retailers' value averaged 53 percent during the 2002–2007 period, more than double the roughly 22 percent of value that they claimed between 1990 and 2001 (Baud and Durand 2012, Figure A2).

Retailers have pursued a variety of complementary strategies to placate their demanding investors. As previously discussed, they took advantage of openings presented by economic liberalization to expand their operations in the global South, thereby widening their customer base while improving their access to cheap inputs, including labor. As retailers have begun to exhaust opportunities for foreign expansion, they have increasingly turned to financial activities to generate new revenue for shareholders. Their diversification into insurance, banking and other financial activities has been a boon to stockowners. Relative

---

[6]Listed in order of their total revenues, the top 10 retailers are Wal-Mart, Carrefour, Metro, Tesco, Kroger, Costco, Target, Home Depot, Sears and Ahold. All but Home Depot and Sears generate a significant portion of their revenues from food sales.

to the previous five-year period, financial assets as a share of total assets increased for seven of the eight food retail corporations analyzed by Baud and Durand (2012, Figure A8) during 2002–2007. On average, financial assets as a share of total assets increased from approximately 20 percent in 1996–2001 to 30 percent in 2002–2007, a 50 percent increase.[7]

Funding for food retailers' financial ventures – and thereby shareholder dividends – has come, in part, at the expense of retail workers and food suppliers. As described above, the food retail sector has become increasingly globalized and concentrated since the 1990s, shifting the balance of power in the agro-food supply chain from food processors to supermarkets. With the ongoing processes of internationalization and mergers and acquisitions, a shrinking number of food retailers are playing an increasingly important role as the gatekeepers between food manufacturers and consumers. The intensification of their oligopsonistic power has enabled food retailers to transfer costs to suppliers, even as they demand lower prices and changes in quality. For example, retailers have imposed 'just in time' inventory management and other changes in their supply chain management, thereby reducing the amount of funds 'immobilized' by inventory and storage costs (Baud and Durand 2012, 256). Similarly, retailers have increased the average time between delivery of food items and payment by as much as 50 percent, thereby freeing up additional funds for their financial activities and/or shareholder dividends (Baud and Durand 2012, Burch and Lawrence 2013). With payment for fresh fruits and vegetables typically delayed 45–60 days, but in some cases as long as 90 (Reardon and Berdegué 2002), the latter practice promises to prove especially challenging to small farmers in search of markets for their produce, a challenge that is only likely to intensify as leading retailers like Wal-Mart seek to ensure fresh produce by cutting out middlemen and contracting directly with farmers.[8]

Just as they have done with their suppliers, food retailers have drawn upon their power over workers to underwrite their financialization. The retail sector is labor-intensive and highly segmented; most workers are low-skilled, hold only part-time positions, are predominantly female, and have low levels of union representation (Baud and Durand 2012). The relatively weak bargaining power of these laborers has been further undermined by globalization and the introduction of labor-saving technologies. Under the pressures of financialization, many workers in the food retail sector have been displaced (Burch and Lawrence 2013); those who have managed to keep their jobs have endured longer and intensified working days even as their compensation remains flat (Rossman 2010, Baud and Durand 2012).

The reorganization of labor within their firms, combined with changes in inventory management, has contributed to improved efficiency within the retail sector since the 1990s. The returns from those efficiencies, however, have not been distributed equitably. Even as their productivity has increased, retail workers' wages have stagnated. Meanwhile,

---

[7]The degree to which food retailers have financialized their assets varies considerably. Among the major food retailers analyzed in Baud and Durand (2012), Carrefour, Ahold, Metro and Target were the most financialized, with financial assets accounting for 30–40 percent of their total asset holdings. Kroger and Wal-Mart were the least financialized with, respectively, 13 percent and 10 percent of their assets in finance. Baud and Durand observe that the most financialized retailers have benefitted the least from the internationalization of the sector and hypothesize that they have tried to appease shareholders by substituting financial revenues for their relatively weak sales abroad.
[8]See Hsu (2013) for a description of Wal-Mart's new produce strategy and Amy Cohen (2013) for a fascinating discussion of why small farmers prefer informal dealings with intermediaries over more direct interactions with large supermarkets.

suppliers have made significant changes to improve inventory management, yet retailers have appropriated the lion's share of the net savings. Flexing their economic power vis-à-vis suppliers and workers, retailers have 'forced' their partners to fund their financialization (Baud and Durand 2012, 256). In so doing, they have helped to reinforce their position of economic power within the agro-food supply chain while rewarding their demanding shareholders with dividends and share buybacks. Moreover, as suggested by the previously described private equity acquisition of Somerfield Supermarkets, the reconfiguration of food retailing to the objectives of financial capital will likely have detrimental impacts on agricultural producers. As will be discussed in the following section, those farmers are likely to feel doubly pinched as the financialization of agricultural risk and price setting heightens the uncertainty of their already precarious livelihoods.

The financialization of agricultural risk and price-setting

Agriculture is a risky endeavor. The uncertainty of weather, pests, plant disease and market prices renders farming a precarious occupation that has become even more perilous in the face of climate change and globalization-induced price volatility. Throughout much of human history, rural communities have mitigated agricultural risk through so-called 'moral economies', whereby locally specific agricultural practices and social institutions such as reciprocity and redistribution have helped to ensure farming families' access to sufficient food (Scott 1976, Watts 1983). Moral economies were largely undermined by colonial and imperial practices, significantly heightening the vulnerability of Southern populations to famine (Watts 1983, Davis 2002). In the 1950s and 1960s, many post-colonial states (re)instituted a variety of protections for agricultural producers, including grain purchasing boards, price supports, crop insurance and subsidy programs. These social protections – along with international commodity agreements that helped to stabilize and boost prices for select commodities in international markets – were largely dismantled during the neoliberal restructuring of the 1980s and 1990s, and agricultural risk management was privatized. Rather than relying on public programs that are purportedly inefficient, expensive and susceptible to corruption, contemporary farmers are now expected to manage risk through financial instruments like derivatives and micro-insurance where the probability of agricultural calamity is assigned a monetary value (Breger Bush 2012). Like other links in the contemporary agro-food value chain, financial motives are playing an increasingly important role in the markets where these instruments are sold. Having been privatized, agricultural risk is now becoming financialized.

Agricultural derivatives are financial contracts whose value is derived from the value of an underlying variable, which is typically the agricultural commodity in question, but might also be some other underlier such as the probability of a weather-related event. Futures contracts – which are standardized contracts in which the seller agrees to deliver a specific quantity of a commodity for a specific price at a specific time and location in the future – are the most familiar type of derivative.[9] Farmers sell futures contracts as means of hedging against the probability that the price for their crop will fall *below* a specified price; end users like grain traders and food processors purchase futures contracts as a means of hedging against the probability that agricultural prices will rise *above* a specified price and that they will have sufficient access to the agricultural product in question. In

---

[9]See Breger Bush (2012) for a helpful and particularly clear description of different types of derivatives contracts.

addition to these hedgers who have an interest in the actual physical product, speculators – who look to profit from changes in the price of crop futures but have no interest in the actual physical crop – are active participants in derivatives markets. A purported benefit of speculators is that they ensure the liquidity of crop futures (i.e. they help to create sufficient demand for futures, enabling farmers to lock in a price for their crop rather than bearing the uncertainty of market conditions at harvest time). Trading derivatives is, thus, a means for shifting economic risk; speculators seek to profit from the riskiness of agricultural production and uncertain markets.

One of the purported benefits of derivatives markets is that they facilitate 'price discovery'. That is, traders' collective wisdom about the probabilities of future economic outcomes will manifest itself as a market price that will guide farmers and help them to make better decisions about the types and quantities of crops to cultivate (Breger Bush 2012). Yet, as John Maynard Keynes argued in the wake of the Great Depression, the future is inherently uncertain and financial actors betting on the future are prone to herd behavior that can send the wrong price signals and exacerbate economic volatility. Indeed, it was Keynes' insights that informed the implementation of a suite of New Deal banking policies in the US and beyond, including regulations on agricultural derivatives markets.[10] Chief among these was the Commodity Exchange Act of 1936, which empowered US federal regulators to establish position limits that restricted the number of contracts that speculators could hold in a given derivatives market. The objective was to enable sufficient speculation for liquidity and price discovery, while simultaneously preventing large traders from herding their counterparts towards speculative bubbles and price volatility.

Faced with intense lobbying that included the likes of Alan Greenspan, the US Congress began easing restrictions on agricultural derivatives trade in the 1980s, a process that culminated with the passage of the Commodity Futures Modernization Act (CFMA) in 2000. Driven by the logic that regulations stymie liquidity and the flow of relevant information into futures pricing,[11] CFMA effectively deregulated commodities trading in the US. It removed position limits on speculators and allowed for the opening of unregulated exchange markets. Combined, these developments opened the door, allowing a variety of new investors – including hedge funds, pension funds, insurance companies, sovereign wealth funds and investment banks – to speculate on an array of increasingly complex agricultural derivatives. Speculation on food commodities exploded, increasing 10-fold between 2000 and 2011 (Spratt 2013).[12] Rather than a means for protecting food producers and users from agricultural uncertainty, derivatives markets are increasingly arenas for speculative betting. In 1996, 88 percent of futures contracts were held for hedging purposes; that share plummeted to 40 percent in 2011 (Spratt 2013).

Many of the new investors have deposited their money in commodity index funds (CIF) that amalgamate derivatives for a variety of commodities into a single value. Pioneered by

---

[10]Ellen Russell (2008a) provides a fascinating account of how, despite their purported intention, New Deal banking policies failed to harness finance as the servant, rather than the master, of the real economy. In addition to her general history, see Clapp and Helleiner (2012), Spratt (2013) and Ghosh *et al.* (2012) for informative histories about Keynesian thinking and the regulation of agricultural derivatives in the US.

[11]As discussed in Spratt (2013) and Ghosh *et al.* (2012), this logic is akin to Milton Friedman's 'Efficient Markets Hypothesis'.

[12]Worldwide, the number of futures and options contracts traded on regulated exchanges increased threefold between 2002 and 2008 while the value of non-exchange listed contracts increased more than 14-fold, to US $13 trillion (Mayer 2009).

Goldman Sachs and now offered by a number of investment banks, CIFs have formally transformed agricultural risk into an 'asset class' that has been especially attractive to large-scale investors like pension funds, hedge funds and sovereign wealth funds because they require little knowledge of the actual markets (Kaufman 2011, Murphy *et al.* 2012). Speculation in CIFs exploded following financial deregulation, ballooning from US $13 billion in 2003 to US $317 billion in 2008 (Kaufman 2010).[13] The index funds accounted for the majority of new investments in agricultural derivatives until the 2007–2008 food price crisis: they made up 65–85 percent of total investment between 2006 and 2008 (Spratt 2013). Since then, hedge funds and other 'active' traders who speculate on short-term price movements have come to play a more prominent role, while CIFs – which tend to take a 'long' position by rolling over commodity futures contracts prior to their expiration and reinvesting those proceeds in new contracts – have seen their share of total investments drop to 45 percent (Ghosh 2010, Spratt 2013). Whether they go short or long, the non-commercial investors who have come to dominate the trade of agricultural derivatives since its deregulation have profited from rising and increasingly volatile food prices. One question that emerges is whether their speculation is, in fact, causing the very price increases and volatility from which they benefit.

When food prices vaulted to record levels in early 2008 – the UN Food and Agriculture Organization (FAO)'s index of food prices shot up 45 percent in just nine months – more than 50 million people were effectively priced out of the food market and the global population of malnourished people soared to more than one billion (FAO 2009). Although prices dropped later that year, they have remained at pre-crisis levels and markedly more volatile, a point that was emphatically illustrated when the price index shot up 33 percent to a new record high during the final six months of 2010 (FAO 2013).

Official attempts to explain the contemporary food price crisis have focused upon the so-called 'market fundamentals'. As discussed in Clapp (2009, 2012), the World Bank, the FAO, the International Monetary Fund (IMF) and other influential bodies attributed rising prices to a mismatch between supply and demand. Most analysts agree with some, but not all, aspects of the official narrative. For instance, there is broad agreement that the diversion of food and farmland to the production of agro-fuels has significantly reduced the supply of food (Bello 2009, Akram-Lodhi 2012, Clapp 2012, Ghosh *et al.* 2012), yet several scholars have challenged the popular argument that the transition to higher-protein diets in India and China – which has been limited to the wealthiest consumers – is straining global food stocks (Clapp 2009, Akram-Lodhi 2012, Ghosh *et al.* 2012). Some critics have observed that while it is possible to explain rising food prices in the language market fundamentals, doing so fails to account for the underlying *causes* of supply and demand curve shifts, namely, the neoliberal restructuring that has undermined agricultural autonomy in the global South and rendered the region dependent upon food imports that are controlled by a handful of transnational agro-food corporations (Bello 2009, Clapp 2009). Given the current structure of the global food system, there is a general consensus that changes in supply, if not necessarily demand, account for some of the recent increase in food prices (Clapp 2009, 2012, Ghosh *et al.* 2012, Spratt 2013). At the same time, however, there is a growing awareness that even though market fundamentals may be pressuring food prices upwards, they cannot explain the *magnitude* of the recent spikes in food prices or the volatility of price movements (Clapp 2012, Ghosh *et al.*

---

[13] A study conducted by Lehman Brothers revealed that the volume of speculation on CIFs increased 1900 percent between 2003 and 2008 (De Schutter 2010).

2012). Many blame the financialization of agricultural derivatives for exacerbating under-lying price movements.

Examining data dating from 1990 to 2011, Ghosh *et al.* (2012) show that there was a strong correlation between increased speculation on commodity futures and actual com-modity prices. In particular, they noted that as liquidity for wheat, maize, soybeans and crude oil began to increase in 2003–2004, particularly from 2007 onwards, the prices for those commodities in spot markets moved in tandem. Ghosh *et al.* (2012) also observe that volatility in the spot markets where physical commodities are actually traded increases with speculation in futures markets, but that the severity of the volatility depends upon the way in which it is measured. Based upon data dating from 1990 to 2011, they observe that price volatility beginning 2007 is at least as strong as, if not stronger than, at any other pre-vious time and that the increased volatility is associated with rising liquidity in futures markets (Ghosh *et al.* 2012, 477–8). Spratt (2013) similarly shows that increased financial investment in agricultural commodities coincides with higher commodity prices and increased price volatility, which is punctuated by severe spikes.

Even as they document the association between speculation and rising and increasingly volatile commodity prices, Ghosh *et al.* (2012) and Spratt (2013) caution that correlation does not necessarily mean causation. Indeed, advocates of deregulated futures markets maintain that, rather than causing food prices to rise, investors were simply attracted to increasing prices. Moreover, they maintain, food markets have always been volatile and hedge funds and other speculators are attracted to that volatility. In a highly influential study conducted for the Organisation for Economic Co-operation and Development (OECD), for example, Irwin and Sanders (2010) do not find a statistically significant relationship between activity in agricultural derivatives markets and price volatility, leading them to dismiss claims that that rising speculation has increased volatility in com-modity markets. They maintain that there is no obvious channel through which speculation in futures markets can impact prices in spot markets.

Countering Irwin and Sanders (2010), Spratt (2013) and Ghosh *et al.* (2012) maintain that activity in futures markets can, in fact, impact actual commodity prices by shaping *expectations*. Given that private actors control most food stocks in the neoliberal era and that information regarding those stocks is proprietary, participants in actual grain markets have limited information about market fundamentals and, thus, look to the futures markets for guidance (Spratt 2013). In other words, buyers and sellers in spot markets look to the futures markets when determining prices, thus jumping on the inves-tor-led bandwagon. Indeed, price changes in futures markets are increasingly leading price changes in spot markets for key commodities like wheat, corn and soybeans (Hernan-dez and Torero 2010), suggesting that financial actors in agricultural derivatives markets are directly influencing global food prices.

The costs and benefits of the food price crisis have not been distributed equally. Perhaps not surprisingly, the beneficiaries have been limited to some of the very financial actors that helped to drive up food prices and exacerbate volatility, namely the hedge funds that specu-late on changes in derivatives prices and the exchanges where revenues are based on trading volumes (Spratt 2013). Meanwhile, food consumers and producers, particularly in the global South, have suffered. Even then, however, it is important to note that not all nations have suffered equally. A large state banking sector, capital controls and active fiscal policy, for example, enabled the Chinese government to mitigate the effects of the crisis upon its citizens (Ghosh 2010). Yet in most areas of the global South, the spikes in global commodity prices have translated into stubbornly high food prices (Ghosh 2010). Spending 50–80 percent of their income on food, poor people in developing countries were hard hit, particularly women and children (Clapp 2012, Heltberg *et al.* 2013). In addition to the obvious and physically detrimental impacts of decreased food consumption, rising food prices have been associated with withdrawing children from school, emotional stress, domestic violence, crime, prostitution and longer working hours for the poor (Heltberg *et al.* 2013).

Price volatility has also hurt farmers. Agricultural producers look to market prices when determining which crops to produce and how much to produce. Yet as speculative activity increasingly distorts prices in spot markets, farmers receive the wrong price signals, resulting in oversowing when prices are artificially high and undersowing when prices are artificially low (Breger Bush 2012, Ghosh 2010). On top of this, volatility has exacerbated the uncertainty of agricultural production. Farmers are not only subject to the whims of nature, but to the whims of speculators as well. As with the weather, failure to accurately predict price movements can have particularly dire consequences for small-scale farmers operating at the margins of agricultural production. Thus, by distorting market signals, the financialization of agricultural derivatives undermines the viability of small-scale agriculture and may have a detrimental impact upon the long-term global food supply (Spratt 2013).

Ironically, even as the financialization of agricultural derivatives has intensified the uncertainty and financial hardships faced by agricultural producers, derivatives and other financial instruments have been promoted as a development tool to help farmers cope with that very uncertainty and volatility. Since the early 1990s, the World Bank, the United Nations (UN) Conference on Trade and Development, the IMF and other major development actors have promoted derivatives as a solution to the poverty and price risk faced by small farmers (Breger Bush 2012). It is not surprising, then, that the fastest growing derivatives markets are located in the global South. Between 2003 and 2006, agricultural contracts in non-OECD exchanges increased by 26 percent, compared to 16 percent in OECD countries. With respective growth rates of 50 percent and 43 percent in 2010, the most rapidly growing derivatives markets are located in Latin America and Asia (Breger Bush 2012). Yet, as Breger Bush (2012) illustrates, rather than improving conditions for small-scale farmers, the proliferation of derivatives markets exposes them to greater risk and contributes to widening inequality. By facilitating speculative trading, unregulated derivatives contribute to more volatile agricultural markets. The small farmers who are supposed to manage price risks in derivatives markets, however, face a number of obstacles for doing so and are largely excluded. Their larger counterparts who already have better means for coping with risk are able to do so, though, resulting in uneven opportunities for mitigating risk. Other beneficiaries include agri-business enterprises that are also able to hedge their risks, financial speculators and other financial elites (Breger Bush 2012).

Recognizing small farmers' exclusion from derivatives markets, influential development institutions like the UN Development Programme and the German Society for International Cooperation, in collaboration with financial service enterprises, have promoted micro-insurance as a means of managing the risks of agricultural production. As Da Costa (2013) observes for the Indian context, however, the number of micro-insurance providers has grown rapidly in recent years, yet very few farmers are purchasing their plans. Consequently, insurance brokers and other actors who stand to benefit from the marketization of risk management have engaged in far-reaching discursive and pedagogical interventions aimed at teaching farmers the 'rationality' of insurance and 'structurally adjusting culture', all with the aim of creating effective demand to sop up the (over)supply of micro-insurance funds. As with agricultural derivatives, Da Costa (2013) notes that micro-insurance treats the symptoms of insecurity and risk, without addressing the underlying causes. As initiatives for micro-insurance become more pervasive and small farmers are programmed to demand it, research into how the practice is transforming agricultural practices and the economic well-being of peasant farmers will be necessary.

The financialization of food trade and processing

Although it has received much less attention in the literature, there is significant evidence suggesting that, like other nodes in the agro-food value chain, the trade and processing of food have become financialized in recent years. Characteristic of the financialization process, this has entailed food traders and processors diversifying into – and earning an increasing share of their revenues from – financial activities. At the same time, financial actors are playing an increasingly active and direct role in the manufacturing, storage and distribution of food. That is, there has been a blurring between finance and these inter-mediate links in the agro-food supply chain. Drawing upon the limited reports on this phenomenon, this section documents how the financialization process is benefitting inves-tors at the expense of food workers, food consumers and commodity traders at the expense other actors in the agro-food and financial sectors.

According to the 'shareholder value' thesis, the rise of finance can be attributed to the alignment of corporate managers' and shareholders' interests such that both believe that the sole purpose of the firm is to maximize the value of stock options (Palley 2007, Krippner 2011). Such has been the transformation of food processing in the era of financialization. Food manufacturers have become increasingly beholden to shareholders demanding returns of 20–30 percent (Rossman 2010). Rather than offering healthier food, for example, food processors have opted to produce products laden with salt, sugar and fat that stimulate overeating and thereby maximize dividends for stockowners (Moss 2013). Similarly, pressure to maximize shareholder value has also been linked to massive job loss in the European Union and the US, as food manufacturers have opted to outsource pro-duction to poorly compensated precarious workers in the global South (Rossman 2010).

In addition to pushing unhealthy but profitable food and increasing worker exploitation, food processors have also increased shareholder value by outsourcing research and devel-opment (R&D) (Rossman 2010). Rather than hiring food scientists and investing their resources in innovation, many of the top food manufactures have financialized R&D. Nestlé, Kraft, Unilever, Pespsico and Coca Cola, among others, have established private equity and venture capital subsidiaries that troll for innovative start-up enterprises that they can target for acquisition, a merger or investment (Nestlé 2002, The Kraft Group n.d., Unilever Ventures Limited n.d., Pepsico 2010, McWilliams 2010). Even if they do not acquire the start-up outright, financial partnerships and venture capital help to ensure control over product procurement. 'Multinational corporations', one observer notes, 'treat the start-up world like a research and development department; they seek nimble, innovative companies that are outside the corporate structure' (McDermott 2012). In so doing, they also increase their profits while strengthening their oligopolistic power.

The financialization of food manufacturing has not only entailed processors becoming more beholden to shareholders and establishing subsidiaries to engage in investment banking, it has also consisted of investment banks playing a more active role in food pro-cessing. Goldman Sachs, for instance, has bet that the 'meatification' (Weis 2013) of Chinese diets will increase the profitability of the country's poultry sector. The investment bank recently purchased a 13 percent holding in the country's second-largest meat and poultry processor and invested another US $300 million for the purchase of 10 poultry farms (Burch and Lawrence 2009).

The blurring of finance and food has unfolded not only in food processing, but in commodity trading as well. Murphy *et al.* (2012) provide an insightful discussion of how the world's four largest grain traders – Archer Daniels Midland (ADM), Bunge, Cargill and Louis Dreyfus (or, as they are collectively referred to, the ABCDs) – are engaging in a variety of financial activities. The ABCDs have all established investment vehicles that allow external investors to speculate on agricultural commodities and other dimensions of food production. Perhaps their most successful ventures have facilitated investment in agricultural derivatives markets, where grain traders have a long history of hedging against undesirable price movements but, since the financial deregulation that began in the 1980s, have become increasingly active in speculation. Growing uncertainty about food prices has spurred interest in the ABCDs' investment funds. As noted above, grain storage that was previously public has been privatized under neoliberal restructuring, translating into significant uncertainty about food supplies. Due to their dominance of agricultural trade and their direct contact with food suppliers, the ABCDs are among the first to know about supply conditions, making their financial products particularly attractive to investors wishing to speculate on agricultural derivatives markets. Indeed, operating under the slogan 'monetize our expertise', Louis Dreyfus' hedge fund, the Alpha Fund, expanded rapidly, growing some 20-fold within its first two years and, ultimately, refusing to accept new investors because the fund had grown so large after a mere three years of operation.

Murphy *et al.* (2012) document how, in addition to commodity investment funds, the ABCDs offer a variety of financial products to investors. Louis Dreyfus, for instance, has established a land investment fund, Calyx Agro, that acquires, converts and then sells – one might call it 'flipping' – farmland in Latin American. Indicative of the extent to which the grain trader has become financialized, the Louis Dreyfus subsidiary is managed by former investment bankers from Bear Stearns and Citigroup and has received investments of US $60–70 million from financial giants American International Group (AIG) and KPS Capital Partners (de Lapérouse 2012). Among the ABCDs, Cargill is the most involved in financial activities (Murphy *et al.* 2012). Since 2003, when Cargill began offering financial services to external investors, it has created a number of financial subsidiaries that offer a variety of financial products, including commodity index funds, asset management services, insurance and opportunities to speculate on real estate, commercial credit and energy. In short, Cargill, the world's largest private company, is not only a grain trader, but a financial enterprise as well.

As with other links in the agro-food value chain, the financialization of food distribution has not been one-way. Even as the ABCDs have become active in financial markets, financial actors are increasingly engaged in food trading. This has taken the form of major financial institutions like AIG and Deutsche Bank investing billions of dollars in grain traders' funds. But it has also entailed financial actors becoming directly involved in the physical storage and transport of agricultural commodities. Catering to speculators' interest in price movements rather than actual products, a number of hedge funds have emerged in recent years to facilitate investors' participation in food and agricultural derivatives markets. In addition to purchasing livestock, grain and other agricultural products, these funds have acquired storage facilities and transport vessels, enabling them to buy maturing futures contracts from fellow investors. In addition to charging fees for their services, the funds benefit from their more direct access to information about agricultural supply (Meyer 2009).

While information about the contribution of financial activities to the ABCDs' revenues is not publicly available, evidence suggests that they have been quite profitable.

Murphy *et al.* (2012, Figure 2) have collected data on the profits of ADM, Bunge and Cargill since the late 1990s. In general terms, profits were relatively flat for the ABCDs in the late 1990s, but they began to steadily increase after the Commodity Futures Trading Act of 2000 formally deregulated speculation in agricultural derivatives markets, and have increased significantly since the onset of excessive food price volatility in 2007. Bunge and Cargill explicitly acknowledge that their strong performance in recent years is based, at least in part, on their financial activities (Murphy *et al.* 2012).

No doubt part of the ABCDs' recent profitability can be attributed to higher food prices. Yet their unique access to food suppliers is also the source of information regarding global food stocks, giving them an advantage when hedging and speculating on price movements. Traders are among the first to know when supplies are falling low, giving them an edge in derivatives markets (Meyer 2011). Cargill, for instance, was among the first to speculate on falling wheat prices in 2008 (Murphy *et al.* 2012), a move correlated with a significant increase in profits. The grain traders have touted this knowledge as a reason to invest in their financial products. As questions have emerged over whether the ABCDs are manipulating financial markets for their own gain, however, observers like Murphy *et al.* (2012) have likened their activities to insider trading.

Though they generally try to maintain a low profile, the major grain traders have actively lobbied for their continued ability to speculate in agricultural derivatives markets. In the wake of the 2008 financial crisis, the US and European Union governments have proposed rules to, once again, regulate commodities trading.[14] A key question is whether the new regulations should apply to grain traders who have long relied on agricultural derivatives as a means for hedging against unfavorable price shifts. In letters, testimony before legislators and meetings with regulators, the major grain traders have maintained that they should be exempt from the new rules. Even as they support regulations on banks and other financial players in derivatives markets, the ABCDs maintain that they themselves should be excused from speculative limits since they are, in fact, end users with a commercial interest in the physical agricultural products (Murphy *et al.* 2012). In other words, the ABCDs who actively blurred hedging with speculation during the deregulated era are lobbying for the exclusive privilege to continue doing so in a regulated environment. This, of course, would give them an even greater advantage in derivatives markets and enhance their power vis-à-vis food producers. Their oligopsony power in spot markets for physical commodities would be replicated in the markets for agricultural risk, thereby strengthening their position in the agro-food value chain.

The financialization of agricultural inputs and farmland

Finance has long played a role in the provisioning of agricultural inputs. Since the advent of industrial agriculture and the attendant commercialization of inputs, farmers have relied upon agricultural credit to purchase seeds, agrochemicals and, of course, farming equipment. Has this node in the agro-food value chain become more financialized in recent

---

[14]As discussed in Clapp and Helleiner (2012), the US led this initiative with the 2010 Dodd-Frank Wall Street Reform and Consumer Protection Act, and the EU is following suit, albeit with rules that are likely to be less stringent. Yet even with the passage of the Dodd-Frank bill in the US, there is still significant uncertainty about how the rules will apply to grain traders who, despite their recent speculative activity, are also legitimate end users with an interest in the physical product.

years? Are suppliers earning a greater share of their revenues from financial activities? Are financial actors playing a greater role in the provisioning of inputs? To date, there has been no systematic analysis of the financialization of agricultural inputs. There is, however, reason to believe that the sector is undergoing transformations similar to those elsewhere along the agro-food chain. Since the 2007–2008 spike in food prices, financial actors have made significant investments in enterprises that produce tractors and other farm equipment, seeds and agrochemicals (Ross 2008, The Economist 2009). A review of the financial press suggests that investors are particularly interested in the fertilizer industry. On the logic that farmers will respond to rising food prices by expanding and intensifying their production, private equity groups have invested in fertilizer producers in China, India, Egypt, western Africa and other key nations (ICIS 2007, Ross 2008, Davis 2011, Friedland 2011, IFC 2011).

Similarly, there is reason to believe that the providers of agricultural inputs are earning a greater share of their revenues from financial activities, particularly the provisioning of credit.[15] The neoliberal restructuring that entailed states reducing – and in most cases eliminating – subsidies for agricultural inputs and curtailing support for rural development banks, did not reduce production pressures on agricultural producers. On the contrary, it exacerbated them, spurring many farmers to intensify agricultural production. Paradoxically, this had the aggregate effect of lowering crop prices, a process known as 'immeserizing growth', where the increased production of agricultural commodities puts downward pressure on prices, thereby exacerbating farmer poverty and, ultimately, debt (Vakulabharanam 2004, 2005). In short, neoliberal restructuring has driven many poor farmers to increase their expenditures on seeds, fertilizers and other inputs that have become more expensive with the elimination of state subsidies even as it has driven down the price of output (Isakson 2013). Ironically, the resulting farmer debt is likely a boon to moneylenders and agricultural suppliers who sell inputs on credit. Whether those suppliers are earning a greater share of their revenues from credit is a question deserving further research. Rising debt has forced many farmers to part with their primary asset and, indeed, the very foundation of their livelihoods: their land. Ironically, they are doing so as land is becoming increasingly more coveted by transnational agribusiness and financial investors.

Farmer ownership of land has been an enduring – though not necessarily universal – feature of capitalist agriculture. While there has been significant integration, both horizontally and vertically, within and along agro-food chains, fragmented family farms have predominated agricultural production (Heffernan 2000, Clapp 2012, Wheaton and Kiernan 2012). The persistence of family farms has been attributed to the riskiness of agricultural production and the challenges of appropriating surplus value. The unique ability of family labor to engage in self-exploitation at key times during the agricultural cycle allows it to produce at a lower cost than hired labor while the uncertainty of weather, pests, plant diseases and prices translates into excessive risk for non-farm capital. Thus, rather than assimilating land and farming, agribusiness has generally engaged in more value-generating activities upstream (i.e. provisioning seeds, equipment and other inputs) and downstream (i.e. trading, processing and retailing food) (Selby 2009, Clapp 2012, Cotula 2012). Meanwhile, it has opted to let farmers absorb the risk of agricultural production while controlling

---

[15]The Syngenta Foundation, the nominally philanthropic arm of the giant agro-chemical and seed enterprise, is promoting weather derivatives in Kenya (Breger Bush 2012, 43). The extent to which the suppliers of agricultural inputs are profiting from the sale of such financial instruments is ripe for investigation.

their practices through contracts, product specifications and other arrangements (Watts 1994, Lewontin 2000, Oya 2012).

The contemporary food crisis is restructuring agro-food supply chains. Rising commodity prices and the fear that sourcing crops will become increasingly more competitive have increased the perceived value of agricultural production and land ownership, fueling the so-called 'land rush' (Cotula 2012). Along with a variety of other actors – including agri-food enterprises, governments, sovereign wealth funds and agro-fuels developers – private institutional investors are scrambling to profit from appreciating land values. Deininger *et al*.'s (2011, 53) systematic review of media reports suggests that, while agribusiness and industry account for the majority of land deals, investment funds are key players.[16] Several observers note that financial actors' interest in farmland has increased significantly since the advent of the food, energy and financial crises in 2005 (HighQuest 2010, Daniel 2012, Fairbairn 2014).[17] Financial insiders estimate the current value of private institutional investments at US \$30–40 billion, with the potential to increase some 30-fold to US \$1 trillion (Wheaton and Kiernan 2012).

There are a number of vehicles for investing in farmland. Investors wishing to speculate on its value can deposit their funds in large investment banks, hedge funds, private and publicly traded REITs, or in companies that combine farm management with strategies to acquire agricultural land (HighQuest 2010). Originally, endowments and wealthy individuals/families were the principal depositors in land acquisition funds. Their role is shrinking, however, as hedge funds and large institutional investors, including pension funds, are contributing more to existing funds or, in some instances, establishing their own (HighQuest 2010). Many of these investors are channeling their funds through investment organizations headquartered in tax havens like the Cayman Islands, Panama and Mauritius (Borras *et al*. 2012, Cotula 2012).

There are varying explanations for the financial sector's growing interest in farmland. Analysts of the Marxian persuasion focus upon the systemic dynamics of capitalism. McMichael (2012), for instance, attributes growing interest in land to capitalism's inherent tendency towards crisis. Declining profits in the real sector combined with the convergence of crises in energy, climate and conventional financial markets have driven financial investors to seek safer opportunities for accumulation in land and agricultural production. Given that much – though certainly not all (Borras *et al*. 2012, Cotula 2012) – of the acquired land is located in foreign markets, these land acquisitions can be likened to a 'spatial fix', wherein financial actors seek accumulation opportunities abroad when domestic markets are no longer capable of delivering (Harvey 2010).[18] Fairbairn (2014) questions whether financial actors' newfound interest in land – a physical input – represents an end to the era of financialization, as investors funnel their surplus capital into an illiquid asset that has traditionally been prized for its use value. Ultimately, however, she concludes that investor interest is yet another example of the blurring between finance and economic production. Financial interest in land is based primarily upon its anticipated appreciation in

---

[16]Deininger *et al*. (2011) also observe that only a few sovereign wealth funds are directly involved in land transactions, noting that most funds prefer to channel their funds through private institutional investors.

[17]Despite growing interest, Fairbairn (2014) notes that even the most enthusiastic investors commit no more than one percent of their overall portfolio to farmland.

[18]In an alternative Marxian explanation, Akram-Lodhi (2012) attributes the global land grab to the need to push down food prices, thereby maintaining the real wage of workers even as their rate of exploitation – and thereby capital accumulation – increases.

value, while the value generated by agricultural production is perceived as an added bonus. In other words, financial actors are first and foremost speculators. Their motivations can be contrasted with other actors in the land rush (e.g. agribusiness, government enterprises) who are more concerned with agricultural production but who, like finance capital, are eager to capitalize upon financial and non-financial returns.

Though not necessarily in contrast with Marxian explanations, orthodox understandings of the financialization of land focus upon the so-called 'market fundamentals'. Financial actors themselves do not point to the accumulation imperative of capitalism, but rising commodity prices, increasing consumption of animal protein, the competition for land that is intensified by urbanization and suburban sprawl and declining land quality, all of which are indicators of rising land values and a sound investment strategy (HighQuest 2010, Magnan 2012). Valid or not, the language of supply and demand legitimizes perceptions of appreciating land values while financial actors' claims that there is close to US $1 trillion of untapped 'investible' land create the spectacle of windfall profits, thereby sparking investors' 'animal spirits' and attracting massive quantities of capital in financial funds (Li 2012, Wheaton and Kiernan 2012).

In addition to high returns, investors are also attracted to land as a means of managing risk. Speculators' flight to quality during economic downturns is not novel, but the emergence of farmland as an 'asset class' compels investors with the dual promise of high yet stable returns (Wheaton and Kiernan 2012, Cotula 2012). In a survey conducted for the OECD, private institutional investors identified inflation hedging – or the expectation that land will retain its value in the face of rising prices – as their primary motivation for investing in farmland. Moreover, they noted, the returns on land have a low correlation with equity markets and are thus a desirable means for diversifying risk in asset portfolios (HighQuest 2010).

Historically, private institutional investors held the majority of their farmland in Europe, North America and Australia, where risks – but also returns – were low. In recent years, however, they have turned their attention to South America, particularly Brazil, and Africa (HighQuest 2010). While these 'frontier' lands have the potential of higher returns, they also come with higher risks. Those risks have been mitigated somewhat in recent years, though, as the governments in host countries and development finance institutions have sought to facilitate the land rush. For their part, states that are enticed by promises of development and spurred by the imperatives of capital accumulation have helped to identify 'idle lands' and enable their acquisition (HighQuest 2010, Borras et al. 2012, Cotula 2012). Meanwhile, the World Bank, through its member bodies, has smoothed private institutional investors' participation in these transactions. In her insightful account, Daniel (2012) describes how the Bank promotes private equity markets as a catalyst for development. Its Multilateral Investment Guarantee Agency (MIGA), for example, provides contracts that guarantee foreign direct investment against a number of risks, thereby enabling fund managers to attract funds from investors who want to insure themselves against non-commercial (i.e. political) risks. Similarly, the Bank's International Finance Corporation (IFC) has backed a number of private equity funds that invest in agriculture, and supports networking among institutional investors. The IFC has also launched a US $500-million fund that provides investors with an exit option from funds operating in emerging markets, thereby making their investments more liquid. Combined, these initiatives have facilitated financial actors' acquisition of low-priced farmland in the South while reducing the risk of doing so, no doubt spurring their participation in the recent land rush.

Analyzing the financialization of land through the political economy framework requires consideration of how financial investment changes land-based social relations

(Bernstein 2010, Borras and Franco 2010, White and Dasgupta 2010). Doing so, however, can be problematic. As noted earlier, private financial institutions are one of several types of actors participating in the land rush. Not all land deals are undertaken by financial actors nor are they necessarily driven by financial motives. Traditional agribusiness and many governments, for instance, are more concerned about their access to food than the appreciation of land values (Murphy et al. 2012). Thus, one should be careful not to conflate the general outcomes of the land rush with the outcomes that are specifically rooted in financialization. How, if at all, do farmland acquisitions by financial actors/motives differ from those by non-financial actors/motives? Do outcomes differ among different types of financial actors? Answering these questions requires one to consider the pre-existing structural and institutional conditions that shape land acquisitions. It also requires an appreciation for the blurring between financial actors (who are increasingly recognizing the productive value of land) and non-financial actors (who are increasingly using land as a financial asset) (Fairbairn 2014).[19]

As Li (2012) has observed, in order for land to become an object of finance, it must be inscribed as such. Historically rooted practices, meanings and relationships with the earth must be abstracted, eliminating qualitative differences and contextual understandings, if land is to become a financial asset with a quantifiable price. Inscribing land with titles, measurements and other instruments allows for comparison and the assignment of value. This abstraction of land into market metrics allows for it to be understood as a 'bundle of assets' that can be de-bundled and re-bundled in novel ways so as to 'unlock' and 'create' value for investors. Such is the case with REITs (that aggregate income streams from land-based investments into a fund that investors can buy into) and opco-propco schemes (whereby an enterprise divides itself into distinct entities, one that owns the land and leases it to another that operates it). The monetization of newly unlocked forms of value accentuates the fictitious commodification of land, thereby heightening its vulnerability to abuse, and creates opportunities for wealthy actors to appropriate land-based values to the detriment of others.

Although media reports tend to focus upon instances of outright land purchases, most deals in the land rush consist of long-term leases (Borras et al. 2012, Cotula 2012). If financial actors are primarily interested in the exchange value of land, one would suspect that they would be more inclined towards land purchases, while agribusiness and other actors who are concerned with agricultural output would have few reservations with long-term leases. To date, no study has identified how, if at all, the types of deals vary among the different types of actors. Given the blurring between financial and non-financial actors, the distinction may not be particularly sharp. Fairbairn (2014), for instance, identifies three types of investor strategies: 'own-operate, 'lease-operate' and 'own-lease out'. Both the 'own-operate' and the 'lease operate' models are indicative of the investors' interest in agricultural production. Indeed, those pursuing the 'lease-operate' strategy are not speculating on the value of land, but on the value of crops, and represent a significant break from traditional financial activity.

The 'own-lease out' model is most in line with understandings of land as a financial asset (Fairbairn 2014). Under this strategy, investors simply acquire land and then rent it

---

[19]It is also worth clarifying that financial control over land is not novel. Credit providers for agricultural inputs and land purchases have long been able to exercise control over agricultural production, often determining which crops will be grown, the conditions under which they will be grown and terms and conditions of their exchange.

to tenant farmers; their interest is in the rental profits and the land's ability to store value, not agricultural production per se. Many investors who pursue this strategy rent the land back to the very farmers who cultivated it before the transaction (HighQuest 2010), suggesting that in some instances the financialization of farmland is associated with a fundamental change in land-based social relations where independent farmers are transformed into tenant farmers. To the extent that they had a choice, producers opting into such an arrangement were likely motivated by debt or some other hardship.

Among the private institutional investors who operate their landholdings – or hire a farm management enterprise to do so on their behalf – many are opting to cultivate flex crops that can be used for multiple purposes. More than three-quarters of the farmland acquired by financial actors is cultivated with major row crops like oilseeds, corn and wheat, while sugarcane is also widely grown (HighQuest 2010). These crops have multiple uses – as food, animal feed, agro-fuels or for industrial purposes – and therefore have a relatively reliable demand. The diversified uses of flex crops are believed to substitute for a diversified product portfolio: price shocks are muffled by redirecting large quantities of versatile crops to where they are in greatest demand. Given their versatility, including their perceived ability to provide 'green energy', flex crops are often touted as a solution to the contemporary food, fuel and climate crises (Borras *et al.* 2012, McMichael 2012). Such logic suggests that the cultivation of flex crops is not only a means for financial actors to mitigate risk, but that it also contributes to social welfare.

Private institutional investors are proud of their farmland activities and proclaim a variety of development-related entailments. In return for lease concessions and opportunities for land purchase, investors have helped to finance schools, healthcare centers and cultural activities like sports teams (HighQuest 2010, Magnan 2012). They also maintain that their projects create jobs, improve public infrastructure, generate taxes and improve market access for farmers neighboring their agricultural projects (High-Quest 2010). They are, perhaps, most proud of their contributions to agricultural productivity.

Based upon the logic that small family farms are backward and inefficient, financial actors frame their activities as socially necessary investments that will modernize agriculture and solve the contemporary food crisis. 'Given the small scale of the average farm globally and the challenges for such businesses accessing capital', two financial insiders write, 'the *scope and need for institutional capital* to be deployed in agriculture in order to improve efficiencies and generate higher returns is significant' (Wheaton and Kiernan 2012, 1, emphasis added). Moreover, they consider the concentration of family farms into larger industrial operations as a solution to Malthusian scarcity and a positive contribution to society:

> What is striking is the proportion of land owned and operated by family farmers, resulting in a very fragmented industry. One of the attractive features from an investment perspective is the opportunity for consolidation given the importance of scale in driving returns from agriculture. (Wheaton and Kiernan 2012, 4)

Whether the consolidation and industrialization of family farms will improve agricultural efficiency is debatable.[20] Whatever the case, the logic that farmers must be displaced

---

[20]Given the inequitable distribution of food and the growing allocation of crops to the agro-fuels red herring, one should also question the very need to increase agricultural productivity.

in order to improve productivity, and that private institutional investors are the best positioned to do so, represents a failure to consider alternative strategies for agrarian development. As the UN's Special Rapporteur on the Right to Food, Olivier de Schutter (2011), has observed, state investment in rural infrastructure and support for land and water rights would also improve agricultural productivity, but it would do so in a way that protects the rights of poor farmers and improves their livelihoods.[21]

Whether private institutional investment in land will generate rural employment is also questionable. Many investment funds boast that their projects will create quality jobs for local actors (HighQuest 2010); some even make promises to do so during their negotiations (Magnan 2012). As several scholars of the land rush have observed, however, investors are primarily interested in land; if it is not functional to their operations, labor will be expelled (Li 2011, Daniel 2012, Levien 2012). In instances where local workers are employed, they are often done so on adverse terms as contract farmers or plantation workers (Alonso-Fradejas 2012, Borras *et al.* 2012).

In short, the financialization of farmland has blurred the line between land as a productive asset and land as an object of speculation. Driven by the Malthusian notion that increased agricultural production is the solution to the contemporary food crisis, the financial sector is betting that the principal yet quantitatively limited input, land, will appreciate in value. In seeking land, private institutional investors are not only fueling the land rush, they are also spurring the development of land markets. Elsewhere along the agro-food supply chain, finance is implicitly pushing farmers off the land through rising input costs (and debt), declining bargaining power vis-à-vis buyers, and increasingly uncertain market conditions. Combined with the pull of appreciating farmland values, many small farmers may be enticed to sell their land, only to find that there are few alternatives elsewhere for their labor. The result will be increasingly precarious livelihoods, rising poverty and a growing concentration of landholdings (cf. De Shutter 2011, Li 2011). Moreover, as finance exerts greater control over how land and labor enter agricultural production, it will undoubtedly prioritize financial dividends over food security, the welfare of workers and the maintenance of farmland. A systematic study of how financial participation in the land rush differs from that of other actors would help to evaluate these claims.

Conclusions and directions for future research

Characteristic of the financialization process, the line between finance and food provisioning has become increasingly blurred in recent decades. Financial actors are seeking to profit from various activities in food and agriculture, while traditional actors within the agro-food sector are increasingly orienting their activities towards financial objectives. This is a process that has unfolded within all of the major links of the agro-food supply chain, from the most upstream node of farmland control to the most downstream node of food retailing. Financialization has transformed power relations within and along the supply chain, allowing empowered actors to appropriate a greater share of agro-food value. Along the chain, financialization has strengthened the position of food retailers as the most powerful set of actors within the agro-food system. Capitalizing upon their unique

---

[21]It is also worth noting that investors are rarely interested in acquiring the most marginal, unproductive lands. Instead, they have a preference for the best quality land in terms of soil fertility, proximity to markets, water availability and irrigation potential, and the availability of other infrastructure (Cotula 2012).

position within the heavily concentrated global food economy, retailers have been able to force food manufacturers, distributors and workers to fund their financial activities, thereby improving the profitability of supermarkets and reinforcing their position of power.

Further upstream, grain traders also appear to be major beneficiaries of the financial transformation of food. With their unique knowledge about grain stocks and the general food supply, the ABCDs and other major traders are well positioned to capitalize on finance-induced food price volatility. All four of the principal grain purveyors have established investment instruments in recent years: they are no longer just hedgers in agricultural derivatives markets, but speculators as well. Evidence suggests that their speculative activities have been particularly profitable. If new financial regulations restrict other financial actors' – but not the major grain traders' – participation in derivatives markets, the profitability of their speculative ventures will undoubtedly increase. So will their ability to shape the prices of futures and the food products that they buy and sell. This would fortify their market power, allowing them to appropriate additional value from downstream food processors who are also getting squeezed by retailers.

Like other enterprises in the agro-food system, food retailers are also heightening their demands on food workers. Having succumbed to the 'shareholder revolution', the managers of agro-food enterprises have realigned their interests with shareholder values, intensifying the workload of laborers even as their wages have remained stagnant and their positions more precarious. Small-scale agricultural producers have not fared well either. In addition to the growing competition for farmland, farmers' position vis-à-vis input suppliers and buyers of various stripes (commodity traders, food retailers and food manufacturers) has weakened as a result of changes in the agro-food system. The vulnerability of small farmers has been exacerbated by the financialization of agricultural risk, which has accentuated the volatility and uncertainty of agricultural prices and market conditions. Faced with the growing threat of dispossession, declining terms of trade and increasingly unstable markets for their output, small farmers are arguably the biggest losers of the financial transformation of food provisioning.

Even as financialization has put tremendous pressure upon small farmers and their livelihoods, it has also generated profitable opportunities for other actors in the agro-food system. This is certainly the case with hedge funds and other investors who speculate on food price volatility. It is also evident with the de-bundling and re-bundling of assets so as to unlock, securitize and market new types of value. The unlocking and ultimate monetization of asset values renders them vulnerable for appropriation by wealthy investors. New investment instruments for land and agricultural risk have indeed proven quite profitable for agribusiness and financial actors while largely excluding the small farmers whose livelihoods are directly impacted by the transformations.

While the wealth of literature on the topic has enabled me to describe key impacts of the financialization upon a generalized agro-food supply chain, there are also many avenues that are ripe for further investigation. Considering the principal nodes of the chain, some have received more attention than others. To date, most of the literature on the financialization of food has focused upon the growing role of financial speculators in agricultural derivatives markets. Other nodes – particularly the provisioning of agricultural inputs and food processing – have received scant attention in the academic literature. Murphy et al.'s (2012) report on the major grain traders is impressive, but it remains alone and it does not appear that any scholars have investigated whether the trade of other agricultural products (e.g. fresh fruits and vegetables) has been transformed by the rise of finance. In terms of the food retail sector, Burch and Lawrence's work (2009, 2013) is an important contribution, but it is largely based upon a single case study and leaves one wishing for a more systematic investigation of the relationship between finance and supermarkets. Meanwhile, there has been a rush to analyze the land rush, yet only a handful of scholars (e.g. Cotula 2012, Daniel 2012, Fairbairn 2014) have focused upon the role of financial actors in land deals and none appears to consider whether the outcomes of transactions involving financial actors/motives differ from those involving non-financial actors/ motives. Regardless of where it is unfolding along the agro-food supply chain, the impacts of financialization upon peasants and other small-scale agricultural producers has received scant attention and is deserving of systematic analysis.

As noted, the financialization of seeds, fertilizers and other agricultural inputs is arguably the most understudied node in the agro-food supply chain. Whether the sector is indeed becoming financialized remains unclear and, assuming that it is, one can only speculate on the outcomes. As I have done in this paper, one could hypothesize that such a process would exacerbate the 'price-cost squeeze' that has indebted many small-scale producers, thereby intensifying their dispossession of farmland. One might further speculate that such a phenomenon would fuel the land rush, but there is a paucity of empirical support for this thesis.

Similarly, the relationship between the financialization of agricultural derivatives and the contemporary land rush is also unexplored. Several studies have observed that rising food prices and food price volatility are spurring the demand for land (HighQuest 2010, Akram-Lodhi 2012, Clapp 2012), while others claim that food price volatility places additional stress upon marginalized small farmers (Breger Bush 2012, Clapp 2012, Spratt 2013). One wonders, however, whether that volatility has pressured farmers to part with their land and, if so, whether financial investors participating in the land rush have subsequently acquired it. Such an outcome would be both plausible and ironic and is deserving of further exploration.

Lastly, given the regressive impact that financialization has had upon the distribution of wealth and value in the agro-food value chain, one wonders about the possibility of alternative arrangements for financing food. Might it be possible to institute a more just and environmentally sustainable food system that is not subject to financial actors' quest for profits? Certainly finance can play an important role in facilitating access to resources and investment in food and agriculture. But, paraphrasing Russell (2008b), might finance be reoriented as a servant to the food economy rather than its master? If so, what might that financial arrangement look like? Instituting this vision might very well require a radical restructuring of the role of finance in food provisioning, but the injustice of billions of malnourished and food-insecure people demands that we consider radical solutions.

References

Akram-Lodhi, A.H. 2012. Contextualizing land grabbing: contemporary land deals, the global subsistence crisis and the world food system. *Canadian Journal of Development Studies*, 33(2), 119–142.

Alonso-Fradejas, A. 2012. Land control-grabbing in guatemala: the political economy of contemporary agrarian change. *Canadian Journal of Development Studies*, 33(4), 509–528.

Arrighi, G. 1994. *The long twentieth century: money, power, and the origins of our times.* New York: Verso.

Baud, C. and C. Durand. 2012. Financialization, globalization, and the making of profits by leading retailers. *Socio-Economic Review*, 10, 241–266.

Bello, W. 2009. *The food wars.* London: Verso.

Bernstein, H. 2010. *Class dynamics of agrarian change: agrarian change and peasants.* Halifax, Nova Scotia: Fernwood Publishing.

Borras, S.M. Jr., C. Kay, S. Gómez, and J. Wilkinson. 2012. Land grabbing and global capitalist accumulation- key features in Latin America. *Canadian Journal of Development Studies*, 33 (4), 402–416.

Borras, S.M. Jr. and J.C. Franco. 2010. From threat to opportunity? Problems with the idea of a 'code of conduct' for Land Grabbing. *Yale Human Rights and Development Law Journal*, 13(2), 507–523.

Breger Bush, S. 2012. *Derivatives and development: a political economy of global finance, farming, and poverty*. New York: Palgrave Macmillan.

Burch, D. and G. Lawrence. 2009. Towards a third food regime: behind the transformation. *Agriculture and Human Values*, 26(4), 267–279.

Burch, D. and G. Lawrence. 2013. Financialization in agri-food supply chains: private equity and the transformation of the retail sector. *Agriculture and Human Values* 30(2), 247–258.

Busch, L. and C. Bain. 2004. New! improved? the transformation of the global agrifood system. *Rural Sociology*, 69(3), 321–346.

Clapp, J. 2009. Food price volatility and vulnerability in the Global South: considering the global economic context. *Third World Quarterly*, 30(6), 1183–1196.

Clapp, J. 2012. *Food*. Malden, MA: Polity Press.

Clapp, J. 2014. Financialization, distance, and global food politics. *The Journal of Peasant Studies*. DOI: 10.1080/03066150.2013.875536

Clapp, J. and E. Helleiner. 2012. Troubled futures? the global food crisis and the politics of agricultural derivatives regulation. *Review of International Political Economy*, 19(2), 181–207.

Cohen, Amy. 2013. Supermarkets in India: struggles over the transformation of agricultural markets and food supply chains. *University of Miami Law Review* 68(1), 19–86.

Cotula, L. 2012. The international political economy of the global land rush: a critical appraisal of trends, scale, geography, and drivers. *The Journal of Peasant Studies*, 39(3–4), 649–680.

Crotty, J. 2009. Structural causes of the global financial crisis: a critical assessment of the 'new financial architecture'. *Cambridge Journal of Economics*, 33, 563–580.

Da Costa, D. 2013. The 'rule of experts' in making dynamic micro-insurance industry in India. *The Journal of Peasant Studies*. 40(5), 845–865.

Daniel, S. 2012. Situating private equity capital in the land grab debate. *Journal of Peasant Studies*, 39 (3–4), 703–729.

Davis, M. 2002. *Late Victorian holocausts: el niño famines and the making of the third world*. London: Vero.

Davis, A. 2011. Primavera takes stake in Chinese fertilizer maker. *Asian Venture Capital Journal*, November 28, 2011. Available from: http://www.avcj.com/avcj/news/2127979/primavera-takes-stake-chinese-fertilizer-maker [Accessed December 26 2013].

De Schutter, O. 2010. *Food commodities speculation and food price crises: regulation to reduce the risks of price volatility – briefing note 02*. Rome: Food and Agriculture Organization of the United Nations.

De Schutter, O. 2011. How not to think about land grabbing: three critiques of large-scale investments in farmland. *The Journal of Peasant Studies*, 38(2), 249–279.

Deininger, K., D. Byerlee, J. Lindsay, A. Norton, H. Selod, and M. Stickler. 2011. *Rising global interest in farmland: can it yield sustainable and equitable benefits?* Washington, DC: World Bank.

Epstien, G. 2001. *Financialization, rentier interests, and central bank policy (version 1.2, June 2002)*. Paper prepared for PERI Conference on the 'Financialization of the World Economy', December 7–8, 2001, University of Massachusetts Amherst.

Epstein, G. 2005. Introduction: financialization and the world economy. In: G. Epstein, ed. *Financialization and the world economy*. Cheltenham, UK: Edward Elgar Publishing, pp. 3–16.

Epstein, G. and A. Jayadev. 2005. The rise of rentier incomes in oecd countries: financialization, central bank policy and labor solidarity. In: G. Epstein, ed. *Financialization and the world economy*. Cheltenham, UK: Edward Elgar Publishing, pp. 46–58.

FAO. 2009. *The state of food insecurity in the world: economic crises – impacts and lessons learned, 2009*. Rome: Food and Agriculture Organization of the United Nations.

FAO. 2013. FAOSTAT. Available from: http://faostat3.fao.org/home/index.html#DOWNLOAD [Accessed 20 June 2013].

Fairbairn, M. 2014. 'Like gold with yield': evolving intersections between farmland and finance. *The Journal of Peasant Studies*, forthcoming. DOI: 10.1080/03066150.2013.873977

Friedland, J. 2011. Morgan Stanley closes $50 million investment in Chinese fertilizer producer Yongye International. Big Emerging Economies (BEEs), June 16, 2011. Available from: http://bigemergingeconomies.wordpress.com/2011/06/16/morgan-stanley-closes-50-million-investment-in-chinese-fertilizer-producer-yongye-international/ [Accessed 10 July 2013].

Ghosh, J. 2010. The unnatural coupling: food and global finance. *Journal of Agrarian Change*, 10(1), 72–86.

Ghosh, J., J. Heintz, and R. Pollin. 2012. Speculation on commodities futures markets and destabilization of global food prices: exploring the connections. *International Journal of Health Sciences*, 42(3), 465–483.

Harvey, D. 2010. *The enigma of capital: and the crises of capitalism*. Oxford, UK: Oxford University Press.

Heltberg, R., N. Hossain, A. Reva, and C. Turk. 2013. Coping and resilience during the food, fuel, and financial crisis. *The Journal of Development Studies*, 49(5), 705–718.

Heffernan, William. 2000. Concentration of ownership and control in agriculture. In: F. Magdoff, J. Bellamy Foster, and F.H. Buttel, eds. *Hungry for profit: the agribusiness threat to farmers, food and the environment*. New York: Monthly Review Press, pp. 61–76.

Hernandez, M. and M. Torero. 2010. *Examining the dynamic relationship between spot and future prices of agricultural commodities*. IFPRI Discussion Paper 00988. Washington, DC: International Food Policy Research Institute.

HighQuest Partners. 2010. *Private Financial Sector Investment in Farmland and Agricultural Infrastructure*. OECD Food, Agriculture, and Fisheries Papers, No. 33, OECD Publishing.

Hsu, T. 2013. Wal-Mart's fresh produce promise: fewer middlemen, faster groceries. *Los Angeles Times*, June 3, 2013.

Hobson, J.A. 2010. *Imperialism: a study*. Cambridge: Cambridge University Press.

ICIS. 2007. Abraaj Capital buys Egyptian fertilizer firm EFC. ICIS., June 11, 2007. Available from: http://www.icis.com/Articles/2007/06/08/4503061/abraaj+capital+buys+egyptian+fertilizer+firm+efc.html [Accessed 10 July 2013].

International Finance Corporation (IFC). 2011. African entrepreneur, backed by IFC private equity, wins award from Ernst and Young. Available from: http://www.ifc.org/wps/wcm/connect/region__ext_content/regions/sub-saharan+africa/news/african_entrepreneur_backed_by_ifc_private_equity_wins_award_from_ernst_young [Accessed 10 July 2013].

Irwin, S.H. and D.R. Sanders. 2010. *The impact of index and swap futures on commodity markets: preliminary results*. OECD Food, Agriculture and Fisheries Working Papers, No. 27, OECD Publishing.

Isakson, S.R. 2013. Maize diversity and the political economy of agrarian restructuring in Guatemala. *Journal of Agrarian Change*, published online: 12 JUL 2013 | DOI:10.1111/joac.12023.

Juhn, T. 2007. Wal-Mart: Mexico's new bank. *Latin Business Chronicle*, July 23, 2007.

Kaufman, F. 2010. The food bubble: how Wall Street starved millions and got away with it. *Harper's Magazine*, 32(July 2010), 27–34.

Kaufman, F. 2011. How Goldman Sachs created the food crisis. *Foreign Policy*, April 27, 2011. Available from http://www.foreignpolicy.com/articles/2011/04/27/royal_flush [Accessed 7 July 2013].

Kraft Group n.d. http://www.thekraftgroup.com/private_equity_investment/

Krippner, G. 2005. The financialization of the American economy. *Socio-Economic Review*, 3(2), 173–208.

Krippner, G. 2011. *Capitalizing on crisis: The political origins of the rise of finance*. Cambridge: Harvard University Press.

de Lapérouse, P. 2012. *Case studies on private investment in farmland and agricultural infrastructure*. Boston, MA: HighQuest Partners.

Lenin, V.I. 1974. *Imperialism: the highest stage of capitalism*. New York: International Publishers.

Lewontin, Richard. 2000. The maturing of capitalist agriculture: farmer as proletarian. In: F. Magdoff, J. Bellamy Foster, and F.H. Buttel, eds. *Hungry for profit: the agribusiness threat to farmers, food and the environment*. New York: Monthly Review Press, pp. 93–106.

Levien, M. 2012. The land question: special economic zones and the political economy of dispossession in India. *The Journal of Peasant Studies*, 39(3–4), 933–970.

Li, T.M. 2011. Centering labor in the land grab debate. *The Journal of Peasant Studies*, 38(2), 281–298.

Li, T.M. 2012. What is land? Anthropological perspectives on the global land rush. Paper presented at the International Conference on Global Land Grabbing II, Oct. 17-19, 2012. Cornell University.

Magnan, A. 2012. New avenues of farm corporatization in the prairie grains sector: farm family entrepreneurs and the case of One Earth Farms. *Agriculture and Human Values*, 29, 161–175.

Mayer, J. 2009. *The growing interdependence between financial and commodity markets*. UNCTAD Discussion Paper No. 195. Geneva: United Nations Trade and Development Conference.

Meyer, G. 2009. Commodity funds get physical. *Financial Times*. December 5, 2009.

Meyer, G. 2011. Commodity traders hit back at planned US futures curbs. *Financial Times*. June 14, 2011.

McDermott, J. 2012. Multinational companies venture into start-up world. *Inc*. September 4, 2012.

McMichael, P. 2012. The land grab and corporate food regime restructuring. *The Journal of Peasant Studies*, 39(3–4), 681–702.

McWilliams, J. 2010. Coca-Cola searches the globe for the next billion dollar brand. *The Atlanta Journal Constitution*, August 22, 2010. Available from: http://www.inc.com/john-mcdermott/multinational-companies-venture-into-start-up-world.html [Accessed 5 July 2013].

Moss, M. 2013. *Salt, sugar, fat: how the food giants hooked us*. New York: Random House.

Murphy, S., David Burch, and Jennifer Clapp. 2012. *Cereal Secrets: The World's Largest Grain Traders and Global Agriculture*. Oxfam Research Reports. Oxford: Oxfam International.

Nestlé. 2002. Nestlé: venture capital fund established. Available from: http://www.nestle.com/media/pressreleases/allpressreleases/venturecapitalfund-5mar02 [Accessed on 5 July 2013].

Orhangazi, Ö. 2008. Financialization and capital accumulation in the non-financial corporate sector: a theoretical and empirical investigation on the US economy: 1973–2003. *Cambridge Journal of Economics*, 32, 863–886.

Oya, C. 2012. Contract farming in Sub-Saharan Africa: a survey of approaches, debates, and issues. *Journal of Agrarian Change*, 12(1), 1–33.

Palley, T. 2007. *Financialization: what it is and why it matters*. The Levy Economics Institute, Working Paper No. 525.

PepsiCo. 2010. Pepsico goes innovation hunting with 20 tech, media start-ups during two-day PepsiCo10 Summit. Available from: http://www.pepsico.com/PressRelease/PepsiCo-Goes-Innovation-Hunting-With-20-Tech-Media-Start-Ups-During-Two-Day-Peps07272010.html [Accessed 5 July 2013].

Reardon, T. and J.A. Berdegué. 2002. The rapid rise of supermarkets in Latin America: challenges and opportunities for development. *Development Policy Review*, 20(4), 371–388.

Reardon, T., C.B. Barrett, J.A. Berdegué, and J.F.M. Swinnen. 2009. Agrifood industry transformation and small farmers in developing countries. *World Development*, 37(11), 1717–1727.

Ross, A. 2008. Profiting from food crisis can be hard to stomach. *Financial Times*, April 25, 2008.

Rossman, P. 2010. What 'financialiation' means for food workers. *Seedling* (January 2010), 21–5.

Russell, E.D. 2008a. *New Deal banking reforms and Keynesian welfare state capitalism*. New York: Routledge.

Russell, E.D. 2008b. Finance as servant? lessons from New Deal financial reform. *Review of Radical Political Economics*, 40(3), 250–257.

Scott, J.C. 1976. *The moral economy of the peasant: rebellion and resistance in Southeast Asia*. New Haven: Yale University Press.

Selby, A. 2009. Institutional investment into agricultural activities: potential benefits and pitfalls. Paper presented at the conference 'Land governance in support of the MDGs: responding to new challenges'. Washington, DC, World Bank, March 9-10.

Spratt, S. 2013. *Food price volatility and financial speculation*. Future Agricultures Consortium, Working Paper 047.

Stockhammer, E. 2004. Financialization and the slowdown of accumulation. *Cambridge Journal of Economics*, 28, 719–41.

The Economist. 2009. Green shoots – no matter how bad things get, people still need to eat. *The Economist*, May 19, 2009. Available from: http://www.economist.com/node/13331189 [Accessed 6 July 2013].

Unilever Ventures Limited. n.d. Home page. Available from: http://www.unileventures.com/ [Accessed 5 July 2013].

Vakulabharanam, V. 2004. *Immiserizing growth: globalization and agrarian change in Telangana, south India between 1985 and 2000*. Ph.D. dissertation, University of Massachusetts, Amherst, Department of Economics.

Vakulabharanam, V. 2005. Growth and distress in a south Indian peasant economy during the era of economic liberalization. *The Journal of Development Studies*, 41(6), 971–997.

Wal-Mart. 2005. *Wal-Mart established private equity fund to drive women and minority-owned business growth.* (Press release.) October 18, 2005.

Watts, M.J. 1983. *Silent violence: food, famine, and peasantry in northern Nigeria.* Berkeley: University of California Press.

Watts, M.J. 1994. Life under contract: contract farming, agrarian restructuring, and flexible accumulation. In: P.D. Little and M.J. Watts, eds. *Living under contract: contract farming and agrarian transformation in Sub-Saharan Africa.* Madison, WI: University of Wisconsin Press, pp. 21–77.

Weis, T. 2013. The meat of the global food crisis. *The Journal of Peasant Studies*, 40(1), 65–85.

Werdigier, J. 2009. British supermarket chain places bets on banking. *The New York Times*, April 30, 2009.

Wheaton, B. and W.J. Kiernan. 2012. *Farmland: an untapped asset class? quantifying the opportunity to invest in agriculture.* Global AgInvesting: Research and Insight, December 2012.

White, B. and Anirban Dasgupta. 2010. Agrofuels capitalism: a view from political economy. *The Journal of Peasant Studies*, 37(4), 593–607.

Wolff, R.D. 2013. *Capitalism hits the fan: the global economic meltdown and what to do about it.* Northampton, MA: Interlink Publishing.

'Like gold with yield': evolving intersections between farmland and finance

Madeleine Fairbairn*

Since 2007, capital markets have acquired a newfound interest in agricultural land as a portfolio investment. This phenomenon is examined through the theoretical lens of financialization. On the surface the trend resembles a sort of financialization in reverse – many new investments involve agricultural production in addition to land ownership. Farmland also fits well into current financial discourses, which emphasize getting the right kind of exposure to long-term agricultural trends and 'value investing' in genuinely productive companies. However, capital markets' current affinity for farmland also represents significant continuity with the financialization era, particularly in the treatment of land as a financial asset. Capital gains are central to current farmland investments, both as a source of inflation hedging growth and of potentially large speculative profits. New types of farmland investment management organizations (FIMOs) are emerging, including from among large farmland operators that formerly valued land primarily as a productive asset. Finally, the first tentative steps toward the securitization of farmland demonstrate the potential for a much more complete financialization of farmland in the future.

Introduction

At the turn of the twenty-first century, farmland was still considered an investment backwater by most of the financial sector. Although some insurance companies have had farmland holdings for years, most institutional investors found farmland, and agricultural investment in general, unappealing compared to the much higher returns to be made in financial markets. However, this began to shift around 2007 as the prices of agricultural commodities started to climb. The recession that began with the bursting of the US housing bubble in 2008 caused the sector to suffer a momentary dip but also added fuel to the fire, as investors sought alternative, and more secure, places to put their money. The effects of the resulting farmland investment boom can be seen in both the Global South and the Global North. The large 'land grabs' (GRAIN 2008) taking place in developing countries have their parallel in roaring land prices in countries with more developed

This article is based on work supported by a National Science Foundation (NSF) Graduate Research Fellowship under grant number DGE-1256259, as well as by a Social Science Research Council (SSRC) International Dissertation Research Fellowship and a Louis and Elsa Thomsen Wisconsin Distinguished Graduate Fellowship. I would like to thank Geoffrey Lawrence, Philip McMichael, Jack Kloppenburg, Katharine Legun, Zenia Kish, and the anonymous reviewers for feedback on earlier drafts.

land markets (Knight Frank 2011), which have led to speculation about a possible land price bubble (Abbott 2011).[1]

Whether or not farmland markets are dangerously overheated, they are certainly hot. Celebrity investors like George Soros are known to be investing in farmland (O'Keefe 2009), and agricultural investment conferences, which provide opportunities for fund managers and farmland operators to network with end investors, have exploded in popularity. Farmland is drawing investment from 'high net worth individuals' as well as institutional investors such as pension funds, hedge funds, university endowments, private foundations, life insurance companies and sovereign wealth funds. While sovereign wealth funds generally have strategic motivations for their farmland investments, private institutional investors are flocking to farmland both for the respectable returns it delivers and for the role that farmland can play in an investment portfolio. Because farmland values have a high correlation to inflation but a low correlation to other investments, it is touted as an inflation hedge and an excellent way to reduce portfolio risk through diversification (HighQuest Partners 2010). Investors generally acquire farmland through an asset management company or operating company.[2] Asset management companies have responded to this sudden investor interest by creating a lavish buffet of new investment vehicles aimed at acquiring farmland. The extent of capital markets' interest in farmland is still relatively minor; even those institutional investors that have most enthusiastically embraced farmland generally commit less than one percent of their portfolios to this uncertain 'new' asset class (Carter 2010), and estimates of total institutional investment in farmland range between US$30 and US$40 billion globally (Wheaton and Kiernan 2012). However, it is undeniable that since 2007, global farmland real estate has undergone a makeover to become a desirable alternative asset class.

In October of 2010, the muckraking financial blog Zero Hedge (2010) wrote about a two-billion-dollar allocation to agricultural land made by the giant pension fund Teachers Insurance and Annuity Association - College Retirement Equities Fund (TIAA-CREF). The many reader comments that follow the post capture the irony of financial markets' sudden affinity for farms. One reader jokes that a farmland bubble is emerging which will culminate with the appearance of a new reality TV show, 'Farm Flippers, Thursdays this fall on HGTV' and even envisions some fake content: 'of course [we] put in all stainless steel & granite feed troughs and watering buckets. We project we'll make a 300 percent profit when we sell next month'. Another reader asks whether the turn to real assets is a 'Sign of Wall Street's fake paper going the way of the dodo? Or, more fake paper?' In this contribution I do not attempt to answer the interesting question of whether or not farmland is in a bubble, but I do take seriously the second reader's question. Slightly rephrased, the question might read: does the turn to farmland, among other real assets, signal a shift away from financialization? Or does it simply indicate that farmland itself is increasingly being treated as a financial asset?

'Financialization' is something of a catch-all term. Here I primarily follow Krippner (2011, 4) in using it to describe 'the tendency for profit making in the economy to occur increasingly through financial channels rather than through productive activities', though

---

[1]Financial sector demand for farmland is only partially responsible for steep land prices. For instance, existing farmers represented 72 percent of Iowa farmland sales in 2009, while investors were responsible for only 23 percent (Duffy 2009).
[2]Confusingly, most institutional investors are actually asset managers themselves, while the real end investors are the pensioners or insurees whose money they manage. However, for clarity's sake I will refer to these institutions as 'investors'.

I also draw on other aspects of the diverse financialization literature. The case of farmland is interesting because the posited distinction between 'productive' and 'financial' sources of profit is not always easy to discern. Land plays two different economic roles; it is an essential factor of production, but it also acts as a reserve of value and creates wealth through passive appreciation. In other words, it is a productive asset that moonlights as a financial asset. I argue that the current wave of farmland investment combines a renewed interest in productive, real assets with an underlying adherence to the logic of financialization. Though farmland's financial qualities have always held some appeal for speculators, the financialization of the global economy since the 1970s has opened up new possibilities for the incorporation of farmland into financial circuits. These new farmland investments are occurring in ways that prioritize capital gains and other financial returns but are not necessarily divorced from productive use.

The relationship between farmland acquisitions and global finance is only just beginning to receive academic scrutiny within the literature on global land grabs. McMichael (2012) provides a useful theoretical framework by situating land grabbing in the context of global food regime restructuring (see also Burch and Lawrence 2009). The current land rush, he argues, signals the deepening contradictions of the corporate food regime. It is part of the response to a crisis precipitated by both rising costs of production (energy prices) and social reproduction (food prices). Finance plays an enabling role in this salvage mission, by increasing the fungibility of land and opening up new frontiers for investment. Harvey (2010) sees the land grab as a way to sop up excess capital; when opportunities for investment at home are limited, new parts of the global economy are brought into capitalism's embrace, providing a 'spatial fix' for the crisis. On an empirical level, Daniel (2012) explores the rise of private equity funds operating in African land markets and the ways that development finance institutions facilitate this trend. The present paper contributes to this nascent interest area with a theoretical examination of the evolving interface between farmland and finance globally. It identifies broad trends in farmland investing with the potential to affect countries in both North and South.

This contribution draws from over 40 interviews with actors along the farmland investment chain – end investors, asset managers and farmland operators – as well as from participant observation at farmland investment conferences. The following section provides the paper's theoretical framework, which combines two areas of political economy: work on financialization, primarily from economic sociology, and work on the treatment of property as a financial asset, primarily from critical geography. The third section describes the ways in which the current wave of farmland investment deviates from the norm of financialization; many investors acquire farmland as part of a productive agricultural operation, and the trend is bolstered by broader discourses that stress the use value of farmland. The fourth section, however, argues that the new farmland investment boom nonetheless represents significant continuity with the financialization era. Capital gains, a mainstay of financialization, are central to even the most productive farmland investments, both as a source of inflation hedging growth and of potentially large speculative profits. The emergence of new types of farmland investment management organizations (FIMOs) also suggests that the desire to profit from farmland as a financial asset exists not only among financial actors but also among commercial actors who have typically invested in farmland primarily as a means of production. Finally, steps toward the securitization of farmland (i.e. the sale of shares in the pooled income stream from various farm properties) represent the frontier of farmland financialization. The conclusion considers possible social and environmental implications of Wall Street's emerging love affair with agriculture.

Financialization and land as a financial asset

Financialization: macro-level and institutional approaches

Epstein (2005, 3) captures the breadth of the financialization literature in his blanket definition of financialization as 'the increasing role of financial motives, financial markets, financial actors and financial institutions in the operation of domestic and international economies'. On a macro level, many theorists with roots in Marxist or World Systems analysis see financialization as a response to the systemic problem of capitalist over accumulation. For Arrighi (1994), financialization is a historically recurring phenomenon in which, midway through a 'cycle of accumulation', capitalist accumulation shifts its emphasis from commodity production and trade to finance (see also Krippner 2011). The US-led cycle of accumulation that occurred in the twentieth century, he argues, shifted into a phase of financial expansion in the early 1970s. The US government, working to maintain its hegemony, facilitated this shift through the abandonment of gold convertibility for floating exchange rates, the adoption of tight monetary policy and high interest rates, and the deregulation of the banking sector. Harvey (2010), like Arrighi, attributes the turn to financialization in the early 1970s to a capitalist crisis of overaccumulation, though he sees it as more historically specific to the political and ideological rise of neoliberalism. The actual mechanism by which financialization is able to postpone a crisis of accumulation is best addressed by literature on market bubbles (Kindleberger and Aliber 2005). During a financial bubble, skyrocketing expectations remove the limit on asset prices, allowing for far higher returns than are available in the stagnating real economy (Arrighi 2009).

The distinction between 'real' and financial sources of profit is a central element of this literature. Following in the Arrighian tradition, Krippner (2005) argues that the financialization of the US economy can be seen as occurring on two fronts. 'Sectoral' financialization describes the fact that the financial sector is playing an increasingly large role in the economy as a whole relative to other sectors; the profits made by banks, asset managers and other providers of financial services have been steadily gaining on those made in other lines of business. 'Non-sectoral' financialization describes the growing importance of financial income in the form of earned interest, dividends and capital gains on investments to non-financial firms; rather than just selling cars and plane tickets, auto companies and airlines increasingly make money from financing car loans or investing in energy derivatives.

Shifting economic institutions have contributed to the financialization process. The growing concentration of investment power in the hands of institutional investors (Useem 1996), the corporate takeover movement of the 1980s, and the emergence of 'shareholder value' as a principle of corporate governance (Fligstein 2001) have all played a role. These trends put increasing pressure on non-financial companies to demonstrate impressive performance for capital markets and to prioritize high returns to investors (Davis 2009). This has led to concern that firms may be sacrificing long-term investment in productive capacity in order to meet the short-term demands of institutional investors (Lazonick and O'Sullivan 2000). Competitive pressure within the business of investment management has also led managers toward shorter and shorter investment horizons (Parenteau 2005).

Another institutional shift associated with financialization, is the proliferation and growing complexity of financial securities in recent years (Davis 2009). Securitization is the aggregation of income streams from a pool of underlying assets to form a new financial instrument in which investors buy shares. It turns illiquid assets liquid and spreads the risk of the underlying enterprise among many investors. Leyshon and Thrift (2007, 100) argue that a key dimension of financialization is the search for ever more unorthodox asset streams to use as a means for raising investor capital. They stress, however, that though the growing complexity of securities has increasingly obscured the relationship between financial assets and the real income streams upon which they are based, this connection, however tenuous, cannot be severed entirely.

Some scholars have recently suggested that the current wave of financialization is close to running its course. Arrighi (2009) argues that the crisis of 1973 was the 'signal crisis' which set off the phase of financial expansion, while the crisis of 2008 was the 'terminal crisis' which indicated that this wave of financialization could no longer sustain itself. For Krippner (2011), the financialization of the US economy was the unintended consequence of government policies aimed at avoiding the thorny distributional questions of the 1970s by turning decisions over to the market. Now, however, she suggests that 'the limits of financialization as a strategy for deferring social and political conflicts appear to have been reached' (137), raising the question of what comes next. On an institutional level, Fligstein (2005) has also hinted that financialization may have reached its limits. He argues that, thanks to such corporate accounting scandals as Enron, 'Financialization in the pursuit of increasing shareholder value has been given a bad name from which it is unlikely to recover' (Fligstein 2005, 223).

The current farmland investment boom can shed some light on the future of financialization. Investor interest in such a tangible, productive asset could lend support to the idea that financialization is 'going the way of the dodo' as the Zero Hedge reader suggested. Land's second economic role as a financial asset, however, complicates this picture.

Land as a financial asset

The distinction between the real and the financial economies becomes somewhat tenuous when applied to farmland. This fuzzy boundary arises from land's double function as productive and financial asset. Harvey (1982), building on Marx, delves into the source of this ambiguity and in doing so lays the groundwork of a theoretical framework for the financialization of land. He argues that the distinction between landlords, who collect ground-rent based on their monopoly control of a natural resource, and capitalists, who collect interest on invested capital via the use of land as a means of production, is increasingly becoming blurred. Rather than consisting of two separate social classes, he argues, capitalist investors are now buying the land themselves and viewing it as a claim on anticipated future revenues – in this case the stream of rental payments – just like any other interest-bearing investment. In other words, property is increasingly being treated as a 'pure financial asset'. 'The land becomes a form of fictitious capital, and the land market functions simply as a particular branch – albeit with some special characteristics – of the circulation of interest-bearing capital' (Harvey 1982, 347).

Research in urban European property markets in the 1980s and 1990s demonstrated that it was no longer just financial investors who had come to see real estate as a financial asset. Haila (1988) and Coakley (1994) both take as their starting point the Marxian view that property has both 'use value' – those qualities which help it to fulfill human needs – and 'exchange value' – what it can acquire on the market. Both researchers found that urban property was being increasingly prized for its exchange value, not only by financial actors, but also by non-financial actors – an observation that recalls Krippner's discussion of 'non-sectoral' financialization. Non-financial firms had 'begun to require maximum profitability also from their real property which has until now served as a framework for activity' (Haila 1988, 92), while even residential property owners took advantage of property booms to flip their homes (Coakley 1994). However, while Harvey and Haila argue that land is becoming a pure financial asset, Coakley (1994) contends that the unique qualities of property – its imperfect substitutability, its illiquidity and its limited divisibility – mean that it is only a 'quasi-financial asset' in which rent and interest remain analytically distinct.

Awareness of land's dual role as productive asset and financial asset can be seen in the economic school of thought arguing that value may lie 'hidden' in property investments, making it possible to 'unlock' this value through institutional arrangements that increase liquidity. The classic version of this theory comes from de Soto (2000), who argues that formalizing property ownership for the poor allows them to release value by using property titles as collateral on loans. A corporate version of this thesis appears in the 'opco-propco' schemes whose premise is that property-owning corporations can raise more investment capital by splitting themselves into two distinct entities: an operating company that runs the business and a property company that owns the property and collects (potentially secur-itizable) rental payments from the operating company (Christophers 2010, Burch and Lawr-ence 2013). Christophers (2010) argues that both the de Soto and opco-propco models rest on a 'mystification' which alleges first that property itself contains a source of value outside of the activities it houses, and second that it is possible to disentangle the exchange and use values of a property and market them separately.

The literature on the treatment of property as a financial asset has tended to focus on urban real estate, with less written about agricultural land. Indeed the financialization of farmland seems to present unique challenges – use and exchange value are particularly dif-ficult to disentangle given that the property itself acts as an essential substrate for the value-producing economic activity, rather than just the location for those activities. However, research on British farmland markets during the 1970s foreshadowed some of the trends seen today. Massey and Catalano (1978) found that financial investors were buying British farmland and leasing it out to tenant farmers, motivated by the rental income and, increasingly, by the potential for property value appreciation. They contrasted this behavior with that of agricultural producers who valued farmland only as a productive asset (i.e. for its use value) and raised concerns that these investors were inflating land prices and outbid-ding 'owner-occupier' farmers. Whatmore (1986, 114), however, rejects this rigid distinc-tion, arguing that 'owner occupiers are active (and not always unwitting) participants in the speculative rise in land prices, rather than the passive victims of outside speculators or of a land market with a mind of its own'. She nonetheless argues that institutional investors do have the effect of importing volatility into land markets. Because they treat land as fictitious capital, their decision to keep or sell it is influenced not just by alterations to the agricultural use value of the land, but by alterations in the wider financial environment, including changes in inflation, interest rates and the profitability of other assets.

The farmland investment boom: a return to the real ...

Taking an Arrighian understanding of financialization as increasing accumulation through financial channels as opposed to productive ones, several aspects of the current farmland investment boom break with the trend. Most importantly, many of the farmland investments that have been initiated since 2007 are functional agricultural projects, not just land pur-chases. Investors looking to get exposure to farmland have two basic investment strategies at their disposal – I will call them 'own-lease out' and 'own-operate'.[3] The own-lease out

---

[3]Investors interested in agricultural production but not farmland ownership could also adopt a third approach, 'lease-operate', in which they produce on rented land giving them the highest risk-return of the three approaches.

approach is the more conservative. The investor simply acquires the land, finds a tenant operator, sits back and begins receiving an income stream in the form of rental payments, as well as capital gains from appreciation. The land acquisition and leasing is often done via an external asset manager, who in turn takes a cut of the profits. This strategy fits Harvey's view of the treatment of land as a pure financial asset. It is attractive to investors who view land as a relatively long-term source of stable returns, portfolio diversification and inflation hedging, including many institutional investors. In the own-operate approach, on the other hand, the investor is financially involved in both the purchase of the land and the agricultural production that takes place on it. Again, the investment is generally undertaken via an investment management organization (discussed in more detail below). In this case, however, the investor is exposed to the higher risks and returns associated with engagement in agricultural production itself, making it particularly popular among those drawn to agricultural investment for the potentially high profits.

In the current farmland rush, many investors are taking an own-operate approach. As a means of production, land has acquired renewed importance over the last few years due to a constellation of factors: population growth, increasing meat consumption in developing countries, biofuel policies that divert grain into energy markets, over-taxed water resources and climate change (Cotula 2012). For many investors, the agricultural commodity bonanza that results from all of this man-made scarcity is simply too good to pass by without investing in commodity production itself. Therefore even some institutional investors, whose long-term liabilities to pensioners or insurees match well with the steady flow of income from rental payments, are opting for a more active strategy involving production income.

In addition, whether or not investors put capital into agricultural operation, the discourses they draw from indicate a view of farmland that is uncharacteristic of financialization. Two current financial perspectives, in particular, support this turn to land and agriculture. First of all, investors who are drawn to farmland are often motivated by a desire to get the right kind of exposure to long-term trends or extreme events that would alter the political economy of global agriculture. Among the new farmland investors, the most common iteration of this perspective is a focus on global population growth and increasing resource scarcity. The influential investor Jeremy Grantham, for instance, espouses an unapologetically neo-Malthusian view (Grantham 2011), which leads him to conclude that farmland and forestry will outperform other assets over the long term (Kolesnikova 2011). Meanwhile, the Hamburg-based investment firm Aquila Capital, which manages funds for agriculture and other real assets, has Dennis Meadows, the former director of the Club of Rome think tank and co-author of *The limits to growth*, on its board of directors (McIntosh 2011). At one major agricultural investment conference, all of the attendees were given a DVD of the documentary film *Last supper for Malthus*, which examines the global food crisis. The film ends on a note of technological optimism, but not before hitting home Malthusian arguments about population growth and resource scarcity. This discursive emphasis on resource scarcity is a reminder that land's productive qualities are far from incidental to the logic of investment.

A second influential financial perspective comes from advocates of 'value investing'. This deceptively simple investment paradigm, popularized by Warren Buffett, emphasizes choosing investments based on their intrinsic value and long-term fundamentals, thereby providing some degree of insulation from the vagaries of investor sentiment. When asked in an interview for his view on gold, Buffett contrasted it unfavorably with farmland, emphasizing productive capacity. He said that if he had a choice between all of the gold in the world, worth US$7 trillion, or an equivalent value in productive assets, he would choose the latter:

> [If] you offered me the choice of looking at some 67-foot cube of gold and ... fondling it occasionally, you know, and then saying, you know, 'Do something for me', and it says, 'I don't do anything. I just stand here and look pretty'. And the alternative to that was to have all the farmland of the country, everything, cotton, corn, soybeans, [and] seven Exxon Mobils ... call me crazy but I'll take the farmland and the Exxon Mobils. (Crippen 2011)

For investors like Buffett, farmland's productive capacity is key to its value as an investment, regardless of whether the investment is in production or just the land itself.

Since 2007, this perspective on farmland has gained adherents due to increased investor distrust of markets. Unlike many financial products, the source of farmland's value is appealingly transparent. One of the farmland fund managers interviewed explained that many of his investors were searching for more concrete investment options:

> They say 'I don't want to have any more derivative operations and I don't have any idea what they are doing at the end of the line. I can see and I can understand soybean production, a sugar mill operation. I can see, I can test, touch, and I can understand all the numbers, so I want to put at least part of the money in this kind of investment'.

For investors motivated by this logic, a direct investment in farmland is significantly different from investments in financial assets based on agriculture. As the keynote speaker at one mid-Western farmland investment conference put it, 'You don't invest in commodities, you speculate or hedge with commodities. You invest in something like land' (Dotzour 2012).

The approach to agricultural investment that emerges from these two interconnected perspectives deviates from the *modus operandi* of the financialization era. At least in theory, it takes a relatively long-term view of farmland ownership and prizes it for its use value. A prominent investor speaking at a recent agricultural investment conference could almost have been paraphrasing Arrighi (2009) on the 'terminal crisis' of financialization:

> The world is changing dramatically. You know, for many periods in world history it was the financial centers that were in charge, and then for many periods it was the people who produced real goods – the oilmen, farmers, the miners – and then you had long periods when the finance people were in charge again. This is a huge change that is taking place, which unfortunately most people don't see ... I mean, finance is a terrible place to go right now. It's over competitive. Huge leverage ...

He concluded that direct involvement in agricultural production or mining was the best way to stay on the right side of this historical shift away from finance.

Of course, an investor who chooses an own-lease out strategy on the thesis that agricultural production will be increasingly vital in years to come is still treating land as Harvey's pure financial asset. However, investor motivations are not entirely inconsequential in that they seem to reveal at least a partial break with financialization construed more broadly. They indicate that, at least among a sub-section of capital market investors, investment in production or in the means of production has a renewed appeal. The discourses and investor rationales that characterize the current turn to farmland investing evince disillusionment with accumulation via financial channels and a desire, albeit partial and perhaps temporary, to return to the real economy.

The farmland investment boom … . or finance as usual?

Concurrent with this movement to make productive investments in agriculture and other natural resources, however, is a contradictory trend in which land is increasingly governed by the logic and tools that emerged with financialization. From this perspective, as TIAA-CREF's Head of Natural Resources and Infrastructure Investments put it, farmland 'is just another asset class that has the potential of going the route that real estate, private equity, [and] hedge funds did in the past' (McFarlane 2010). Rather than being treated as a *pure* financial asset as Harvey suggests, however, I will argue that the new farmland investments are premised on land's profitability as both a productive and a financial asset.

This section discusses three aspects of the ongoing financialization of farmland. First, I point out that even the productive, own-operate investments discussed above place a heavy emphasis on the profits to be made from land appreciation. Second, the emergence of new farmland management entities from within both the financial sector and the agribusiness sector demonstrates that this treatment of land as a financial asset goes beyond capital markets to those who have traditionally been interested in land for its use value alone. Finally, the emergence of farmland securitization schemes illustrates an extreme case of farmland financialization in which the profit streams from agricultural land are used as the basis to construct an actual financial asset.

Cultivating capital gains

The farmland investments initiated since 2007 place a heavy emphasis on capital gains, a type of financial return. The cash returns to the productive use of farmland are generally in the range of 3–7 percent (Allison 2005). This is a profoundly uninspiring figure to institutional investors, who are often accustomed to double-digit returns and who, in the case of pension funds, frequently base estimates of future obligations to retirees on a return expectation of at least 8 percent (Reilly 2010). Under these circumstances, modest farmland has largely managed to capture the eye of capital markets because of its potential to appreciate. Of the farmland fund managers interviewed for this study, almost all expected at least 50 percent of their fund's total internal rate of return (IRR) to come from land appreciation, and some expected substantially more. Here I discuss the importance of capital gains to farmland's appeal as both an inflation hedge and as a real estate speculation.

Many investors are drawn to farmland primarily because it is widely believed to act as an inflation hedge, preserving the value of invested capital better than most financial assets. These hedgers may be quite conservative; they often seek a long-term ownership stake in developed country farmland (such as North American or Australian) and lease out their land. Farmland's desirability as a store of value and inflation hedge is perhaps best illustrated by the comparisons between farmland and gold that have proliferated over the last few years. Like gold, farmland is limited in quantity, appreciates over time and benefits from the 'flight to quality' during economic downturns. Unlike gold, however, farmland is also a means of production, a fact that – Warren Buffet's example notwithstanding – sometimes gets lost in the metaphor. In media and investment publications, farmland is frequently referred to as 'black gold' (Cole 2012), as 'like gold with yield' (Koven 2012) or 'gold with a coupon' (Land Commodities 2009). At one investment conference, a South American agricultural fund manager took this analogy even further, arguing that if Brazilian and Argentine row crop farmland is like gold, then a more niche investment in Chilean vineyards or orchards is like investing in diamonds, emeralds and rubies. Such expressions are telling because they imply that farmland's primary appeal is its ability to store and even increase in value (leading to capital gains), while the fact that it also comes 'with yield' in the form of operating returns or rent is just the icing on the cake. These comparisons imply that it is a store of value first and foremost and a means of production only as an afterthought.

For many other investors, however, farmland's inflation hedging properties alone do not constitute sufficient motivation to invest. As a manager at one university endowment put it,

> farmland competes for every investment dollar like any other asset class would. That said we look for certain diversification, but we are not willing to accept a lower yield on the thesis of food prices going up or keeping an inflation hedge.

This quotation reflects Harvey's image of land being treated as an investment like any other and demonstrates that many capital market investors are extremely reluctant to temper their high return expectations to accommodate farmland. For these more aggressive investors and the asset managers who work for them, the potential for large capital gains takes on an even more prominent role. They invest in regions that are undergoing particularly fast appreciation whether due to policy changes, infrastructural improvements or simply growing investor interest. This speculative approach means that timing is important; according to one investment publication, 'their focus is a carefully timed purchase and subsequent disposal' (InvestAg Savills 2011).

Although passive appreciation is often key to these more aggressive farmland investments, it is not the only source of capital gains. Many farmland managers also actively cultivate appreciation by employing a 'transformative' approach that seeks to add value to the property. The methods for adding value range from simply formalizing legal titles to the wholesale transformation of forested land into farmland. Other common transformations include the addition of irrigation or transportation infrastructure and the consolidation of a number of smaller properties. In addition, operation itself is often a route to obtaining capital gains. When I asked one European pension fund manager why he preferred an own-operate approach to farmland investment, his answer was simple: 'if you participate in the operating part of the business you have a better control over the land appreciation' since land value is based largely on productivity. Once these sources of appreciation are added to the operating returns, the IRR envisioned by asset managers can easily surpass 20 percent for transformative investment strategies on marginal land in Latin America, Africa and Eastern Europe.

The point I wish to make here is that, due to land's dual nature as a productive and a financial asset, it is possible to use the land productively while simultaneously speculating on financial returns from its appreciation. The ongoing centrality of capital gains, for both hedgers and speculators, indicates that the farmland investment boom has not deviated much from the reliance on financial profits that Arrighi, Krippner and others associate with financialization. The use of land as a financial asset is obvious among those investors who adopt an own-lease out approach, as their returns are constituted by rental income and capital gains on appreciation (itself just rent capitalized into the value of the land). However, among those who adopt an own-operate approach, about half the returns still take the form of capital gains. Coakley's assessment of property as a quasi-financial asset appears particularly apt in this case. Contrary to simplistic portrayals of recent large-scale farmland acquisitions as *either* productive *or* speculative, this demonstrates that they can be, and frequently are, both at the same time.

The new FIMOs

Structural changes within the farmland investment sector indicate that land is being used as a financial asset by two different sets of actors; new farmland managers are emerging from within the financial sector but also from within agribusiness itself. Just as Haila (1988) and Coakley (1994) observed in the case of the booming urban property markets of the 1980s, farmland is being treated as a financial asset not only by financial companies but also by non-financial companies that previously saw it primarily as a source of use value.

In certain ways, the shifts occurring within farmland investing mirror those that have already occurred in US timberlands. The economic transformations that began in the 1970s – the increasing size and power of institutional investors and the corporate takeover movement – contributed to a financialization of US timberland beginning in the 1980s (Gunnoe and Gellert 2010). Vertically integrated US timber companies, facing increasing market pressure, began to view their land holdings as deadweight on their balance sheets, and ownership was gradually transferred to institutional investors. The land was either included in a real estate investment trust (REIT) or was managed on behalf of institutional investors by a Timberland Investment Management Organization (TIMO). This section considers the emergence of new asset managers entirely or partially dedicated to farmland, referred to here as FIMOs (farmland REITs are discussed in the following section). In the US, three major FIMOs – Hancock Agricultural Investment Group (HAIG), Prudential Agricultural Investments and UBS Agrivest – have existed since the 1980s or 1990s, and the former two share parent companies with major TIMOs. Like TIMOs, these management firms assemble a portfolio of land tailored to fit the client's investment thesis and appetite for risk in exchange for a management fee. They generally have a minimum investment of US$50 million and so are accessible only to institutions and extremely wealthy individuals. They also tend to take a relatively long-term view of farmland assets in which land is held for years or decades as a source of rental income and a store of value. In recent years, however, the farmland investment landscape has changed with the emergence of two new types of FIMOs.

The first type of FIMO has its origins in the financial sector. The years since 2007 have seen the advent and proliferation of farmland private equity funds (Bergdolt and Mittal 2012, Daniel 2012)[4]. While private equity funds are generally associated with the purchase, upgrading and resale of companies, the new farmland funds may acquire farmland-owning agribusinesses or simply invest directly in a portfolio of land. Like typical private equity funds, however, they are usually set up as limited partnerships, operate for a fixed term of seven or 10 years, and have management fees and carried interest on the order of 2 and 20 percent respectively. While this fee structure is likely higher than that of the managed accounts offered by traditional FIMOs, these funds also have much lower barriers to entry, sometimes available to investors with as little as US$200,000 to put into farmland. They therefore offer investors exposure to a portfolio of farmland that is generally at least somewhat geographically diversified – and therefore less risky – for an amount of capital that would otherwise have barely been sufficient for the down payment on one US farm

---

[4]The new farmland investment vehicles actually include private equity funds, hedge funds, venture capital and specialized farmland funds operated by more mainstream asset managers. However, because the traditional distinctions between these vehicles do not always apply in the case of farmland (Bergdolt and Mittal 2012), and because the majority of new farmland funds have what would generally be considered a private equity-like structure, I will focus my discussion on the role of private equity.

property. Investors now encounter a wide range of options for making private equity investments in global farmland, from NCH Capital's Agribusiness Partners Fund, which boasts 700,000 ha of farmland in the former Soviet Union and Baltic States (Bergdolt and Mittal 2012), to Emergent Asset Management's African Agri-Land Fund, which focuses on sub-Saharan Africa (Daniel 2012).

In order to return capital to investors after the term of the fund is complete and receive their own compensation, the fund managers must have some kind of exit strategy. The most common exit strategies are taking the entire fund public via an initial public offering (IPO) on the stock market, selling off the properties to a strategic buyer or rolling them over into a new fund. This last option would allow investors to keep the farmland assets even after the fund's term ended. Because most of these funds are only in their third or fourth year of operation, it is not yet possible to know the form that most of these exits will take. Although many of the funds produce on, and often make improvements to, the land they acquire, they treat their portfolio of farmland much like any other investment portfolio in terms of expected profits and time frame of investment.

A second type of FIMO which has emerged since 2007 has its lineage in large agricultural operators, some of which are seeking to capitalize on high rates of farmland appreciation by spinning off a part of their farmland portfolio into a separate asset management business. This is particularly the case in South America where a concentrated land ownership structure has made it possible for operators to own hundreds of thousands of hectares of land. The case of the publicly traded, Brazilian agribusiness SLC Agrícola is illustrative. This cotton, corn and soy producer has recently created a separate agricultural property company called SLC LandCo. In order to construct LandCo., SLC took 60,000 ha of its existing 200,000 ha land portfolio and used it to raise ~US$240 million from British asset management firm Valliance in exchange for a share of just under 50 percent in LandCo. (SLC Agrícola 2012). These funds will be used to purchase additional agricultural land with potential for rapid appreciation, all of which will be operated by SLC. The creation of this land-focused fund in addition to SLC's normal operations signals the current appeal of capital gains from land appreciation. The company underscores this point in the title of their current investor presentation: 'SLC Agrícola: value from both farm and land' (SLC Agrícola 2012). Another public Brazilian agribusiness, the sugar-alcohol sector company Cosan, has adopted a similar model. In 2008, Cosan collaborated with TIAA-CREF to create Radar Propriedades Agrícolas, a rural real estate business. As the Cosan website explains, Radar aims to 'capitalize on new business opportunities in the Brazilian rural real estate market, purchasing properties with significant potential for appreciation and leasing them to major agricultural producers. After they reach their target value, the properties are put on the market' (Cosan 2012).

The examples of LandCo and Radar demonstrate that, in a booming land market, agricultural operators are increasingly aware of the exchange value of their land base. High-Quest Partners (2010, 9) explain that the type of restructuring they have undertaken serves to 'create a platform for raising capital from a larger universe of investors which maintains a preference for land ownership (a hard asset) over investing in farm management operations'. These new FIMOs make use of the same logic that Christophers (2010) observes in opco-propco restructurings. Although the parent companies are still primarily commercial operators and the land is still used as a productive asset, these firms are taking steps to more effectively profit from farmland appreciation. While treatment of land as a financial asset is perhaps to be expected in the case of the new farmland private equity funds, whose roots are in the financial sector, it is more telling in the case of the FIMOs that have emerged from within commercial agriculture itself.

Increasing land liquidity through farmland securitization

Securitization represents the frontier of farmland financialization. It would transform farm-land from a notoriously illiquid asset to an extremely liquid one. Securitization of residen-tial real estate is, of course, widespread and was intimately connected with the crash of the US housing market in 2008.[5] Securitization of farmland real estate, however, is only in its initial stages. It would likely mean the aggregation of the rental payments made by tenant farmers on several properties into a single income stream that investors could then buy into, probably in the form of stock in a publicly listed farmland fund.

The securitization of farmland is all the more significant because it actually poses some serious difficulties not present in the securitization of other types of real estate. Land's ability to store value and appreciate over time, which makes it desirable to many investors, also makes it a weight on public company balance sheets. Buildings, equipment and most other capital assets are classified as depreciable by the Generally Accepted Accounting Principles (GAAP). Farmland, however, is not. Asset depreciation allows a company to declare the initial capital outlay for the asset as a tax-deductible expense over the years that follow. For publicly listed companies with a large amount of their fixed assets in farm-land, the inability to depreciate sets them at a disadvantage relative to other public compa-nies. In the shareholder value era, when stock price largely depends on company financial statements, farmland can therefore pose something of a liability in public markets.

Until recently, North American retail investors and those who wanted a more liquid investment could only invest in farmland indirectly by buying stock in a landowning public company, such as the South America-based agribusiness giants AdecoAgro, Cosan and Cresud, all of which own hundreds of thousands of hectares of land and are traded on the New York Stock Exchange. In 2007, investors gained a second investment option with the advent of the agribusiness exchange-traded fund (ETF). ETFs, such as the Market Vectors Agribusiness Fund, hold securities for publicly traded agribusinesses, and shares in the fund are themselves traded like stocks. Because many of the agribusi-nesses whose stocks are included in these ETFs own farmland, they give investors some indirect exposure to farmland.

The most obvious way for the securitization of farmland to occur is via a REIT. Estab-lished in the US with the Real Estate Investment Trust Act of 1960, a REIT is a corporate entity that is exempt from paying corporate taxes by virtue of the fact that it distributes 90 percent of its income directly to investors. The US has several timberland REITs, as men-tioned above, while Australia and Malaysia boast public REITs focusing on timber and palm oil production, respectively. However, the international leader in farmland securitiza-tion is, strangely enough, Bulgaria. Bulgarian REITs, known as Special Purpose Investment Companies (SPICs), were made possible with the passage of a 2003 act that exempted these entities from corporate tax provided they, like US REITs, distribute 90 percent of income to investors (DTT 2005). At least five public REITs were created in 2005 and 2006 with the view of profiting from inevitable land price increases when Bulgaria joined the European Union (EU) in 2007. They also aim to profit from the improved rent to be gained by con-solidating the fragmented plots that resulted from the distribution of former state farms to their previous owners during the transition from communism.

---

[5]The term 'securitization' is most often used to refer to the bundling of debt obligations, as in mort-gage-backed securities. However, as Leyshon and Thrift (2007) observe, almost any income stream can be turned into a security. Urban real estate has also been securitized through equity REITs, in which the income stream comes from rental payments.

Until a few years ago, North America did not even have a single *private* farmland REIT, but now there are several, and a few companies are racing to take farmland public. The first through the gate is Gladstone Land Corporation, a farmland-focused real estate firm based in Virginia that raised US$50 million in a January 2013 IPO (NASDAQ symbol: LAND). Gladstone Land's parent company, Gladstone Investment Corporation, already runs a public REIT composed of commercial real estate, and Gladstone Land intends to apply for REIT tax status for the 2013 tax year (NASDAQ 2013). Gladstone Land owns 14 farms in California, Florida, Michigan and Oregon, comprising 1950 acres (Gladstone Land 2013b). The company takes no part in farm operation, and its profits come from leasing the farm properties out to corporate and independent farmer tenants. It acquires land, in part, through sale-leaseback deals, in which the farmer sells land to the company in return for a long-term lease to continue as the farm operator (Gladstone Land 2013a).

Another firm that has expressed interest in taking farmland public is the Canadian farmland investment company Bonnefield Financial. In January of 2012, Bonnefield announced that it had applied to the Canadian security regulatory authority to launch a C$100 million initial public offering of a farmland ETF on the Toronto Stock Exchange (Canada Newswire 2012). Bonnefield already owns around 7000 acres of Canadian farmland which, like Gladstone Land, it acquires, in part, through sale-leaseback deals (Bonnefield 2012). In Saskatchewan and Manitoba, where corporate ownership restrictions prohibit public companies from owning land, Bonnefield intends to buy farmland mortgages instead of the land itself (Koven 2012) – a disconcerting idea given the role that mortgage-backed securities played in the financial crisis.

Turning farmland into a public security can have the unintended consequence of allowing use values and exchange values to become further detached. Although labor-using agricultural production remains the source of value in farmland investments (Harvey 1982) and securities depend upon such real income streams for their worth (Leyshon and Thrift 2007), they allow for an increasing divergence between the two. In an interview, an executive at one of the Bulgarian REITs told me that the crisis had increased this divergence in his company: 'After the financial crisis there is a big difference between the book value of the share and the market value because the price on the stock exchange is not so closely connected to our profits and activities'. The issue of share prices diverging from assessed land price is not specific to REITs, but is also a trait of farmland-owning public companies more generally. For instance, analysts often comment that shares in the South American farmland operator AdecoAgro trade well below their net asset value (see for instance Orihuela 2012). This divergence may relate back to the unique challenges of taking farmland public mentioned above.

However, public farmland funds are not the only unusual financial vehicles aimed at increasing the liquidity of land. A new 'crowdfunding' company called Fquare, launched in August of 2012, is in the business of selling *private* farmland securities. Crowdfunding, best known for donation-based web sites like Kickstarter, is no longer just about supporting artists and charities. In April of 2012, the Jumpstart Our Business Startups (JOBS) Act was signed into law, reducing the securities regulations that apply to crowdfunding (Cortese 2013). While crowdfunded companies could previously only compensate their 'investors' with gifts like t-shirts and signed CDs, investors can now receive company debt or equity in return for their investment. In short, investment crowdfunding has become a new type of private market, which is easily accessible over the internet and not highly regulated (Rattner 2013). So far Fquare accepts only accredited US investors – individuals with a relatively high level of wealth and financial sophistication – but once the Securities and Exchange Commission (SEC) fully implements the JOBS Act, its founders plan to accept all retail investors. An investment in Fquare buys an ownership stake in an operational Corn Belt grain farm acquired via sale-leaseback. Investor profits come from farm lease payments and take the form of quarterly dividends in the range of 3–6 percent. Investors are able to select which farm properties they hold equity in, and both investment periods (3, 5 or 7 years), and minimum investments (as low as US$5000) vary between investment properties (Fquare 2013). Perhaps most significantly, Fquare hopes eventually to establish a secondary market in which investors can buy and sell their farmland ownership shares to other Fquare investors, essentially rendering farmland liquid.

Although farmland has always had appeal as a financial asset, the amount of fixed capital it involves and its illiquidity have acted as barriers to investment. By securitizing farmland, Gladstone, Fquare and other companies like them are attempting to dismantle these barriers.

Conclusion

Occurring in the wake of a global financial crisis and in the midst of global environmental shifts that have brought renewed prominence to natural resources, the current farmland investment boom could be seen to indicate a deviation from the process of financialization. Farmland investors often draw from discourses that stress the profitability of long-term, productive investments, and frequently choose an own-operate approach that involves investment in agricultural production as well as the land itself. In many ways, however, this trend represents a continuation of financialization into new territories. Many farmland investors are eager to get exposure to agricultural production, but their investment calculus is also heavily dependent on the potential for capital gains from land appreciation. These investments depend on both the use- and exchange-value aspects of land. Meanwhile, new farmland investment vehicles, from private equity funds to public securities, are making farmland more liquid and accessible to a wider range of investors. FIMOs are emerging both from within the financial sector and from agribusiness itself, indicating that the use of land as a financial asset is not restricted to professional investors. Instead, the sector is characterized by crossover; financiers are using land as a productive asset, while operators are using it as a financial asset. Rather than a situation in which land is treated as a pure financial asset, land's financial qualities are increasingly valued but not necessarily divorced from its productive qualities. We may be seeing the emergence of a new type of financialization for an era of growing resource scarcity – one in which farmland's role as a quasi-financial asset will be even more prominent. As McMichael (2012, 686) observes, the restructuring of the corporate food regime involves the opening of new investment opportunities for capital with the result that 'the so-called rational planning of planetary resources such as land (and water) is driven as much by financial goals as by material considerations'.

Several excellent macro-level overviews of land grabbing (Cotula 2012, McMichael 2012, White *et al.* 2012) have signaled the importance of financial processes but have not elaborated much on their role. This contribution has aimed to put some further meat on the bones of the land grab-financialization connection. It adds to existing empirical research (GRAIN 2011, Bergdolt and Mittal 2012, Daniel 2012) by exploring how developments in the relationship between farmland and finance extend beyond the popular image of investors snapping up farmland in the Global South, to trends as diverse as Latin American agribusiness restructuring and soaring US land prices.

I have given relatively short shrift to the potential consequences of these developments. This is partly because the many worrying social and environmental implications associated with the global rush for farmland – among them peasant dispossession, deepening food insecurity and sweeping conversion to industrial agriculture – have already been well rehearsed. However, there are several implications of increasing interest in land as a financial asset that deserve special mention. First, to the extent that investors use an own-lease out approach, they contribute to the separation of ownership and control in land markets. The sale-leaseback arrangements pursued by Gladstone, Bonnefield and others can provide farmers with much needed financing, but they also transfer ownership away from the person farming the land. Aside from the obvious impact this has on the structure of agriculture, it also reduces the farmer's incentive to use sustainable practices by removing his or her stake in future productivity.

Some of the ways that investors 'add value' to farmland before re-selling could also reduce access to land for smallholders. Many companies, like the Bulgarian REITs mentioned above, see consolidation of small properties as an integral part of their strategy of land transformation. Their reasoning is that larger plots will be more attractive to agribusinesses and other strategic buyers that could potentially serve as their exit. In addition, some companies claim to add value by clarifying legal title where it was previously murky. In many parts of the Global South, an ironclad property title, lease or other use right will come at the expense of local residents whose legally flimsy claim lies only in years or generations of life rooted in that location.

There is also a danger of importing the short-termism of finance into land markets. This concern relates particularly to the more speculative investments being pursued by private equity funds. If capital gains are to be realized, rather than just serving the purpose of value storage, then the land (or the company that owns the land) must eventually be sold. For this reason, the new farmland private equity funds generally have seven- or 10-year time horizons. Fund managers need an exit to get paid, and an exit usually implies a sale. The idea of entering into land ownership with an 'exit strategy' in place would thoroughly confound most of the world's farmers, for whom hanging on to their land is a primary objective. For the investors involved, however, seven years actually is a long-term commitment, given that they can drop an unprofitable stock in an instant. Although many private equity fund managers argue that their short tenure as landowner will involve soil quality or other property improvements as a means to increase profit on re-sale, it seems equally likely that such a short-term view could lead to careless treatment of soil and water resources.

The financialization of farmland could also alter land market dynamics. If attempts at farmland securitization progress, it would become possible to buy or sell farmland almost instantaneously and for retail investors to acquire land as a financial asset. The increasing liquidity and volume of investment associated with securitization could greatly increase the volatility of farmland markets. Though increased volatility translates into the possibility of higher profits for speculators, it would not necessarily be welcome to those more staid farmland investors that were drawn to the sector for the steady, predictable returns. However, these investors – many of the pension funds and others employing an own-lease out strategy – could also contribute to changing land market dynamics. Global pension funds alone manage over US$20 trillion in assets (Hua 2012). If all allocated just 1 percent of their portfolios to farmland investments, there would be US$200 billion of pension money competing in global land markets. Many commentators have argued that the increasing participation of index funds in agricultural commodity markets has contributed to soaring global grain prices (Wahl 2009), and this could potentially have a similar effect. This amount of capital could raise the floor of land prices, putting it out of reach of small farmers, especially if it is concentrated it a handful of attractive markets.

Increasing financial interest in farmland may prove to be a transient phenomenon. The farmland bubble, if indeed one exists, may soon burst or simply deflate, particularly given that the appeal of land as a financial asset is highly dependent on interest rates. If, however, powerful institutional investors and financial companies continue to embrace farmland as a financial asset, it could have lasting effects on land ownership and farming worldwide.

References

Abbott, C. 2011. U.S. farmland boom may carry long-term risk: FDIC. *Reuters*, 10 March. Available from: http://www.reuters.com/article/2011/03/10/us-fdic-farmland-idUSTRE72968T20110310 [Accessed on 22 March 2012].

Allison, K. 2005. Investors dabble with living off the land. *Financial Times*, 5 April.

Arrighi, G. 1994. *The long twentieth century: money, power, and the origins of our times*. New York: Verso.

Arrighi, G. 2009. The winding paths of capital: interview by David Harvey. *New Left Review*, 56, 61–96.

Bergdolt, C. and A. Mittal. 2012. Betting on world agriculture: U.S. private equity managers eye agricultural returns. Oakland, CA: The Oakland Institute. Available from: http://www.oaklandinstitute.org/sites/oaklandinstitute.org/files/OI_report_Betting_on_World_Agriculture.pdf [Accessed on 9 July 2013].

Bonnefield. 2012. Bonnefield Canadian farmland LP1. Available from: http://www.bonnefield.com/uploads/pdfs/Bonnefield%20LP%20I%20facts.pdf [Accessed on 29 March 2012].

Burch, D. and G. Lawrence. 2009. Towards a third food regime: behind the transformation. *Agriculture and Human Values*, 26(4), 267–279.

Burch, D. and G. Lawrence. 2013. Financialization in agri-food supply chains: private equity and the transformation of the retail sector. Agriculture and Human Values, 30(2), 247–258.

Canada Newswire. 2012. Bonnefield Canadian farmland corp. announces filing of preliminary prospectus for initial public offering. *Canada Newswire*, 18 January. Available from: http://cnw.ca/4BI5 [Accessed on 29 March 2012].

Carter, D. 2010. Fertile ground for investment. *Pensions & Investments*, 19 April.

Christophers, B. 2010. On voodoo economics: theorising relations of property, value and contemporary capitalism. *Transactions of the Institute of British Geographers*, 35, 94–108.

Coakley, J. 1994. The integration of property and financial markets. *Environment and Planning A*, 26, 697–713.

Cole, R. 2012. The new black gold: U.S. farmland. *The Globe and Mail*, 22 March.

Cortese, A. 2013. The crowdfunding crowd is anxious. *The New York Times*, 5 January.

Cosan. 2012. Radar. São Paulo: Cosan. Available from: http://www.cosan.com.br/cosan2009/web/conteudo_eni.asp?idioma=1&conta=46&tipo=26752 [Accessed on 23 May 2012].

Cotula, L. 2012. The international political economy of the global land rush: a critical appraisal of trends, scale, geography and drivers. *Journal of Peasant Studies*, 39(3–4), 649–680.

Crippen, A. 2011. CNBC buffet transcript part 2: the 'zebra' that got away. CNBC. Available from: http://www.cnbc.com/id/41867379/CNBC_Buffett_Transcript_Part_2_The_Zebra_That_Got_Away [Accessed on 25 May 2012].

Daniel, S. 2012. Situating private equity capital in the land grab debate. *Journal of Peasant Studies*, 39 (3–4), 703–729.

Davis, G. 2009. *Managed by the markets: how finance reshaped America*. Oxford, UK: Oxford University Press.

De Soto, H. 2000. *The mystery of capital: why capitalism triumphs in the West and fails everywhere else*. New York: Basic Books.

Dotzour, M. 2012. Economic outlook for investors and business decision makers. Presentation at the Land Investment Expo, 20 January, Des Moine, Iowa. Available from: http://www.youtube.com/watch?v=41lzJqt5xiA&context=C362f9bcADOEgsToPDskK_AhG2gCLboOeBTiBvV3vF [Accessed on 1 June 2012].

DTT. 2005. Bulgaria: brief overview of the Bulgarian legal framework for the funds industry. Available from: http://www.dtt-lawoffice.com/new/downloads/Mutual_Funds_report_Bulgaria_05_12_05.pdf [Accessed on 23 May 2012].

Duffy, M. 2009. 2009 Iowa land value survey: overview. Available from: http://www.extension.iastate.edu/landvalue/lvs2009/background09.html [Accessed on 6 July 2012].

Epstein, G. 2005. Introduction: financialization and the world economy. *In*: G. Epstein, ed. *Financialization and the world economy*. Cheltenham, UK: Edward Elgar Publishing, pp. 3–16.

Fligstein, N. 2001. *The architecture of markets: an economic sociology of twenty-first-century capitalist societies*. Princeton, NJ: Princeton University Press.

Fligstein, N. 2005. The end of (shareholder value) ideology? *Political Power and Social Theory*, 17, 223–228.

Fquare. 2013. Frequently asked questions. Available from: https://fquare.com/faqs.aspx [Accessed on 10 July 2013].

Gladstone Land. 2013a. Overview. Available from: http://gladstoneland.investorroom.com/overview [Accessed on 10 July 2013].

Gladstone Land. 2013b. Portfolio. Available from: http://gladstoneland.investorroom.com/portfolio [Accessed on 10 July 2013].

GRAIN. 2008. *SEIZED! The 2008 land grab for food and financial security*. Barcelona: GRAIN. Available from: http://www.grain.org/article/entries/93-seized-the-2008-landgrab-for-food-and-financial-security [Accessed on 14 June 2012].

GRAIN. 2011. Pension funds: key players in the global farmland grab. *Against the Grain*, June. Available from: http://www.grain.org/article/entries/4287-pension-funds-key-players-in-the-global-farmland-grab [Accessed on 19 July 2013].

Grantham, J. 2011. GMO quarterly letter, April 2011. Available from: http://www.scribd.com/doc/54681895/Jeremy-Grantham-Investor-Letter-1Q-2011 [Accessed on 30 June 2012].

Gunnoe, A. and P. Gellert. 2010. Financialization, shareholder value, and the transformation of timberland ownership in the U.S. *Critical Sociology*, 37(3), 265–284.

Haila, A. 1988. Land as a financial asset: the theory of urban rent as a mirror of economic transformation. *Antipode*, 20(2), 79–101.

Harvey, D. 1982. *The limits to capital*. Oxford, UK: Blackwell.

Harvey, D. 2010. *The enigma of capital: and the crises of capitalism*. Oxford, UK: Oxford University Press.

HighQuest Partners. 2010. Private financial sector investment in farmland and agricultural infrastructure. Paris: Organisation for Economic Co-operation and Development (OECD). Available from: http://www.oecd.org/officialdocuments/publicdisplaydocumentpdf/?cote=TAD/CA/APM/WP (2010)11/FINAL&docLanguage=En [Accessed on 14 June 2012].

Hua, T. 2012. OECD countries' pension assets surpass $20 trillion. *Pensions and Investments*, 21 September.

InvestAg Savills. 2011. International farmland market bulletin. Available from: http://www.investag.co.uk/Bulletin2011.pdf [Accessed on 24 May 2012].

Kindleberger, C. and R. Aliber. 2005. *Manias, panics, and crashes: a history of financial crises*. 5th ed. Hoboken, NJ: Wiley and Sons.

Knight Frank. 2011. How the land lies: review of the international farmland market. *The Wealth Report 2011*. Available from: http://www.knightfrank.com/wealthreport/2011/international-farmland-market/ [Accessed on 22 May 2012].

Kolesnikova, M. 2011. Grantham says farmland will outperform all global assets. *Bloomberg.com*, 10 August. Available from: http://www.bloomberg.com/news/2011-08-10/grantham-says-farmland-will-outperform-all-global-assets-1-.html [Accessed on 30 June 2012].

Koven, P. 2012. ETF may stand for exchange-traded farmland. *Financial Post*, 19 January.

Krippner, G. 2005. The financialization of the American economy. *Socio-Economic Review*, 3(2), 173–208.

Krippner, G. 2011. *Capitalizing on crisis: the political origins of the rise of finance*. Cambridge, MA: Harvard University Press.

Land Commodities. 2009. The land commodities global agriculture and farmland investment report 2009. Available from: http://www.landcommodities.com/ [Accessed on 23 May 2012].

Lazonick, W. and M. O'Sullivan. 2000. Maximizing shareholder value: a new ideology for corporate governance. *Economy and Society*, 29(1), 13–35.

Leyshon, A. and N. Thrift. 2007. The capitalization of almost everything: the future of finance and capitalism. *Theory, Culture & Society*, 24(7-8), 97–115.

Massey, D. and A. Catalano. 1978. *Capital and land: landownership by capital in Great Britain*. London: Edward Arnold.

McFarlane, S. 2010. Pension funds to bulk up farmland investments. *Reuters*, 29 June. Available from: http://uk.reuters.com/article/2010/06/29/uk-pensions-farmland-idUKLNE65S01K20100 629 [Accessed on 24 July 2013].

McIntosh, B. 2010. Aquila Capital: absolute return and real asset strategies from Germany. *The Hedge Fund Journal*, Dec 2010/Jan 2011.

McMichael, P. 2012. The land grab and corporate food regime restructuring. *Journal of Peasant Studies*, 39(3-4), 681–701.

NASDAQ. 2013. Gladstone Land Corp (LAND) IPO. Available from: http://www.nasdaq.com/markets/ipos/company/gladstone-land-corp-834473-70799?tab=news [Accessed on: 10 July 2013].

O'Keefe, B. 2009. Betting the farm. *Fortune Magazine*, 16 June.

Orihuela, R. 2012. Soros-backed farms ripe for bid at 36% discount: real M&A. *Bloomberg.com*, 14 June. Available from: http://www.bloomberg.com/news/2012-06-14/soros-backed-farms-ripe-for-bid-at-36-discount-real-m-a.html [Accessed on 30 June 2012].

Parenteau, R. 2005. The late 1990s' bubble: financialization in the extreme. *In*: G. Epstein, ed. *Financialization and the world economy*. Cheltenham, UK: Edward Elgar Publishing, pp. 111–148.

Rattner, S. 2013. A sneaky way to deregulate. *The New York Times*, 3 March.

Reilly, D. 2010. Pension gaps loom larger. *The Wall Street Journal*, 18 September.

SLC Agrícola. 2012. SLC Agrícola: value from both farm and land. Available from: http://www.mzweb.com.br/slcagricola2009/web/arquivos/SLCE3_PresentationInstitutional_201205_ENG.pdf [Accessed on 30 June 2012].

Useem, M. 1996. *Investor capitalism: how money managers are changing the face of corporate America*. New York: Basic Books.

Wahl, P. 2009. *Food speculation: the main factor of the price bubble of 2008*. Berlin: World Economy, Ecology and Development (WEED).

Whatmore, S. 1986. Landownership relations and the development of modern British agriculture. *In*: G. Cox, P. Lowe, and M. Winter, eds. *Agriculture: people and policies*. Amsterdam: Springer Netherlands, pp. 105–125.

Wheaton, B. and W. Kiernan. 2012. Farmland: an untapped asset class? *Food for Thought*, December 2012. Sydney, AU: Macquarie Agricultural Funds Management. Available from: http://www.macquarie.com/dafiles/Internet/mgl/com/agriculture/docs/food-for-thought/food-for-thought-dec2012-anz.pdf [Accessed on 24 July 2013].

White, B., S. Borras Jr., R. Hall, I. Scoones, and W. Wolford. 2012. The new enclosures: critical perspectives on corporate land deals. *Journal of Peasant Studies*, 39(3–4), 619–647.

Zero Hedge. 2010. Is TIAA-CREF investing in farmland a harbinger of the next asset bubble? Available from: http://www.zerohedge.com/article/tiaa-cref-investing-farmland-harbinger-next-asset-bubble [Accessed on 18 March 2012].

Financialization, distance and global food politics

Jennifer Clapp

This paper provides a new perspective on the political implications of intensified financialization in the global food system. There has been a growing recognition of the role of finance in the global food system, in particular the way in which financial markets have become a mode of accumulation for large transnational agribusiness players within the current food regime. This paper highlights a further political implication of agrifood system financialization, namely how it fosters 'distancing' in the food system and how that distance shapes the broader context of global food politics. Specifically, the paper advances two interrelated arguments. First, a new kind of distancing has emerged within the global food system as a result of financialization that has (a) increased the number of the number and type of actors involved in global agrifood commodity chains and (b) abstracted food from its physical form into highly complex agricultural commodity derivatives. Second, this distancing has obscured the links between financial actors and food system outcomes in ways that make the political context for opposition to financialization especially challenging.

Introduction

Recent decades have seen phenomenal growth in the sale and purchase of financial products linked to agricultural commodities and farmland by banks, agricultural commodity trading firms and investment funds. This trend is taking place alongside a larger process of financialization within the global economy, which has seen financial markets play an increasingly important role in investment decisions and outcomes in a variety of sectors. The food studies literature is only just beginning to examine what this greater role for financial actors in the global economy means for political dynamics within the contemporary global food system (e.g. Burch and Lawrence 2009, Daniel 2012). A growing number of studies have situated financialization within the context of food regimes, showing how new financial instruments are widely used by transnational agrifood corporations as yet another mode of accumulation that further solidifies their dominant role in the global food regime (Burch and Lawrence 2009, McMichael 2012, 2013). This work has been important in developing a deeper understanding of how changing dynamics of capitalism in the global economic context drive corporate investment decisions with repercussions throughout the food system.

For helpful comments and advice, the author would like to thank Kimberly Burnett, Rachael Chong, Taarini Chopra, Eric Helleiner, Sarah Martin, Sophia Murphy, and Jan Aart Scholte, and three anonymous reviewers. Funding for this research was provided by the Social Sciences and Humanities Research Council of Canada.

This paper highlights an additional political implication of the increased role of financial actors in the food system. At the same time that financialization opens new opportunities for corporate accumulation in the global food system, some of its specific dynamics shape the political context for resistance. In particular, both the increased activity of financial actors and the growing range of specific agriculture-based financial investment tools that they utilize have contributed to a new kind of 'distancing' within the food system. Distance – which includes the geographical expanse from farm to plate along global commodity chains, as well as knowledge gaps about the social and environmental impacts of food production – affects the distribution of power and influence over the governance of the food system (Kneen 1995, Princen 2002, Clapp 2012). Heightened financialization in the global food system contributes to distancing in two ways. First, it increases the number of actors involved in global agrifood commodity chains and, second, it abstracts food from its physical form into highly complex agricultural commodity derivatives that are difficult to understand for all but seasoned financial traders.

Both of these kinds of distancing tend to obscure the role that financial actors play in the food system, making it difficult to link them to the social and ecological consequences of financial investment activities on the ground. Financial derivatives, such as index funds that derive their value from changes in an index tracking the prices of commodities, farmland and agrifood firm shares, have been prominent forms of financial investment in the sector, and have been popular with financial speculators. These financial investments, in turn, affect food prices and provide the capital for firms to make investments in the productive sector, often resulting in negative side effects, including higher and more volatile food prices as well as a host of environmental and social problems associated with large-scale farmland acquisitions. But distancing associated with financialization means that the role that financial actors play in fuelling those problems is not always transparent. This lack of transparency about which actors are involved in driving these trends creates space for competing narratives – often advanced by the financial actors themselves – that point to other explanations for negative social and environmental outcomes. Distance thus shapes the political context for groups that seek to oppose financialization, complicating attempts by civil society to challenge the dominant norms around the role of finance in the food system.

The next section of this paper maps out the theoretical literatures on the rise of financialization, its interpretation thus far in the food studies literature and the ways in which the concept of distancing can help to further highlight the implications of financialization for food system politics. The subsequent section then illustrates how new kinds of distancing have emerged with the financialization of the food system by tracing the various new actors involved and the types of financial investments they employ that have abstracted food from its physical form and add new decision points within commodity chains. The final section shows that distancing brought about by financialization in the food system has worked to obscure the role of financial actors, and the costs imposed by their activities, in ways that complicate the efforts of civil society groups that seek to reduce the impact of financial actors on food price volatility and hunger.

Finance, food and distance

The growing significance of financial actors in the world economy in recent decades has given rise to a burgeoning literature on 'financialization' and its implications (Epstein 2005, Engelen 2008, Montgomerie 2008). Financialization, as defined by Epstein, refers to the 'increasing importance of financial markets, financial motives, financial institutions, and financial elites in the operation of the economy and its governing institutions, both at the national and international levels' (Epstein 2005, 3). The literature on financialization has sought to explain how and why financial markets and the associated actors, institutions and structures have emerged as such a powerful force in the global economy as well as their implications. These works have pointed to various political factors – from states' embrace of neoliberal economic ideology to competitive pressures and lobbying by specific private interests – that have driven a dramatic deregulation and globalization of the world's financial markets. Financialization is widely seen to be a response to the exhaustion of the Fordist economic growth model, where financial capital has replaced productive capital in the quest for new profits. This process has seen the rise to dominance of a finance-led form of capitalism in which the ownership of financial assets drives investment decisions and allows for new modes of accumulation (see Montgomerie 2008). The financialization literature has also examined the rise of new relationships between financial markets, firms, individuals and the broader economy (French *et al.* 2011, Hall 2012).

Financial actors have long had a role within the food system via futures markets, but this role has grown remarkably in recent decades. Prior to the 2007–2008 food crisis, few scholars of financialization had focused their attention specifically on the food sector (Clapp and Helleiner 2012a). But rapidly rising and volatile food prices in that crisis encouraged a number of scholars to begin to note the growing role of financial speculation on agricultural commodity futures markets as one of the contributing causes of those price trends (Clapp 2009, Mittal 2009, Ghosh 2010). Looking at the structural causes and consequences of the rise of financial actors in the food system, several scholars have sought to explain the rise of financialization in the food sector as part of the broader trend toward a finance-led capitalism within the corporate food regime (Burch and Lawrence 2009, McMichael 2012). Mirroring the broader financialization literature, these works make the case that financialization provides an additional accumulation strategy for major agrifood firms within a transformed global food regime. This work shows that profit-seeking financial actors in turn channel speculative investments into the sector, enabling transnational corporations, including large agribusiness and supermarket chains, further opportunity for profit accumulation (Burch and Lawrence 2009, McMichael 2005, 2009, 2013). Some have also pointed to the role of financial speculation in land grabs, as financial capital is drawn to the fixed asset of land and the revenues that flow from the offshoring of agriculture, which also serves corporate accumulation strategies (Cotula 2012, Daniel 2012, McMichael 2012). These recent studies have been important in identifying some of the key forces and actors involved in this process, and signal the need for further investigation of the role of financial actors in the food system to tease out its specific implications.

In addition to shaping the food regime by providing new avenues for accumulation, the specific dynamics of financialization – the types of actors involved and the characteristics of the new financial tools they utilize – also affect food politics in other ways by shaping the political context for resistance. By adding new and highly complex financial relationships within the food regime, the precise role of financial actors and the consequences of their investments on the ground have become more opaque, making political action to address it more challenging. The concept of distance can help to shed light on light on this impact of financialization. Distance in global commodity chains refers to the separation between production and consumption decisions (Friedmann 1994, Kneen 1995, Princen 1997, 2002, Clapp 2012). As Freidmann notes, distance has become a dominant norm as food systems have become more globalized, which in turn has suppressed the particularities of both time and place in agriculture as well as diets, with important implications for politics (Friedmann 1994, 379). Princen shows that distance in global commodity chains can occur along several dimensions, including geography (physical distance), culture (knowledge about the conditions of production), bargaining power (ability to drive decisions) and agency (the number of middlepersons in a commodity chain) (Princen 2002). Kneen notes that technological change also expands distance by separating raw food from the final product (Kneen 1995). Greater distance between the point of production and consumption tends to constrain information feedback concerning the social and ecological implications of production processes and economic relationships at different points along commodity chains (Kneen 1995, 25, Princen 1997, Dauvergne 2008). With multiple middlepersons typically present within commodity chains that span across greater geographical distances, for example, opportunities are opened up for some agents to wield more bargaining power and withhold information that ultimately severs feedback loops which would normally make clear the lines of responsibility to protect a resource (Princen 2002, 122–3).

The way in which distancing constrains information feedback is important for the political context. When distance expands as more actors enter the chain, powerful actors are also more likely to externalize or shade costs, rendering them 'invisible' to others (Kneen 1995, Princen 2002). Environmental problems can then be easily displaced and consumers are left largely unaware of the full ecological and social consequences of their own consumption (Princen 1997, 2001, 2002, 108–15, Dauvergne 1997, 2008). The shading of costs also blurs the lines of responsibility for addressing them. Dauvergne argues that elongated commodity chains in the current era of economic globalization are particularly problematic. He makes the case that uneven global trade, investment and finance relationships fuel myriad environmental and social consequences in different locations around the world, with the costs typically falling disproportionately on the world's poor (Dauvergne 2008, 10). In such cases, the politics of resisting these forces and addressing the costs are fraught with challenges (Princen 2002, 123–30, Dauvergne 2008, 210). Corporate actors can exploit the obscured nature of their own role in externalizing costs by developing competing discourses that aim to shape or reshape the way people view their activities in a positive light (see Fuchs 2007, Clapp and Fuchs 2009).

These conceptual insights from the broader literature on distancing help to illuminate political dynamics emerging within a more financialized food system. To start, it can be noted that financialization drives distancing within the food system in two key ways, each of which shapes the broader political context. First, new actors and their vast financial resources are brought into the food system, changing the boundaries of agrifood commodity chains as more decision points are added from production to consumption. Financial investors may not be directly involved in production and marketing of actual food products, but they have gained bargaining power around the edges of the physical agrifood commodity chains through the provision of funding, the extraction of profits and their broader influence over market and price trends (Burch and Lawrence 2009). Although financial investors have had a long relationship with agricultural commodity markets, financial deregulation in the sector has enabled large numbers of investors, operating through banks and financial subsidiaries of trader firms, to enter these markets en masse. Because of their sheer size, investors now wield enormous influence over market conditions, and ultimately prices, of food commodities.

Second, financialization promotes a new kind of distancing by encouraging a greater abstraction of agricultural commodities from their physical form. As the rules governing agricultural commodity markets became more relaxed (see below), the use of new and more intricate financial derivatives proliferated (Burch and Lawrence 2009). These products are designed as financial investments, to give investors an opportunity to gain exposure to commodities in order to diversify their financial portfolios, but without having to purchase the physical commodities. For these investors, agricultural derivatives are attractive for the financial opportunities that they offer. The virtual dimension of the product takes on its own value and can generate profits, even though the investor does not own, or have any need for, the physical commodity with which it is associated. In this way, investment in the commodity is separated from its physical form. Yet, at the same time, these new agricultural derivatives are locked to physical markets in new and complex ways, and trends in one market affect trends in the other (Russi 2013).

These new kinds of distancing associated with financialization – involving both an increased number of players in the commodity chain as well the abstraction of physical commodities to a new financial form – in turn have important implications for the politics of food. These forms of distancing in the food system, not unlike greater distancing in other commodity chains, have an obscuring effect that shapes the political context (Princen 2002). In this case, financial investments fuel activities that often result in negative social and environmental effects on a global scale, yet those costs are often externalized, or shaded. The increase in the number of financial agents involved in and around agrifood commodity chains constrains information feedback, which makes cost externalization easier and more likely. As a result, the ways in which the global financial system is linked to outcomes in the food system – in particular when negative social and environmental costs occur – is not always clear. The activities of financial institutions and investors are often taking place virtually in financial centers concentrated in the world's wealthiest countries, often well before commodities are grown or delivered, and often at a great physical and cultural distance from the point of production. Banks and financial investors are buying and selling products based on agricultural commodities, and influencing the global food system in ways that are not always visible to the general public. This lack of clarity creates opportunities for competing narratives, or discourses, to emerge that seek to explain the negative outcomes in ways that portray financial actors as providers of solutions rather than sources of problems. The result of both new forms of distancing and the emergence of competing discourses presents a challenging political context for civil society organizations that are seeking to oppose financialization in the global food system.

Agriculture as financial investment

The link between financial investors and agricultural commodity trade has existed for centuries (Bryan and Rafferty 2006). Futures exchanges for agricultural commodities were established in London, for example, in the eighteenth century. Futures markets provided a means by which farmers and grain merchants could purchase and sell agricultural commodities for delivery at a future date. The ability to make deals in the 'future' enabled both sellers and buyers to lock in prices and hedge their risks in a sector that is highly uncertain due to weather fluctuations and the perishability of foodstuffs. As such, these markets have played an important role in 'price discovery' for commodities. Financial investors speculating on price movements in these early futures markets provided liquidity in cases where farmers and end users did not find equal matches for their needs. More institutionalized commodity futures trading markets emerged in other UK cities and in the United States by the mid-1800s and the practice of commodities futures trading became widespread (Cronon 1991). The ability to hedge their positions in the grain markets enabled large grain trading companies to expand their scope and size in the latter part of the nineteenth century (Morgan 1979).

The possibility that speculators might manipulate markets by taking large positions was recognized early on. Because of this risk, agricultural futures markets in the United

States – home to the largest agricultural commodity futures exchange, the Chicago Mercantile Exchange Group – have been tightly regulated since the early 1900s. The 1922 Grain Futures Act mandated that all futures trading had to take place on approved exchanges that outlawed price manipulation. Since 1923, daily reporting on trading by market traders in the markets was required. The 1936 US Commodity Exchange Act gave US federal regulators the authority to establish 'position limits' on those traders who were deemed to be 'non-commercial'; that is, those not involved in the business side of the commodity as either farmers, grain elevator operators or end users such as commodity firms and food processors. These non-commercial traders were not seen as bona fide hedgers in the markets. Rather, they were viewed as financial speculators, and the number of futures contracts there were legally allowed to hold at any time was strictly controlled. The aim of the legislation was not to outlaw speculation, but rather to prevent 'excessive' speculation that might result in market manipulation and sudden sharp price shifts (US Senate 2009, Clapp and Helleiner 2012a). Since 1974, the Commodity Futures Trading Commission (CFTC) has maintained regulatory oversight of commodity futures markets in the US, including monitoring of position limits.

Banks see opportunity in agriculture

The above regulations were put in place to prevent market manipulation and sharp price shifts. But those regulations began to be relaxed in the 1980s and 1990s (Ghosh 2010). In response to pressure from some large banks to relax the tight position limits for non-commercial operators, the CFTC issued what were referred to as 'no action letters'. These letters enabled specific banks that requested them to exceed position limits on the grounds that their positions in commodity markets were hedges against real risks they faced in financial markets. Banks needed to hedge their own financial risks by engaging in physical markets in large quantities because they began to sell financial derivative products to investors that were based on agricultural commodity markets (Clapp and Helleiner 2012a).

A common financial derivative that banks began to sell is known as a 'commodity index fund' (CIF). These derivative products track changes in the prices of a bundle of different types of commodities as an index. A general commodity index is made up of the prices of agricultural commodities, minerals, livestock and petroleum products. Agricultural products typically make up around one third of the value of these indices. What the CIF offers investors is an opportunity to gain exposure to commodity markets without being required to purchase the actual commodities on exchanges. The most popular of these index funds are the Standard and Poor's Goldman Sachs Commodity Index (GSCI) and the Dow Jones American International Group (AIG) Index (IATP 2008, De Schutter 2010). In selling these products, the banks act as middle operators, providing a financial derivative product *based on* commodity markets to investors 'over the counter' (OTC), meaning that it is a non-standardized derivative contract and thus cannot be traded on an open exchange, but rather is arranged informally (Russi 2013, 47). At the same time, the sale of these financial products posed a financial risk for banks that sold them. If commodity prices in the index rose, they would have to pay out returns to investors. To hedge these new financial risks, the banks began to purchase actual commodity futures contracts on commodity exchanges, so that they would actually gain financially if prices rose, and thus be able to make the payments to investors. This need to invest in the commodity futures markets was precisely why these banks pressed for a relaxation of position limits (Clapp and Helleiner 2012a). The new rules enabled them to expand the sale of OTC agricultural derivatives products to investors who themselves were unable to participate directly in commodity exchanges. But the OTC products and commodity exchanges became linked because banks operated in commodity exchanges to hedge their risks in the OTC market.

In 2000, the relaxation of regulations was codied with the passage of the Commodity Futures Modernization Act (CFMA) in the US (Ghosh 2010). This law exempted OTC derivative trading from oversight by the CFTC (Tett 2009). In effect, the sale of OTC derivatives products was not regulated, and purely speculative trading in these types of derivatives products was allowed. This deregulation in the United States, the most tightly regulated commodity futures markets, brought it more into line with markets in other countries. The EU, for example, had only light regulations on its commodity futures market, and prior to 2008 placed no regulations on OTC derivatives trading (Vander Stichele 2011, 1).

Other kinds of financial investment products linked to the agricultural sector also began to be offered by banks after 2000, including funds that invested not just in commodities, but also farmland and agriculture-based firms (Burch and Lawrence 2009, 271–2, Daniel 2012, McMichael 2012, 688–91, White *et al.* 2012). Buying a stake in farmland provides financial investors exposure to the agricultural production that underlies commodity prices. And investment of this kind has been made both more attractive, and easier, by the increased financialization of food and agriculture. It is attractive because the exposure to farmland expands investors' opportunities to earn returns from the production of food and biofuel crops in a context where both financial and commodity markets (including fossil and biomass energy) are increasingly volatile and uncertain, but also inextricably linked (White and Dasgupta 2010, McMichael 2012, Clapp and Helleiner 2012b).

The involvement of financial investors in farmland has also been made easier by the development of financial derivatives that are based on land investment (Burch and Lawrence 2009, McMichael 2012). Some new agriculture funds specialize in farmland acquisition, and some 66 funds now include land in their portfolio (Buxton *et al.* 2012, 1). BlackRock, for example, an investment firm that is the world's largest manager of assets, established a World Agriculture Fund in 2010 that invests in a range of agriculture-based assets, including commodity futures, farmland, agricultural input firms, and food processing and trading companies, which it bundles into an index (BlackRock 2012). Agriculture-based exchange-trade funds (ETFs) have also emerged, including the DaxGlobal Agribusiness Index and the Dow Jones Global Equity Agriculture Index. These funds track the performance of the largest agricultural firms and sell shares on the stock exchange. According the International Institute for Environment and Development, around 190 private equity firms are acquiring land and other agricultural assets on behalf of their investors (Buxton *et al.* 2012).

Commodity trading firms deepen the linkage

Banks were not the only financial actors to capitalize on the changing face of commodity futures markets that resulted from changes to regulations. Tapping into rising investor demand for commodity derivative financial products, the large agricultural commodity trading firms also began to get into the business. Archer Daniels Midland (ADM), Bunge, Cargill and Louis Dreyfus – known in the business as the ABCD firms – are heavily engaged in the agricultural derivatives market. Each of these firms has a long history: Bunge dates back to the early 1800s, Cargill to the 1860s, Dreyfus to the 1880s and ADM to the early 1900s (Morgan 1979). The ABCD firms operate under a complex business model that involves dealing in bulk commodities and trading high volumes at typically low margins. Each of these firms is intimately linked to the world of complex agricultural commodity chains, with different aspects of their business touching all aspects of those chains from production to consumption. And each has privileged access to information that has helped them to maintain advantage over their competitors (Murphy *et al.* 2012).

The commodity trading firms have long used their information advantage to manage their own business risks by purchasing and selling agricultural commodity futures contracts on commodity exchanges (see Morgan 1979, Kneen 2002). In some cases, these firms are engaged in hedging of their own business operations. But it is virtually impossible to tell when these firms are instead making purely speculative investments based on their own inside knowledge of agricultural commodity markets. As the *Wall Street Journal* noted in 2009, 'In contrast to stocks, commodities trading is the only major US market where companies are allowed to act on inside information to manage risks others might not know about' (David 2009). Commodity trading firms are often the first to become aware of crop shortages or other interruptions to agricultural trade, giving them an information advantage in the futures markets (Meyer 2011).

Commodity trading firms were able to capitalize on their specialized knowledge of the sector and in the past two decades have made financial and risk management a major part of their business structure (Murphy *et al.* 2012). In this period, each of the ABCD companies established financial subsidiary firms that specialized in this task. The ABCD financial services firms gradually began to meet not just their own risk management needs, but also those of third-party investors. The financial arms of these firms became very active in selling OTC derivatives, much like the banks were doing. Cargill, for example, founded Cargill Risk Management (CRM) in 1994, explicitly to sell individualized OTC products for its own purposes and for third-party customers. In 2003, Cargill established another independently managed subsidiary, Black River Asset Management, which started to manage the funds of third-party investors in 2004. ADM operates ADM Investor Services (ADMIS), a subsidiary which also sells OTC agricultural derivatives to third parties and which has two separate financial subsidiaries that operate under it, selling investment products to business partners and third-party institutional investors. Louis Dreyfus established two agricultural hedge funds: the Alpha Fund that specializes in agricultural commodities, and Calyx Agro that specializes in farmland investments in Latin America. Bunge has two financial divisions, Bunge Global Markets and Bunge Limited Finance Corporation (Murphy *et al.* 2012).

The distinction between banks and commodity trading firms has become increasingly blurred since the mid 1990s as both sets of actors became actively engaged in selling OTC agricultural commodity derivatives products such as commodity index funds (Burch and Lawrence 2009, 277). The market for these products grew rapidly after 2000 when the CMFA came into place, and the total assets of financial speculators in agricultural commodity markets increased from US$65 billion in 2006 to some US$126 billion by early 2011 (Worthy 2011, 13). In the US wheat futures market, for example, financial speculators' share of the trade increased from 12 percent in the mid-1990s to 61 percent in 2011 (Worthy 2011, 13).

Large-scale third-party investors

In this period, agriculture-based financial investment products became hugely popular among investors, including large-scale institutional investment funds and wealthy individuals. Institutional investors include insurance companies, pension funds, mutual funds, sovereign wealth funds, hedge funds, and university and foundation endowments. These investment funds essentially pool the resources of those that invest in them, enabling them to expand and diversify their investment options while sharing transaction costs (Burch and Lawrence 2009, 272–3, Buxton *et al.* 2012). According to the Bank for International Settlements, in 2005, insurance companies, pension funds and mutual funds together managed US$46 trillion (BIS 2007). These large-scale investors have some unique features, one of which is that they tend to be passive investors with a significant proportion of their investments. With large amounts of money to invest, they tend to make long-term investment decisions that do not require active management, and members do not always have detailed knowledge of their own investments.

Large-scale financial investors increasingly began to seek exposure to commodities as an asset class after 2000, as commodity prices in general were rising in this period (Burch and Lawrence 2009, 273). Food and energy commodities in particular saw important upward price shifts, as these sectors became increasingly locked together through the rise of biofuels as an alternative energy source and financial derivatives that bundled food and energy commodities together (McMichael 2010, Clapp and Helleiner 2012b). Investors sought to purchase financial products from large banks and the financial arms of commodity trading firms who offered exposure to commodities and farmland through CIFs and other kinds of agriculture-based financial investments that were becoming more readily available to them. Some estimates put overall commodity investments of institutional investors at around US$320 billion, which is up significantly from the US$6 billion they held in investments in this sector in 2002 (Buxton *et al.* 2012, 2).

Institutional investors are also increasing their exposure to farmland via new derivative products on offer from large investment banks and commodity firms (Daniel 2012). Farmland is especially attractive to institutional investors because it not only holds value in itself due to its limited supply, but also produces a flow of income from the production that arises from it. These traits suit investors with long time horizons, such as pension funds, that seek to capitalize on rising demand for food and biofuels on a fixed land base. BlackRock's World Agriculture Fund noted above, for example, markets to institutional investors and includes agricultural land in its fund (BlackRock 2012). Pension funds have been actively adding farmland to their asset mix, holding approximately US$5–15 billion in farmland assets (GRAIN 2011). The Calyx Agro fund, noted above, explicitly seeks to identify, acquire, develop, concert and sell farmland for large institutional investment funds such as AIG, with a focus on Latin America (Calyx Agro no date, Henriques 2008). Private equity funds EmVest and SilverStreet Capital have established funds that target institutional investors and focus specifically on African farmland (Daniel 2012, 705–6).

This closer look at the regulatory trends and the actors involved in financialization in the food system illustrates the ways in which distancing has expanded. Deregulation has enabled the entry of new actors, or middlepersons, into agrifood commodity chains. Banks, institutional investors and new financial investment arms of the ABCDs are now active participants in agrifood supply chains, bringing with them considerable decision-making power based on their financial weight. Along with new actors, new types of relationships between actors, in the form of abstract commodity derivatives, have also changed the shape and culture of those commodity chains. Financial investors are not necessarily buying agricultural commodities in their physical form, but rather are seeking profits in and around the commodity chain, through new kinds of financial derivatives that are abstractions from the physical commodity. CIFs, in particular, have been especially attractive for the needs of these large-scale third-party investors. They are easy to obtain because they are sold OTC, and investors can hold them for long periods of time, waiting to reap profits as commodity prices climb. The pooling of financial resources of multiple investors and the availability of new financial products also enable institutional investors to buy a stake in farmland and other land-based derivatives. These changes brought by financialization have important implications for the politics of food, as discussed below.

Financialization and the political context for opposition

Increased distancing that has resulted from financialization has worked to obscure the role that financial investors play in driving certain outcomes in the global food system. The increased number of financial actors in the agrifood sector and the introduction of new and complex agricultural derivatives have made the boundaries around commodity chains, and the lines of responsibility within them, less clear. This new context complicates political action around food. Some analysts have linked rising financial speculation in agricultural commodities and land to food price volatility and speculative land grabs, each of which imposes social and ecological costs in the world's poorest countries (e.g. Dauvergne and Neville 2010, Ghosh 2010, Daniel 2012, McMichael 2012, White et al. 2012, Russi 2013). But distancing, as outlined above, has made the precise impact of each financial investment hard to pinpoint with certainty because the flow of information is inhibited by greater distance that financialization has brought to the food system. Financial investors' funds are often pooled and managed by others, which in turn makes drawing a causal link between particular financial investments and their potential negative outcomes, such as hunger or environmental degradation in specific locations, extremely difficult (Clapp and Dauvergne 2011, 217). The new forms of distancing, in other words, have expanded the opportunity for powerful agents within commodity chains to shade costs.

The difficulty of drawing a direct causal line from financial investments to outcomes in the food system has made the political context much more challenging for civil society organizations campaigning on these issues. Although there are growing pressures for progressive reform in the food system in the face of crisis (e.g. Giménez and Shattuck 2011), the lack of certainty about the specific impact of financial actors caused by distancing creates room for competing narratives that feature other, non-financial explanations for hunger, displacement from land and environmental degradation. Instead of seeing the rise in financial investment in agriculture as a potential problem, other narratives in fact often portray financialization as a solution to problems in the food system. These narratives typically emanate from financial actors themselves as well as powerful organizations that support them. The World Bank, for example, has played a key role in advocating for more commodity exchanges in developing countries at the same time that it has portrayed large-scale land acquisitions as an important development opportunity (World Bank 2007, 2010, Bush 2012).

Both the new forms of distancing, and the competing discourses they give rise to, complicate efforts of civil society organizations that are seeking to curb the effects of financial actors in the food system. In this political context, the typical response of states has been to adopt weak regulations. These trends can be illustrated through a brief overview of the recent politics surrounding food price volatility.

Uncertainty and debate over role of finance in driving food price volatility

It is widely understood that financialization in the food sector has exposed agricultural prices to broader trends in financial markets (Russi 2013). As noted above, financial market turmoil after 2006 contributed to disruptions in food markets that encouraged investors to capitalize on commodities that were at that time widely advertised to be a more stable and higher-return investment than other kinds of financial derivatives. As money poured into commodities in large amounts in this period, food prices began to climb. In the 2006–2008 period, average world prices for rice rose by 217 percent, wheat by 136 percent, maize by 125 percent and soybeans by 107 percent (WRI 2008). But, by the end of 2008, prices fell just as dramatically as financial markets collapsed and investors retreated. Volatility and renewed price spikes then returned in 2010–2012. Fluctuating food prices on world markets in this period have caused enormous disruptions to food security of the world's poorest people (Daviron *et al.* 2011, FAO 2011, IFPRI 2011). Most analysts agree that there is some link between the recent rise of financial speculation in commodity markets and food price trends. But because distancing has obscured the precise lines of cause and effect, the extent to which financial speculation is viewed as central cause of food price volatility or a marginal force is highly contested. In other words, distancing has fuelled uncertainty about cause, effect and responsibility, giving rise to competing discourses about the issue.

When food prices rose steeply after 2006, a number of analysts and civil society organizations pointed to speculative finance as a driving factor. Bolstered by the Food and Agriculture Organization (FAO)'s analysis that a significant portion of the price volatility was well beyond what would be explained by underlying supply and demand for food (FAO 2008), these analysts made the case that the fundamentals of supply and demand alone could not explain the food price spikes (Mittal 2009). They argued that speculation was the only plausible explanation, and that it drove up food prices and made them more volatile as financial speculators moved in and out of commodity derivatives in order to seek profits as financial market conditions changed (e.g. IATP 2008, Ghosh 2010, Russi 2013). Speculative investment in CIFs was blamed in particular for pushing up food prices. Even though they are only indices of prices on commodity markets, large movements of speculative capital into these particular financial products can cause severe disruptions to commodity markets (Masters 2008). As institutional and other investors purchase index funds, banks and commodity trading firms hedge the risks associated with the sale of those products on commodity futures markets, in effect linking the physical market with the index. And although they are betting huge sums on the physical commodity futures market, these transactions are in the hands of relatively few traders. At the height of the food price rises of 2008, just six traders held up to 60 percent of the Chicago wheat futures contracts that were linked to index funds (US Senate 2009). In this context, critics argued, even very small changes in how investment portfolios of thousands of investors are managed can result in sharp changes in agricultural prices. Several recent studies that have presented sophisticated models teasing out the different factors that have affected commodity price volatility point to financial speculation as a significant force (Lagi *et al.* 2011, Basak and Pavlova 2013).

Others have taken issue with these lines of argument, inserting a competing narrative into the space obscured by distancing. The World Bank and the Organisation for Economic Co-operation and Development (OECD), for example, have presented highly technical arguments about the functioning of futures markets that downplay the idea that financial speculation fuels food price volatility (World Bank 2008, Irwin and Sanders 2010). Supporters of this view have made the case that there is little evidence to prove that causal link and they suggest that critics of financial speculation have misunderstood the technicalities of how the markets work (Irwin *et al.* 2009). For these analysts, commodity futures markets have provided liquidity and capacity to absorb risk to global food markets in ways that have instead helped to stabilize food prices (Irwin *et al.* 2009, Sanders and Irwin 2010). From this standpoint, the primary causes of food price volatility can be found elsewhere, namely in the fundamentals of supply and demand in food markets themselves (Irwin *et al.* 2009).

The debate over the impact of commodity speculation on food prices continues. By 2010, it was clear that food price volatility was persisting, and pressure grew to uncover the cause. The G20 announced that it was taking up the issue under France's leadership the following year. Around this time, several international organizations began to stress that speculation in agricultural commodity futures markets and financial derivatives was a factor in at least exacerbating price volatility trends. The Bank for International Settlements noted, for example, that financialization influences commodity prices, especially in the short term. Several United Nations (UN) reports also came to a similar conclusion (De Schutter 2010, BIS 2011, UNCTAD 2011). An UNCTAD (United Nations Conference on Trade and Development) report explains that investors often act in a herd-like fashion, following each other due to the lack of perfect information. This herd behaviour can make prices swing up and down more dramatically than they otherwise would have done (UNCTAD 2011). Although these reports provided some fuel to those seeking greater regulation of financial markets to stem excessive commodity speculation, progress on this front has been slow and uneven, indicating that the debate on the role of finance in food price volatility is far from settled (Clapp and Murphy 2013).

Civil society campaigns and financial sector backlash

A number of civil society groups concerned about the impact of food price volatility on hunger have sought to challenge the new role of financial actors in the food system by calling for divestment from speculative activities that drive up food prices and for stronger state-based regulation to curb speculation on food. The new types of distancing brought about by financialization, however, have made the political context within which these groups operate more difficult in several ways. First, these groups have had to engage with a large number of new actors in the agrifood commodity chain, namely financial institutions, which they previously had not dealt with in any extensive way. Second, the abstract nature of agricultural financial derivatives has complicated the general public's understanding of the issue and meant that campaigners have to become knowledgeable about a highly technical sector in a very short period of time. And third, because these types of distancing created room for the emergence of competing narratives about the causes of food price volatility (as outlined above), civil society groups face the challenge not only of explaining their concerns to the general public, but also of constantly engaging with these alternative narratives in order to ensure that their message is not undermined.

Although distancing brought by financialization introduced these new political challenges, civil society organizations were not deterred from launching campaigns that sought to shine a light on the role of financial actors in fuelling food price volatility. As the G20 took up the issue of food price volatility in 2010–2011, a number of organizations based in Europe, including World Development Movement (WDM), Oxfam, Friends of the Earth, and Foodwatch, launched campaigns that sought to stop banks' participation in agricultural commodity speculation and to improve financial sector regulation (Jones 2010, Foodwatch 2011, Herman *et al.* 2011, Worthy 2011, Friends of the Earth Europe 2012). In the US, a coalition of groups launched the Stop Gambling on Hunger campaign, including the Institute for Agriculture and Trade Policy, Better Markets and Commodity Markets Oversight Coalition (CMOC) (Stop Gambling on Hunger n.d.). An open letter posted on the World Development Movement website has been signed by over 100 civil society groups which allied in a broader global campaign to 'stop gambling on food and hunger' (WDM 2011). Many of these groups targeted large banks familiar to the general public that deal in agricultural commodity derivatives. The World Development Movement, for example, has targeted Barclays Bank and Goldman Sachs, while Oxfam France and Foodwatch focused their attention on major banks in France and Germany, respectively (e.g. Foodwatch 2011, Scott 2011, Oxfam France 2013).

Some successes have been won by these groups, but given the complex political environment in which these campaigns were launched, the overall outcome has been mixed and the successes are fragile. In the first half of 2012, several German and Austrian banks, including Deutsche Bank, announced that they would be removing agricultural products from their index fund products. In early 2013, several major French banks pulled out of commodity speculation in direct response to an Oxfam France report that targeted them (Oxfam International 2013). And in the UK, Barclays announced in early 2013 that it too would be withdrawing from speculation on food prices (Treanor 2013). By mid-2013, the WDM reported that some 11 European banks had pulled out of financial investment in agricultural commodities (Haigh 2013).

But the WDM also warned that these banks could change course at any time. Indeed, one major financial institution that had halted agricultural commodity speculation in 2012, Deutsche Bank, has already reversed its decision, announcing it was reinstating its agricultural commodity investments. On making this announcement, the bank invoked the competing narratives cited above, arguing that it found 'no convincing evidence' that financial speculation in agricultural commodity markets had driven up food prices or made them more volatile (Deutsche Bank 2013). Lobbyists for the banks have also become more aggressive in trying to shape the public debate on the issue by taking advantage of the uncertainties about the cause and effect generated by distancing. The International Swaps and Derivatives Association (ISDA), an association representing over 800 financial institutions including Barclays, Deutsche Bank, JP Morgan, Goldman Sachs and Morgan Stanley, among others, launched a website in early 2013 that counters growing criticism of commodity speculation by selectively reproducing quotations from reports and media appearances of analysts who argue that financial speculation is not a cause of price volatility and that financial reform to address it is not needed (ISDA n.d.). Civil society groups in turn have coordinated their own efforts more tightly and have filled their own websites not just with their own studies, but those of other groups, academics and policymakers that agree with their assessment of the role of speculation in causing food price volatility. But it has often been challenging to present to the general public what are often quite technical debates about the impacts of the abstract financial products involved.

In addition to targeting specific financial institutions to demand divestment, civil society organizations have also worked to strengthen financial regulation to stem commodity speculation. To engage in these debates, these groups had to learn the highly technical details of abstract financial derivatives as well as the complex regulatory context. The outcomes, once again, have been mixed. The 2008 food and financial crises generated significant political momentum to reform its financial market regulations, including demands from civil society lobby groups such as the CMOC for tighter position limits and improved reporting in commodity derivative markets (Clapp and Helleiner 2012a). After long and hard-fought struggles, the 2010 Dodd-Frank Wall Street Reform and Consumer Protection Act did back stronger measures to limit speculation in agricultural commodities, calling upon the CFTC to put in place more stringent position limits and more rigorous and transparent reporting of OTC trades, among other things.

But the implementation of Dodd Frank has been slow and uneven, raising concerns about its long-term fate. In late 2011, the CFTC was faced with a legal challenge from ISDA and the Securities Industry and Financial Markets Association over its rules concerning position limits (Protess 2011). These lobby groups made the case that the rules should not be implemented because the CFTC moved ahead to impose position limits without first determining whether those limits were either 'necessary' or 'appropriate'. Because distancing obscures the lines of cause and effect, the implementation of the act was easily contested. Weeks before the rules were due to be implemented in 2012, the court ruled that the CFTC failed to prove that such rules were 'necessary', and as such the rules did not come into place. The CFTC appealed the ruling and subsequently revised the rules to relax some of its requirements while also providing a detailed review of the literature to bolster the case for their need (Protess 2012, Schoenberg 2012, CFTC 2013).

Civil society groups are also lobbying policymakers in Europe to impose position limits and more comprehensive reporting on commodity derivatives markets as part of financial market reforms under the EU Markets in Financial Instruments Directive (MiFID). As in the US case, however, opponents of reform have capitalized on the obscured role of financial actors caused by distancing. Like their US counterparts, these opponents have called into question the causal relationships drawn by the civil society groups, thereby weakening the momentum behind the reform efforts. Civil society groups have expressed strong concern about the direction of the negotiations (Friends of the Earth 2013). Even if the European Union does tighten regulation in this area, implementation of the new rules is very likely to meet with further resistance by the financial industry, as has been the case in the US.

From this brief review of the recent campaigns against agricultural commodity speculation, it is apparent that the results have been mixed, and the political context remains a complicated one. Although some financial institutions have chosen to withdraw from financial speculation on food prices, other banks and lobby groups have fought back and are seeking to shape public discourse toward a more positive view of the impact of agricultural commodity speculation by financial actors. Progress on financial regulatory reforms in the US and EU has also been slow and uneven. The campaigns have required civil society groups to engage with new financial actors and financial regulatory issues. Distancing has also obscured the impact of financial investments on the global food system in ways that complicate the efforts of campaigners to engage the general public and that open the door for competing narratives to undermine campaigners' goals.

Conclusion

Financialization in the food system has added a layer of complexity to agrifood markets in recent years. New financial actors have entered agrifood commodity chains and have acquired significant power within them through the use of new complex financial derivative products. But the influence of these actors and the outcomes of their investments are not always transparent because of the distancing that financialization has also generated. Financialization has contributed to greater distancing within the food system by facilitating the entry of new actors taking profits along, in and around agrifood commodity chains, and by encouraging more abstraction of the commodity from its original form, in this case into a 'virtual' financial derivative product.

By obscuring the cause and effect of financial investment activities from public view, distancing increases opportunities for ecological and social costs to be externalized on a global scale. With many more financial actors now involved in and around the edges of the agrifood supply chain, and with their investments in a mix of abstract financial derivatives and pooled investment funds linked to food commodities, agriculture and farmland, it is difficult to ascertain the precise role of specific investors in driving the negative outcomes as well as the lines of responsibility for addressing them. This lack of clarity on the role of finance has complicated the efforts of civil society groups to engage the general public and has opened spaces for counter-narratives to emerge over the role of financial actors in fuelling the problem. In this challenging political environment, it is not surprising to see only slow and uneven progress.

The analysis in this paper contributes to the broader literature on financialization as well as more recent work on the financialization of food by showing how the specific dynamics of financial markets contribute to distancing in the global food system, which in turn affects the political context for resistance to the growing power of financial actors within that system. The paper also contributes to the conceptual literature on distancing by highlighting the way in which financialization fosters new forms of distancing, in particular through abstraction of food from its physical to a financial form. It also shows that distancing is important to examine not just in terms of its implications for production and consumption decisions, but also how it affects investment decisions that have crucial implications for social and ecological outcomes relating to both production and consumption.

References

Basak, S. and A. Pavlova. 2013. *A model of financialization of commodities*. Available from: http://ssrn.com/abstract=2201600 [Accessed 8 July 2013].

BIS. 2007. *Institutional investors, global savings and asset allocation*. CGFS Paper 27. Committee on the Global Financial System. Basel: Bank for International Settlements.

BIS. 2011. 81st Annual Report. Basel: Bank for International Settlements. Available from: http://www.bis.org/publ/arpdf/ar2011e.pdf [Accessed 8 July 2013].

BlackRock. 2013. Black Rock world agricultural fund fact sheet. May. Available from: http://www.blackrock.com.hk/content/groups/hongkongsite/documents/literature/1111121111.pdf [Accessed 8 July 2013].

Bryan, D. and M. Rafferty. 2006. *Capitalism with derivatives*. Basingstoke: Palgrave Macmillan.

Burch, D. and G. Lawrence. 2009. Towards a third food regime: Behind the transformation. *Agriculture and Human Values*, 26(4), 267–279.

Bush, S.B. 2012. *Derivatives and development*. New York: Palgrave Macmillan.

Buxton, A., M. Campanale, and L. Cotula. 2012. *Farms and funds: Investment funds in the global land rush*. IIED Briefing. January. London: IIED. Available from: http://pubs.iied.org/pdfs/17121IIED.pdf [Accessed 8 July 2013].

Calyx Agro. No date. Company homepage. Available from: http://www.calyxagro.com/company.php [Accessed 8 July 2013].

Clapp, J. 2009. Food price volatility and vulnerability in the Global South: Considering the global economic context. *Third World Quarterly*, 30(6), 1183–1196.

Clapp, J. 2012. *Food*. Cambridge: Polity.

Clapp, J. and D. Fuchs, eds. 2009. *Corporate power in global agrifood governance*. Cambridge, MA: MIT Press.

Clapp, J. and P. Dauvergne. 2011. *Paths to a green world*. Cambridge, MA: MIT Press.

Clapp, J. and E. Helleiner. 2012a. Troubled futures? The global food crisis and the politics of agricultural derivatives regulation. *Review of International Political Economy*, 19(2), 181–207.

Clapp, J. and E. Helleiner. 2012b. International political economy and the environment: Back to the basics? *International Affairs*, 88(3), 485–501.

Clapp, J. and S. Murphy. 2013. The G20 and food security: A mismatch in global governance? *Global Policy*, 4(2), 129–138.

Commodity Futures Trading Commission. 2013. Position limits for derivatives. Available from: http://www.cftc.gov/ucm/groups/public/@newsroom/documents/file/federalregister110513c.pdf [Accessed 8 December 2013].

Cotula, L. 2012. The international political economy of the global land rush: A critical appraisal of trends, scale, geography and drivers. *The Journal of Peasant Studies*, 39(3–4), 649–680.

Cronon, W. 1991. *Nature's metropolis: Chicago and the Great West*. New York: W. W. Norton & Company.

Daniel, S. 2012. Situating private equity capital in the land grab debate. *Journal of Peasant Studies*, 39 (3–4), 703–729.

Dauvergne, P. 1997. *Shadows in the forest: Japan and the politics of timber in Southeast Asia*. Cambridge, MA: MIT Press.

Dauvergne, P. 2008. *The shadows of consumption*. Cambridge, MA: MIT Press.

Dauvergne, P. and K.J. Neville. 2010. Forests, food, and fuel in the tropics: The uneven social and ecological consequences of the emerging political economy of biofuels. *The Journal of Peasant Studies*, 37(4), 631–660.

David, A. 2009. Cargill's inside view helps it buck downturn. *Wall Street Journal*, 14 Jan.

Daviron, B., N.N. Dembele, S. Murphy, and S. Rashid. 2011. Price volatility and food security. A report by the High Level Panel of Experts on Food Security and Nutrition of the Committee on World Food Security. Available from: http://www.fao.org/fileadmin/user_upload/hlpe/hlpe_documents/ HLPE-price-volatility-and-food-security-report-July-2011.pdf [Accessed 22 August 2012].

De Schutter, O. 2010. *Food commodities speculation and food price crises*. UN Special Rapporteur on the Right to Food. Briefing Note 02 – September. Available from: http://www.srfood.org/images/ stories/pdf/otherdocuments/20102309_briefing_note_02_ee_ok.pdf [Accessed 8 July 2013].

Deutsche Bank. 2013. *Questions and answers on investments in agricultural commodities*. Frankfurt: Deutsche Bank. Available from: https://www.db.com/en/content/company/headlines_agricultural_ commodities_faq.htm [Accessed 8 July 2013].

Engelen, E. 2008. The case for financialization. *Competition & Change*, 12(2), 111–119.

Epstein, G. 2005. Introduction: Financialization and the world economy. *In*: G. Epstein, ed. *Financialization and the world economy*. Cheltenham: Edwar Elgar, 3–16.

FAO. 2008. *Food outlook*. June. Rome: Food and Agriculture Organization. Available from: ftp://ftp. fao.org/docrep/fao/010/ai466e/ai466e00.pdf [Accessed 8 July 2013].

FAO. 2011. *The state of food insecurity in the world 2011*. Rome: Food and Agriculture Organization.

Foodwatch. 2011. *The Hunger-Makers: How Deutsche Bank, Goldman Sachs and other financial institutions are speculating with food at the expense of the poorest*. Berlin: Foodwatch.

French, S., A. Leyshon, and T. Wainwright. 2011. Financializing space, spacing financialization. *Progress in Human Geography*, 35(6), 798–819.

Friedmann, H. 1994. Distance and durability: Shaky foundations of the world food economy. *In*: P. McMichael, ed. *The global restructuring of agro-food systems*. Ithaca: Cornell University Press, 258–276.

Friends of the Earth Europe. 2012. *Farming money: How European banks and private finance profit from food speculation and land grabs*. Available from: https://www.foeeurope.org/farming-money-Jan2012 [Accessed 23 December 2013].

Friends of the Earth. 2013. *MiFID2: Set to fail on food speculation*. Policy briefing. April. Available from: https://www.foeeurope.org/sites/default/files/makefinancework_mifid_loopholes_june2013. pdf [Accessed 8 July 2013].

Fuchs, D. 2007. *Business power in global governance*. Boulder, CO: Lynne Rienner Publishers.

Ghosh, J. 2010. The unnatural coupling: Food and global finance. *Journal of Agrarian Change*, 10(1), 72–86.

Giménez, E.H. and A. Shattuck. 2011. Food crises, food regimes and food movements: Rumblings of reform or tides of transformation? *Journal of Peasant Studies*, 38(1), 109–144.

GRAIN. 2011. Pension funds: Key players in the global farmland grab. Available from: http://www. grain.org/article/entries/4287-pension-funds-key-players-in-the-global-farmland-grab [Accessed 8 July 2013].

Haigh, C. 2013. DZ Bank switches sides. World Development Movement. 29 May. Available from: http://www.wdm.org.uk/food-and-hunger/switching-sides [Accessed 8 July 2013].

Hall, S. 2012. Geographies of money and finance III: Financial circuits and the 'real economy'. *Progress in Human Geography*, 37(2), 285–292.

Henriques, D.B. 2008. Food is gold, and investors pour billions into farming. *New York Times*, 5 June. Available from: http://www.nytimes.com/2008/06/05/business/05farm.html?_r=1&dlbk [Accessed May 2012].

Herman, M., R. Kelly, and R. Nash. 2011. *Not a game: Speculation vs. food security*. Oxfam Issue Briefing. October. Available from: http://www.oxfam.org/en/grow/policy/not-game-specul ation-vs-food-security [Accessed 8 July 2013].

IATP. (2008). *Commodities market speculation: The risk to food security and agriculture*. Minneapolis: Institute for Agriculture and Trade Policy. Available from: http://www.iatp.org/files/451_2_104414.pdf [Accessed 8 July 2013].

IFPRI. 2011. *Global hunger index: The challenge of hunger: Taming price spikes and excessive food price volatility*. Washington, D.C.: International Food Policy Research Institute. Available from: http://www.ifpri.org/sites/default/files/publications/ghi11.pdf [Accessed 8 July 2013].

International Swaps and Derivatives Association. No date. Commodity fact website. Available from: http://www.commodityfact.org [Accessed 8 July 2013].

Irwin, S.H. and D.R. Sanders. 2010. *The impact of index and swap funds on commodity futures markets: Preliminary results*. OECD Food, Agriculture and Fisheries Papers, No. 27, OECD Publishing. Available from: http://dx.doi.org/10.1787/5kmd40wl1t5f-en [Accessed 8 July 2013].

Irwin, S.H., D.R. Sanders, and R.P. Merrin. 2009. Devil or angel? The role of speculation in the recent commodity price boom (and bust). *Journal of Agricultural and Applied Economics*, 41(2), 377–391.

Jones, T. 2010. *The great hunger lottery: How banking speculation causes food crises*. World Development Movement. July. Available from: http://www.wdm.org.uk/sites/default/files/hunger%20lottery%20report_6.10.pdf [Accessed 8 July 2013].

Kneen, B. 1995. *From land to mouth: Understanding the food system*. Toronto: NC Press.

Kneen, B. 2002. *The invisible giant*. London: Pluto Press.

Lagi, M., Y. Bar-Yam, K.Z. Bertrand, and Y. Bar-Yam. 2011. *The food crises: A quantitative model of food prices including speculators and ethanol conversion*. Cambridge, MA: New England Complex Systems Institute. Available from: http://necsi.edu/research/social/food_prices.pdf [Accessed 8 July 2013].

Masters, M. 2008. *Testimony before US Senate Committee on Homeland Security and Governmental Affairs*. Washington, D.C. May 20.

McMichael, P. 2005. Global development and the corporate food regime. *In*: F.H. Buttel and P. McMichael, eds. *New directions in the sociology of global development*. Oxford: Elsevier Press, pp. 265–299.

McMichael, P. 2009. A food regime geneology. *The Journal of Peasant Studies*, 36(1), 139–169.

McMichael, Philip. 2010. Agrofuels in the food regime. *The Journal of Peasant Studies*, 37(4), 609–629.

McMichael, P. 2012. The land grab and corporate food regime restructuring. *The Journal of Peasant Studies*, 29(3–4), 681–701.

McMichael, P. 2013. Value-chain agriculture and debt relations: Contradictory outcomes. *Third World Quarterly*, 34(4), 671–690.

Meyer, G. 2011. Commodity traders hit back at planned US futures curbs. *Financial Times*, 14 June.

Mittal, A. 2009. The blame game: Understanding structural causes of the food crisis. *In*: J. Clapp and M. Cohen, eds. *The global food crisis: Governance challenges and opportunities*. Ontario: WLU Press, 13–28.

Montgomerie, J. 2008. Bridging the critical divide: Global finance, financialisation and contemporary capitalism. *Contemporary Politics*, 14(3), 233–252.

Morgan, D. 1979. *Merchants of grain: The power and profits of the five giant companies at the center of the world's food supply*. New York: Viking.

Murphy, S., D. Burch, and J. Clapp. 2012. *Cereal secrets: The world's largest grain traders and global agriculture*. Oxfam International. Available from: http://www.oxfam.org/en/grow/policy/cereal-secrets-worlds-largest-grain-traders-global-agriculture [Accessed July 8 2013].

Oxfam France. 2013. Réform bancaire: ces banques françaises qui spéculent sur la faim. Available from: http://www.oxfamfrance.org/IMG/pdf/rapport_oxfam_france_reforme_bancaire_120213.pdf [Accessed 8 July 2013].

Oxfam International. 2013. Key Eurozone banks step back from food speculation. Press release. Available from: http://www.oxfam.org/en/pressroom/pressrelease/2013-02-18/key-eurozone-banks-step-back-food-speculation [Accessed 8 July 2013].

Princen, T. 1997. The shading and distancing of commerce: When internalization is not enough. *Ecological Economics*, 20(3), 235–253.

Princen, T. 2001. Consumption and its externalities: Where economy meets ecology. *Global Environmental Politics*, 1(3), 11–30.

Princen, T. 2002. Distancing: Consumption and the severing of feedback. *In*: T. Princen, M. Maniates, and K. Conca, eds. *Confronting consumption*. Cambridge, MA: MIT Press, pp. 103–131.

Protess, B. 2011. Wall St. groups sue regulator to challenge new trading rule. *New York Times*, 2 Dec. Available from: http://dealbook.nytimes.com/2011/12/02/wall-street-groups-sue-regulator-over-dodd-frank/ [Accessed 8 July 2013].

Protess, B. 2012. Judge strikes down a Dodd-Frank trading rule. *New York Times*, 28 Sept. Available from: http://dealbook.nytimes.com/2012/09/28/judge-strikes-down-dodd-frank-trading-rule/ [Accessed 8 July 2013].

Russi, L. 2013. *Hungry capital: The financialization of food*. Winchester, UK: Zero Books.

Sanders, D.R. and S. Irwin. 2010. A speculative bubble in commodity futures prices? Crosssectional evidence. *Agricultural Economics*, 41(1), 25–32.

Schoenberg, T. 2012. CFTC to appeal ruling rejecting Dodd-Frank trading limits. *Bloomberg News*. *Bloomberg Businessweek*, 15 Nov. Available from: http://www.businessweek.com/news/2012-11-15/cftc-to-appeal-ruling-rejecting-dodd-frank-trading-limits [Accessed 8 July 2013].

Scott, B. 2011. Barclays PLC & Agricultural Commodity Derivatives. *World Development Movement*. Available from: http://www.wdm.org.uk/food-speculation/barclays-plc-and-agricultural-commodity-derivatives-report

Stop Gambling on Hunger. No date. Homepage. Available at: http://stopgamblingonhunger.com [Accessed 8 July 2013].

Tett, G. 2009. *Fool's gold*. New York: The Free Press.

Treanor, J. 2013. Barclays cuts 3700 hobs in overhaul. *The Guardian*, 12 Feb. Available from: http://www.guardian.co.uk/business/2013/feb/12/barclays-cuts-3700-jobs-strategic-review?INTCMP=ILCNETTXT3487 [Accessed 8 July 2013].

UNCTAD. 2011. Price formation in financialized commodity markets: The role of information. Available from: http://www.unctad.org/en/docs/gds20111_en.pdf [Accessed 8 July 2013].

US Senate. 2009. Excessive speculation in the wheat market. Majority and minority staff report. Permanent Subcommittee on Investigations, June 24. Washington, DC.

Vander Stichele, M. 2011. Regulating the commodity and exchange derivatives markets: The case of the EU. UNCTAD Public Symposium 2011. 22–24 June.

White, B. and A. Dasgupta. 2010. Agrofuels capitalism: A view from political economy. *The Journal of Peasant Studies*, 37(4), 593–607.

White, B., S.M. Borras Jr., R. Hall, I. Scoones, and W. Wolford. 2012. The new enclosures: Critical perspectives on corporate land deals. *The Journal of Peasant Studies*, 39(3–4), 619–647.

World Bank. 2007. *World development report 2008: Agriculture for development*, Washington D.C.: The World Bank.

World Bank. 2008. Double jeopardy: Responding to high food and fuel prices. Paper presented at the G8 Hokkaido-Toyako Summit. 2 July. Available at: http://siteresources.worldbank.org/INTPOVERTY/Resources/335642-1210859591030/G8-HL-summit-paper.pdf [Accessed 8 July 2013].

World Bank. 2010. *Rising global interest in farmland: Can it yield sustainable and equitable benefits?* Washington, D.C.: The World Bank.

World Development Movement. 2011. Stop gambling on food & hunger: Call for immediate action on financial speculation on food commodities. Available from: http://www.wdm.org.uk/sites/default/files/Food%20spec%20statement%2002.2011_0.pdf [Accessed 8 July 2013].

World Resources Institute. 2008. Rattling supply chains: The effect of environmental trends on input costs for the fast-moving consumer goods industry. Available from: http://pdf.wri.org/rattling_supply_chains.pdf [Accessed 8 July 2013].

Worthy, M. 2011. *Broken markets: How financial market regulation can help prevent another global food crisis*. World Development Movement.

Jennifer Clapp is a Canada Research Chair in Global Food Security and Sustainability and Professor in Faculty of Environment at the University of Waterloo, Canada and a recent recipient of a Trudeau Fellowship. She has published widely on the global governance of problems that arise at the intersection of the global economy, the environment, and food security. Her most recent books include *Hunger in the balance: the new politics of international food aid* (Cornell University Press, 2012), *Food* (Polity, 2012) and *Corporate power in global agrifood governance* (co-edited with Doris Fuchs, MIT Press, 2009).

Moral economy in a global era: the politics of provisions during contemporary food price spikes

Naomi Hossain and Devangana Kalita

The wave of food riots since 2007 revived interest in why people protest in periods of dearth, yet research has to date failed to make sense of the political cultures of food protests. The concept of the moral economy in European history is explored here to make sense of contemporary political perspectives on how food markets should work in Bangladesh, Indonesia, Kenya and Zambia. The concrete expressions of these moral economies are localized and politically contingent, yet there are broad areas of common ground across settings. As with the moral economies of seventeenth- and eighteenth-century Europe, there is strong popular feeling against speculation and collusion in food markets in times of dearth, and an emphasis on the responsibilities of public authorities to act. But whereas the moral economy in European histories focused on customary paternalistic obligations, the contemporary emphasis is on formal and electoral accountabilities as a means of triggering public action. The paper concludes with a discussion of a research agenda on the moral economy and the politics of provisions in globalised present-day food markets.

Introduction: the importance of popular ideologies of how food markets should work

In March 2011, a young Bangladeshi rickshaw-puller said the following about how food markets should (be made to) work:

> The government should make sure that the businessmen are not storing food items in their store-houses rather they are supplying/selling them in the market. The problem is, when the businessmen raise the price, the government does not tell them anything. There is a reason behind this and that is the government takes bribes from the businessmen. The money taken from them through bribes is very important for the ruling party as they will spend this money in the next election to come back [to] power again.

This paper draws mainly on primary research conducted with Duncan Green, Rizki Fillaili, Ferdous Jahan, Grace Lubaale and Mwila Mulumbi, funded by Oxfam GB in 2011. The authors are very grateful for their permission to reuse this material here. The paper has also benefited from funding from DFID-ESRC grant reference ES/J018317/1 which has enabled literature review and discussions about the moral economy with partners on the Food Riots and Food Rights project, including Luis Brito, Ferdous Jahan, Anuradha Joshi, Celestine Nyamu-Musembi, Biraj Patnaik, Michael Sambo, Dipa Sinha, Patta Scott-Villiers and Alex Shankland. The paper has been greatly improved thanks to feedback from two anonymous reviewers. All errors remain those of the authors alone.

> There is only one way to keep the price of food items under control and that is to hold dem-
> onstrations against price hikes. The fact is, in Bangladesh you cannot achieve anything
> without a show-down, without a demonstration. If the people go to the street and vandalize
> 20 or 30 cars, only then will effective steps be taken, the price will come down and the poor
> will be able to live a happy life.

This fragment of popular political culture came from a focus group discussion in a northern Dhaka slum, as part of research on experiences of food price volatility. In Dhaka in March 2011, the price of coarse rice (the kind eaten by rickshaw-pullers and other people on low-incomes) was BDT 34 per kg (then USD 0.47), almost as high as at its peak during the 2008 food price spike.[1] During the 2008 crisis, 20,000 factory workers had rioted in Dhaka, protesting low wages in a time of high food prices (Schneider 2008). Yet it seemed that retail rice prices could be controlled, when the will existed: in 2009, after sales of cheap grain and other state actions to curb prices, rice had dropped to BDT 20/kg (USD 0.28). This followed Bangladesh's return to multiparty democracy after two years of military-backed rule (2007–2008) and the steep drop in global commodity prices after the global financial crisis struck in 2008.

The rickshaw-puller's statement has many hallmarks of the 'moral economy' studied by E.P. Thompson, the social historian of eighteenth-century English food riots. The moral economy ideas of English rioters prioritised rights to affordable food over those of free markets, and articulated obligations of the public authorities to protect those as matters of custom. These protests typically occurred at crisis moments of economic adjustment, at times when free trade was in the ideological ascendant and its virtues were being asserted. The moral economy provided the popular mandate for public authorities to protect people against food-related shocks, as well as the theory for food rioters in well-documented instances (Thompson 1971, Thompson 1991).

The rickshaw-puller's statement shares a number of elements with the English rioters of the past. It came at a crisis moment of economic adjustment, at a time when global markets in food were rapidly extending and integrating. The rickshaw-puller displays suspicion of the motives and means of actors in the food trade, a belief that government should act to bring high food prices down and faith in direct action as a means of triggering government action on prices. There is no faith here in the view that malfunctioning food markets will correct themselves automatically, and the aim is firmly that 'the poor will be able to live a happy life'. This is a normative position on how food markets should work, in which the market is subordinated to or embedded within society's moral norms.

Despite its generic similarities to eighteenth-century English views on how food markets should be made to work, the rickshaw-puller's moral economy is also localised and politically contingent: political elite corruption is the source of governance failures (Bangladesh continues to be listed as one of the most corrupt political administrations in the world), electoral cycles determine the degree of political will (the two main parties usually alternate in power) and a disgruntled crowd is relatively easy to organise (rapid urbanisation, population density, mass poverty). References to customary or paternalistic obligations that featured in eighteenth-century English political culture were absent in

---

[1] Nearly 27 percent of non-agricultural daily labourer consumption was rice in the period prior to the food price spikes, so rapid price increases would most likely have meant a sharp decline in consumption among rickshaw-pullers and similar occupation groups (World Bank 2013). Rice prices are from the Food and Agricultural Organisation's Global Information and Early Warning System price tool, based on Government of Bangladesh data.

contemporary Dhaka. This view of the moral economy is no utopian vision, but closely attuned to the political possibilities of action.

Comparing moral economies across contexts (space and time) in this way offers the prospect of a richer understanding of the political cultures governing food markets. It also directs attention to the local political possibilities of action, or what Bohstedt has called the 'politics of provisions' (Bohstedt 1992, 2010). It sheds light on what change in food markets means to people affected by it, and the directions of travel of political mobilisation around food. This paper contributes to such understanding through a comparative analysis of perspectives of people, who, like this young rickshaw-puller, find themselves vulnerable to the volatilities of early twenty-first-century global food markets and the vagaries of the official response. It explores views from qualitative research in four developing countries – Bangladesh, Indonesia, Kenya and Zambia – at a time when riots and related protests in many countries have indicated mass discontent with how food markets are functioning. It asks: how are food markets – and the rights to profit from food trade – viewed in these times? How do norms about entitlements to protection against food insecurity inform the domestic politics of food? And how do the political possibilities – the scope for organising and the prospects of a response – inform what people think and do about food markets?

Why raise these questions now? For three reasons. First, the 2007–2008 'food crisis' marked the end of three decades of low world food prices and the start of a period of global food price volatility featuring major spikes in 2008 and 2011 (Clapp 2009, Gilbert and Morgan 2010, Naylor and Falcon 2010, FAO 2011, Von Braun and Gebreyohanes 2012). In 2007–2008, there were incidents labelled 'food riots' in between 25 and 30 countries, and the 2011 uprisings that turned into the Arab Spring were related to food prices (Schneider 2008, Lagi *et al.* 2011). Cross-country research has correlated price spikes with unrest, finding, among other things, that low-income countries and weakly democratic polities are particularly prone to riots, and that the relationship between global food prices and civil unrest has become stronger with market integration and the 'contagion' effect of the internet and social networking (Arezki and Bruckner 2011, Arora *et al.* 2011, Bellemare 2011). However, cross-country work has tended to 'sophisticate and quantify evidence which is only imperfectly understood' (Thompson 1971, 77), because it relies substantially on 'spasmodic' explanations of why people riot that reduce people to bellied bodies. We learn little about the political perspectives on food markets that inform protests (although see Hossain 2009, Patel and McMichael 2009, Bush 2010, Brinkman and Hendrix 2011, O'Brien 2012). For Thompson, exploring the moral economy helps us address the question: '[b]eing hungry … what do people do? How is their behaviour modified by custom, culture, and reason?' (Thompson 1971, 77–8).

The second reason the political culture of food markets merits research now is that ideas about human rights to food, a set of rights with legally enforceable claims and institutional support, are fast gaining ground. This includes in countries like India, where getting a basic meal has been the primary struggle for generations. If claims to food within the moral economy rest on custom or tradition, how might a legally enforceable human right to food change that? To what extent might the institutional and legal basis shape the ideas about how food markets should run?

A third reason to explore contemporary variants of the moral economy is the increasingly global integration of food markets. Do small country governments have the power to protect people against food price changes that arise in global commodity markets? If not, can moral economy ideas survive persistent enduring failures by public authorities? The moral economy in the historical past was fed by its successes; what if food riots fail, or are put down, or become so routine that they are safely ignored? How meaningful is the idea of the moral economy in an era of global food markets?

The paper draws centrally on the concept of the moral economy for the insights it may enable into the determinants of different political cultures. Yet it also pays attention to what the concept may miss in the present day, and to situating it within its local political space. The concept of the moral economy being used here focuses on popular perspectives on how food markets should work, and the associated responsibilities for action. This generic focus on political cultures of food markets makes it possible to compare and contrast across rural and urban, Asian and African, low- and middle-income contexts with different degrees of global market integration.

The paper is organised into four sections. The next section discusses the concept of the moral economy, and specifies how the term is used. Section three presents analysis of original research into popular perspectives on food markets during the 2011 food price spike, and the continuing period of food price volatility (mainly rises) into 2012. The section also describes the methods and data used, and compares and contrasts past and present political cultures of food markets. Section four concludes, with a discussion of the implications of the research for the understanding of the 2007–2008 food riots, and an agenda for research taking into account the effects of globalisation and right to food movements on the moral economy and the politics of provisions.

The many meanings of the moral economy

The idea of a moral economy has appealed to scholars from a wide range of social science disciplines. The concept was revived by E.P. Thompson in *The moral economy of the English crowd in the eighteenth century* (1971) to refer to the popular political culture (views, beliefs, practices) of food markets, founded on a specific model of grain marketing derived from earlier Tudor practices of market regulation. The moral economy tended to be most noticeable during moments of food crises, when demands for a 'just price' could take the form of protests usually labelled 'food riots'. For Thompson, food riots in eighteenth-century England were forms of collective action undertaken by the poor, legitimised by customary expectations that food provisioning took precedence over legal rules in the market-place, in a context where the poor felt threatened by market failures to assure food access at a moment of transition from paternalism to laissez faire. The idea of the moral economy excited a vast literature, some supportive and adulatory, much highly critical and dismissive (often because ideas about the moral economy were a poor fit with neo-liberal sentiments about the virtues of free markets; see in particular Thompson 1991, 266–78).

Scott famously extended the concept to explain peasant risk management institutions in contexts of subsistence and vulnerability in Southeast Asia. It is arguably now Scott's formulation which is most prominent in the study of contemporary societies (see Greenough 1983 for a discussion of Scott's use of the term). By Scott's definition, the moral economy was the Southeast Asian peasant rebels'

> notion of economic justice and their working definition of exploitation – their view of which claims on their product were tolerable and which intolerable ... the central economic and political transformations of the colonial era served to systematically violate the peasantry's vision of social equity [so] a class 'of low classness' came to provide, far more than the proletariat, the shock troops of rebellion and revolution. (Scott 1977, 3–4)

Elements of Scott's definition form the core of a broad conception of the moral economy. These include (1) conceptions of material justice, (2) the idea of clear limits to the tolerance of exploitation and unfairness, and (3) the impacts of sweeping economic and political change on understandings of fairness and equity. The moral economy also provides the theory or ideological justification for protest, although, as Bohstedt (2010) has shown, it is under highly specific political and institutional conditions that these actually occur.

For Scott, as for Thompson, the moral economy is not a static 'pre-modern' or monolithic entity, but involves dynamic processes of negotiation, confrontation and competing ideas of the moral economy (Moore 1983, Gore 1993), the 'rules' of which tend to become particularly visible during times of crisis. In its broader conception of popular ideas and values about material justice and the legitimacy of its defence, the moral economy has been treated as a central explanatory concept across a wide range of scholarship concerned with, among other issues, the ethics of food and food trade (Trentmann 2007, Jackson *et al.* 2009); famine, risk management and subsistence (Scott 1977, Greenough 1982, Keyes 1983, Appadurai 1984, Fafchamps 1992, Adams 1993); rebellion and other modes of contentious politics, including peasant and farmer organisations (Tilly 1971, Coles 1978, Orlove 1997, Edelman 2005, Patel and McMichael 2009), and broader debates about the cultural and political dimensions of markets (Booth 1994, Arnold 2001). In this paper, however, we explore the moral economy as the popular ideology of how food markets should work, remaining close to E.P. Thompson's original sense. This makes sense given how closely the present period appears to parallel the relevant historical period in the experience of food price spikes and related crises of economic transitions. This focus on food markets means less focus on the moral economy of growing, than on that of buying, food. A focus on the moral economy of consumption in food markets is important because in the present food regime, people living in poverty tend to be net food consumers (Ivanic and Martin 2008); because an increasing proportion of the world's poor are urban, suggesting the need to revisit questions of urban bias in food policy (Jones and Corbridge 2010), and because there are good reasons to believe that poor rural farmers are not the main beneficiaries of food price rises (Vanhaute 2011), so that poor rural producers are also unlikely to be celebrating food price rises.[2]

Other aspects of Thompson's approach to the moral economy tend to be less commonly noted, but are relevant to how we use the term here. The first is that 'confrontations in the marketplace' may no longer have precisely the same meaning, as 'market' has come to mean something abstract and metaphorical (and often, facelessly sinister) as often as an actual space in which transactions occur. As Edelman puts it

> [E]ven in the mid-19th century market by itself often referred primarily to a specific physical location where particular types of goods were stored and traded ... Only later did it assume the metaphorical and deterritorialized qualities that increasingly adhere to it. The political sleight-of-hand that accompanied this semantic shift involved making the institutions that actually shaped markets invisible as well as creating the appearance of a separate and autonomous economic domain disembedded from society ... [P]easants' "confrontations in the market-place" now occur more and more in a "market-place" that no longer has a "place" in it, which naturally affects the character of their political responses.
>
> (Edelman 2005, 332)

A question is how 'the market' appears in popular political culture, whether as de-territorialised metaphor or as the place you buy your vegetables, and what that means for the

---

[2]Although there remains considerable debate about why the rural poor should in theory benefit from high food prices (Swinnen 2011) and, indeed, about whether the food price spike of 2008 resulted in a net increase in the numbers of hungry people (Headey 2013).

political culture of food markets. Finally, the present paper aims to avoid romanticising the moral economy. This includes treating as discourse rather than fact the 'selective recon-structions' of past paternalism and the 'superstitions' Thompson noted in the moral economy of the English crowd. Our starting point is not that the moral economy is especially moral, but that in drawing on moral-type arguments, it is a political strategy – sometimes effective – of the relatively weak.

Popular political cultures of food markets during the 2011 global price spike

Contemporary research on experiences of food price spikes

We turn now to an exploration of popular political cultures of food markets in contem-porary developing countries. This draws on original qualitative data from eight case study research sites, one rural and one urban each in Bangladesh, Indonesia, Kenya and Zambia. The data were collected in February–April 2011, just after the 2011 price spike. Staple (rice and maize meal or flour) and other food prices were high and had risen sharply in the past year in Bangladesh, Indonesia and Kenya, but not in Zambia, where maize meal prices had been stable or reduced slightly compared to the previous year. People there believed the elections set for November 2011 meant pressure on the incumbent government to keep prices low.[3] These patterns held both for national price levels (from official sources) and for the prices people reported paying in their local markets.

The data are from 32 focus group discussions, as well as key informant and other inter-views. An average of 10 people participated in each focus group and, with additional inter-views and household case studies, almost 500 people were included in this research (although some groups were more interested and vocal than others). Focus groups were organised by occupation group and gender. The 2011 round was the third round of data col-lection, so a relatively rich picture of these communities' experiences of the global food, fuel and financial shocks after 2008 had been built up.[4] Data from other rounds of the research is used to help situate and explain views expressed in 2011. The data used here were primarily responses to two sets of questions, about:

(1) perceived causes of recent food price rises; using participatory tools, people were asked to list, analyse and rank different causes of recent food price rises;
(2) actions that needed to be taken, by whom, and how, to alleviate the effects of rapid food price changes. Topics included (i) views on what should happen when food prices rise, (ii) possible preventive actions, (iii) actual actions taken, (iv) responsi-bilities for action, (iv) reasons for failures to act, (v) information channels through which policymakers understand food price rise impacts and (vi) means by which citizens can express concerns about price rises.

---

[3]It seems more likely that good maize harvests in 2010–2011 helped explain lower staple prices; Zambia has a very low staple import dependency ratio of only five percent (according to FAO (2013)). However, 2011 electoral politics did play a role in food security: evidence shows that ferti-lizer subsidies were targeted to constituencies who voted for the ruling party in the previous election (Mason et al. 2013).
[4]A number of outputs are available from this data. See Hossain et al. (2010), Hossain and Green (2011), Hossain and McGregor (2011) and Heltberg et al. (2012).

Interviews and focus group notes were translated and coded thematically, using NVivo 10 software. The analysis undertook to identify, group and explain the main emerging themes, but did not seek to quantify these, as this would have given a false sense of their representativeness. However, the paper indicates themes on which there was greater agreement, as well as those which were minority views.

The research sites are diverse places, situated within rural, urban or peri-urban settings in four different national political and economic systems. The selection of sites attempted to gather accounts of the global crises across a range of low-income and/or precarious occupation and social groups, to understand their differentiated impacts. Each locale features social and occupational groups with different relationships to food markets, as well as means and modes of popular political action. The urban sites included sprawling low-income slum neighbourhoods in Dhaka and Nairobi, and a more mixed-income informal area in Lusaka. All included diverse informal sector occupation groups: daily wage labourers and workers in the construction, waste recycling, transport, domestic and sexual service sectors. These groups tend to be particularly exposed to commodity price rises because their earnings are insecure. Urban slums tend to be excluded from most social services in Bangladesh, where policies display a strong rural bias; by contrast, Nairobi slum-dwellers had access to social protection programmes as part of the new constitution of 2010. The Lusaka community comprised relatively well-off small business owners and public sector workers, whose earnings tended to rise with inflation, as well as more precarious small vendors, traders and service-providers, for whom cost-of-living rises were a major peril. The peri-urban site near Jakarta included a large informal sector occupation group, mainly servicing the relatively well-paid but insecure internal migrant industrial workers. Both in the site near Jakarta and the Dhaka slum, people showed great awareness of global markets, as export sector employment was an important part of the local economy. The urban and peri-urban sites were selected to include some formal sector occupation groups, globally-integrated export sectors, and informal and precarious occupations.

In or close to the sites in Dhaka, Nairobi and near Jakarta, there had been food- and/or livelihood-related protests in 2011, in which some of the local community were said to have participated. These ranged from civil society-organised demonstrations for fair price maize flour in Nairobi, to riots of rickshaw-pullers over restricted routes in Dhaka, and trade union activism over weakening workers' rights post-crisis in the industrial area near Jakarta. These recent local events, and the fact that food riots and other related protests took place in each of these countries in the years since 2007, mean that people's views on how government should act were formed within a context in which mass collective action had recently taken place on issues linked to the right to subsistence. It should be noted that these sites were selected in 2009, prior to these local political events having taken place.

The rural communities in coastal Kenya and northern Zambia mainly comprised subsistence farmers. These were selected as broadly typical of poor rural communities, with the researchers had prior links. The Kenyan community had been facing the long drought of the past decade, and climate change was a key concern. The main current challenge facing Zambian farmers was steep agricultural input price rises and their inability to access input subsidies; these were selectively distributed, mainly to organised groups in the locale. The rural community in the northwest of Bangladesh was selected because of its high concentration of people living in extreme poverty. It included rice farmers and landless agricultural wage workers, as well as new non-farm activities such as brick-making and cross-border smuggling, including of fertilizer, to and from India. The rural Indonesian site was a trans-migrant community of rubber farmers and workers in South Kalimantan, but with a significant local mining concession. Rubber farmers and tappers are highly exposed to global commodity price volatility, which is why the site was selected for the research.

Research participants were drawn from across the occupation groups. Interviews with key informants from the professions and official positions, from markets and the food trade were also conducted.

Speculation and corruption

A suspicion that business interests were involved in hoarding and speculation in times of dearth and high prices emerged as an important explanation of sharp food price rises. Cartels and illicit links between politicians and big businessmen were detected in the Bangladesh and Indonesian sites:

> Prices increase if production falls, when production falls and the demand for the good increases, price rises eventually. On the other hand, price falls when production increases beyond the demand. But sometimes, increased production does not cause a decline in price. The reason behind this is – the ill-intentions of the businessmen. They buy rice, stock it, create a fake crisis and then sell rice at a higher price. (male wage labourer in Naogaon, rural northwest Bangladesh)
>
> [T]op businessmen of the country sometimes develop a syndicate among themselves. They store the food items and tell each other not to sell food in the market. They come to a consensus that they should store the food items a little longer, should raise the price and should sell them when the price almost doubles. These people are real criminals. (shopkeeper in a large north Dhaka market)

In Bangladesh, the belief that a 'syndicate' or cartel profiteers from perceived or actual food shortages is widely held, with discussions of such activities in the news media and public discourse. The election manifesto of the Awami League (which won a landslide victory in late 2008) stated that 'hoarding and profiteering syndicates will be eliminated' as part of its effort to tackle the food crisis, echoing the perception that speculation played a large role during the 1974 famine (Islam 2003).

As in Bangladesh, concerns about hoarding to raise prices during the fasting month of Ramadan proved to be a major concern in majority Muslim Indonesia. Fuel shortages and price hikes were closely associated with food price shocks in Indonesia in the 2000s, and subsidy cuts are hotly resisted, including through mass riots in 2012:

> Speculators also play a role. This happened for example in April/May when the businessmen loaded up fuel that caused panic among the people and caused prices of people's needs to soar. In addition, on the eve of the fasting month in 2010, there was also hoarding of staple foods that caused an increase of price. To solve this, the Trade Office raided the market with the police officers to catch these 'naughty' businessmen. (male rubber tapper in Banjar, Kalimantan)

Resonances with the moral economy narratives around the specifics of grain marketing in eighteenth-century Europe were strong in several accounts, including reference to the length of the marketing chain, correct practices of dealers, profit margins, fair prices and suspected collusion between the rich and powerful. One variation on this theme was a perceived link between official corruption and food price rises. In rural and urban Kenya, official corruption was widely blamed for problems in the relief food distribution system. In one focus group of young men in Mukuru in Nairobi, it was argued that 'relief food does not reach the intended or the affected people because of the corrupt local provincial administrators who divert the food to their households and to their cronies'. For Kenyan youth, this is a particular concern because relief food tends to be earmarked for 'vulnerable' groups, which excludes young men. Political elite corruption is seen as part of the generalised malaise in the public sector in public discourse in Kenya, Bangladesh and Indonesia, as with many other developing countries, so it is unsurprising it is linked to weak regulation of food distribution systems.

Ethics and religion, or their failure to curb speculation, were cited in Bangladesh. In Dhaka, a discussion among small shop-keepers revealed that even they believed big wholesalers were corrupt: one man argued that efforts to regulate markets were themselves doomed to failure unless honest people could be found to take action. Another retailer argued that 'businessmen should get some moral teaching. If they were afraid of Allah and conducted business honestly, the situation would improve', highlighting the religious inflection that the moral economy can acquire.

In rural Zambia, corruption and collusion among powerful groups was detected in fertilizer rather than in food markets. The distribution of subsidised fertilizer is seen as crucial for local food security because most people in this poor northern region depend on the maize they grow. Yet cooperatives through which fertilizer was distributed were seen to be politically aligned, and the government was seen to be more supportive of big than subsistence farmers.

Governmental responsibilities

A second generic feature that emerged was a strong emphasis on citizen-state relationships in the regulation of markets. For Thompson's rioters, the responsibilities of public authorities were cast as customary and paternalistic. In the present-day discourse, there appears to be a more formal institutional stress on political or institutional responsibilities, which marks a difference with earlier moral economies. It also highlights the importance of the politics of provisions – the political possibilities afforded by electoral competition and ideas of citizenship – in shaping how responsibilities are framed within the moral economy itself (Bohstedt 2010).

Government responsibilities to protect consumers in the marketplace require that they act to protect food production for domestic consumption:

> The food situation has been caused by long-term factors, which government has failed to address over time. There has never been a time when we felt that the situation had improved, it has either been the same or worse ... The long-term factors include poor terms of trade, poor agricultural policy, poor governance, expensive and inadequate inputs. (Chief, Northern Province in Zambia)

As noted above, the distribution of subsidised fertilizer is a key concern in the food security situation in this part of rural Zambia, and the Chief noted that 'political interference' had been an issue in determining beneficiary selection. The area as a whole had benefited relatively little in the previous few years from subsidised fertilizer. Because the distribution of subsidised inputs is seen as so critical to food security in the area, much of the discussion around state responsibilities with respect to food revolved in this context around the distorted market for subsidies.

In Kenya, which had recently experienced significant tax reforms, Mukuru slum residents argued that the government was to blame for failing to control prices of essential food commodities, and for charging food producers high taxes, making them raise prices. MPs' (Members of Parliament) ongoing efforts to vote themselves large new salary increases were fingered by a young male transport worker:

> The government has been unsuccessful in controlling the prices of food but has instead exacerbated the crisis by increasing the number of ministries and salaries of the ministers. The money used for paying high salaries for ministers and members of parliament could have been used for initiating irrigation projects and for buying enough relief food to cushion citizens against high food prices.

A series of protests highlighted the failures of the Kenyan government to meet its responsibilities to protect rights to food. Consumer group protests against plans to tax all food items forced the government to exempt maize and bread from the tax in 2013 and Mukuru slum residents participated in the 2011 'Unga Revolution', in which street protestors called for a fair price for unga or maize flour.[5]

The way a claim or responsibility was articulated appeared to depend on the existence of an official mandate for action. The government action demanded could be either market and price regulation, or producer or consumer subsidies. Whether or not people felt protest was an option, and to whom they targeted their protests, depended greatly on the highly localised politics of the food system: in rural Zambia, for instance, where no other officials were available for blame, small and subsistence farmers focused on agricultural extension and the distribution of subsidised fertilizers as their most visible connection to food security policies of the state. As a Zambian subsistence farmer explained, the recent situation had been of high input and consumer but low farmgate prices:

> When the food price rises, costs for farm inputs should be reduced. One thing triggers the other, so it would be good to deal with the trigger mechanisms. For instance, government should deal with what triggers the price of fertilizer, because when fertilizer prices go up, this puts us in a crisis. And as you know government is not even willing to offer a good price for the maize we grow. So where is the sense in increasing the price of fertilizer, while at the same time reducing the floor price of maize?

The target for rural Zambian grievances about food security in this context is the agriculture officer and the policies surrounding fertilizer distribution. By contrast, in Kenya the new responsibility assigned to chiefs to provide relief food under the new constitution provided a rallying point; respondents spoke clearly about their new constitutional rights to relief food, and gave accounts of camping out at the local Chief's office until he provided some.

What could be expected sometimes referred to an earlier golden era in which state regulation was visible and active:

> In the olden days, there were price control or price controllers which helped to maintain the affordability of food, especially mealie meal. This used to help to prevent sudden price rises. The current government needs to adopt such past measures.

Where government action to control food prices was taken, it was seen as captured by or used to protect the interests of elite and powerful groups, rather than that of the poor. In

---

[5]These protests have not been well documented to date, possibly because they have had limited, if any, success, in changing policies. See IRIN (n. d.) 'Kenya's Unga Revolution', and Reuters (2013).

Indonesia, a rubber tapper argued that pro-market reforms to reduce fertilizer subsidy were driven by corruption and that fertilizer was distributed to palm oil plantations rather than small producers. A small shopkeeper in Dhaka similarly noted that the government is 'taking action against the "small" shop-owners while not touching the real culprits – the big scale black-marketeers'.

A woman subsistence farmer in Zambia noted simply that 'government does not care about us peasant farmers, they care more about the commercial farmers'.

The protection of elite interests was connected to the sentiment that the state fails to act because it 'does not care' for the poor, either because politicians were in collusion with business interests as noted above, or they were too busy with their own petty concerns. In none of the sites did people believe politicians to lack adequate information about how people were being adversely affected by food price rises, and television, print and other media were widely cited as sources of information. A local politician in rural Bangladesh said, 'the government has completely failed. In this capitalist world, people are simply helpless in the hand of the government and the system. The government really does not think about the ordinary people'. Women subsistence farmers in rural Zambia similarly said:

> Government knows what is going on, they know what we are experiencing, they just choose to turn a blind eye to our situation … We don't have the commitment from our government. Our MP does not even know how we live or survive. There is just no concern.

A striking finding was the common emphasis on specifically *electoral* accountabilities, which was naturally absent in the moral economy of pre-enfranchised Europe. That the contemporary moral economy refers specifically to electoral accountability responds to Appadurai's (1984) addition of 'enfranchisement' to Amarya Sen's theory of entitlement failure: in multiparty democracies, however flawed, people feel some power over policymakers and decision-takers. However, 'enfranchisement' in the form of voting offers very limited scope for voters to participate in making the rules about rights to food and its proper distribution. Respondents were typically cynical about the extent to which electoral accountabilities could ensure acceptable food prices:

> Because we are near [government presidential elections], government has stabilised the floor price of maize, which has helped to reduce the price of mealie meal. The [current ruling party] knows that people will not vote for them when they are hungry, so they have ensured that the price of mealie meal is reduced. But let's wait after the elections this year, the price of most food items, especially mealie meal, will rise. (woman petty trader, in Lusaka)

When asked how governments could be influenced, the common view among young and urban respondents was that informal or unruly means such as demonstrations could convince governments to act. The Arab uprisings were world headlines at the time of the research, and at least one youth took inspiration from this:

> [T]he leadership should change completely … This government is so corrupt how come in Moi's era food accessibility was so good unlike today yet there was no devolvement of funds as today? Maybe it is the high time we go the Egypt way … We need a leadership change! (young transport worker in Nairobi)

In Bangladesh, similarly, people felt that collective action was likely to work, or to be relatively more effective than other means of influencing an official response. However, in

Zambia and Indonesia, few people argued for direct action. In both countries, the view was that influencing local officials or forcing government to pay attention to the issue by raising awareness through the mass media or research (such as we were conducting) were more likely means of triggering action. This signalled both optimism about how information and media could influence action but also pessimism about the power or prospects of unorganised mass protest.

Real and metaphorical markets

Markets could mean real or metaphorical sites of exchange. The term mainly meant physical marketplaces, but it was used in its more metaphorical sense in a handful of cases, usually about distant and unknowable global market forces. This use suggests a sense of uncertainty about whether markets work as they are supposed to, and of intractability with respect to how they can be engaged with.

That actual actions in physical marketplaces could have an impact on the prices faced by consumers was noted in several cases. A woman market stall holder in a busy crossroads market in rural Zambia said:

> Nobody has bothered to address the predicament we find ourselves in when the suppliers increase their prices and at the same time the customers demand lower prices. We complain to the market executive committee, but there is nothing they can also do.

In Dhaka, a young transport worker speaking in a group that included shopkeepers proposed that

> People should come together and decide a price ceiling after having a discussion among themselves. They can hang a signboard in the market which will show how much they will pay at most for a particular food item. This will indicate that if the shopkeepers charge more than this, people will not buy from that shop. For instance, if people decide that for 1 kg of lentil they will not pay more than Tk 40, the shopkeeper will sell lentil at that price. If any shopkeeper charges Tk. 42, people will not buy from him.

Both statements suggest confrontations in actual marketplaces were the concern. However, the workings of markets and the scope for intervention were also seen as mysteriously abstract, as a market trader in rural Zambia explained:

> Government has made the food trading business free and open, so it's a challenge for action to be taken when the people who are producing decide to increase their prices ... There may be other factors that affect the rise in food prices which may be outside the power of the food producers. In such instances there is nothing that anyone can do.

In Bangladesh, people spoke of the need for signboards for approved prices in marketplaces while noting that the workings of markets were opaque, probably due to corruption.

Global markets often seemed abstract and unreal, and, for the small number of respondents who discussed their role in local price-setting, a convenient scapegoat behind which political corruption involving powerful food commodity cartels could be concealed. One young man, a shopper in a small Dhaka market, said

> [T]hose who are engaged in importing food products are really corrupt people. They import the necessary amount of food items and then make a call to these two ministers [the Finance and Commerce Ministers], 'Boss, we have stored the food items in our godowns. We will sell them

in the market just after 10 days. Please allow us to do this and in exchange we will send Tk. 10 crore (10 million, USD 1.3 million) to your Swiss bank account'.

He thought the recent price hike was unrelated to global food prices, and that these were stories put about by political leaders to save face. Elsewhere people seemed to believe that global markets affected not only the prices they were paying but the state of their economies more generally:

> There are external and internal factors that prevent government from taking action when food prices rise. External factors are mainly related to the conditions that IMF [International Monetary Fund] and World Bank put on the government ... We are basically at the mercy of the global economy and those giving us money [aid donors]. What happens at the global level will affect us whether we like it or not, and there is very little that our government can do.
> (woman focus group discussant in Lusaka)

That local food prices could be affected by distant markets was also highlighted by awareness of how local food prices and global markets were linked through fuel prices. More proximate factors such as environmental change, agricultural decline and agricultural input cost rises, population increases and failures to regulate markets on behalf of consumers were more commonly cited as causes. However, food traders, retailers and commodity export sector workers – all groups with occupational imperatives to understand the global economy – appeared to find global markets less abstract and more real. A rubber farmer in Kalimantan in Indonesia argued that although Indonesian government policy was to avoid rice imports in order to protect the price for local rice farmers, it was the rice merchants and commodity speculators who mainly benefited. He noted that officials said that rice price rises had to do to with shortfalls in rice stocks, yet there were ever more people selling rice in his village. His guess was that speculators were 'messing with the price', and that they could only be defeated by allowing imported rice into his village. Despite some nostalgia for locally-produced food, whether respondents approved or disapproved of exports or imports appeared, in general, to be contingent on their understanding of what had shaped current price levels.

Spatial separation and the complexity of present-day commodity price chains appear to mitigate against popular acceptance that local prices may be influenced by global forces. However, as the evidence about the transmission channels affecting local prices is mixed and contingent on a complex range of variables,[6] and there is little agreement on the role of commodity price speculation on food markets,[7] it is hardly surprising to find such a range of contradictory views about the benefits or otherwise of global food trade.

Conclusions and a research agenda

Political theory for protestors?

Findings about the popular political culture of food markets at a moment of rapid food price rises in present-day Bangladesh, Indonesia, Kenya and Zambia highlight the continuing relevance of the concept of the moral economy in understanding popular political cultures around contemporary food markets.

---

[6]For instance, the 2008 food price spike appears to have had a direct and substantial impact on many African countries' food prices, even though few displayed close associations between global and national food prices in the previous period (Minot 2011).
[7]But see Ghosh (2010) and Ghosh et al. (2011).

Similarities in the moral economies of eighteenth-century Devon bread rioters and twenty-first-century Bangladeshi or Kenyan slum-dwellers need not be overstated, yet there is common ground in beliefs that markets are manipulated or fail, chiefly because of elite collusion or greed, that authorities have the power and the responsibility to act, and that direct action may be necessary to correct failures to do so. There are also differences between these perspectives. These lie partly in the nature of the responsibilities assigned to public authorities: customary and paternalistic appear to have been replaced by formal institutional and electoral accountabilities; this may reflect both the rapid decline of effective patronage ties where these once existed, and the enfranchisement Appadurai (1984) discusses in relation to food market politics. They also relate to the political possibilities of action: not all hungry people riot, and for many collective action is a risk.

Whether moral economy ideas actually evolve into 'food riots' or other kinds of protests depends ultimately on the political opportunities and potential, as well as the scope for popular mobilisation. When situated within the local political context of elections, perceptions of corruption and public policies and provisions, it becomes clearer that moral economy ideas are not formed in a vacuum of political practice: instead, ideas about what is wrong with food markets and how they should be amended are shaped firmly by the framework of the politically possible.

The politics of provisions in a global era

This exploration of the political culture or moral economy of food markets has given rise to thinking about how to research the contemporary practice of the politics of provisions. In particular, it suggests the need to extend the understanding of food riots to a fuller understanding of why they succeed, to the extent that they do, thereby creating popular political cultures and repertoires for action which have a long historical shadow.

At the same time that the world has seen a wave of food riot-type protests, there has also been an upsurge in right to food movement campaigning, comprising grassroots mobilisation, civil society activism and public interest litigation and other legal activism. The idea of a legally enforceable human right to food is clearly and compellingly connected to the idea of a moral economy and may, again, mark the shift away from customary and paternalist obligations to far harder, more institutionalised and more actionable claims. The extent to which food riots and right to food movements succeed in generating a response, and how the politics of provisions shapes that response, is the subject of an ongoing research project in Bangladesh, India, Kenya and Mozambique.[8] However, the globalised nature of recent food price shocks raises new questions for the study of the politics of provisions. Prime among these is that of the scope for action by small country governments. While Zambians and Bangladeshis may expect their governments to act when prices rise – they may punish them electorally for failures to do so – their governments may have relatively little room for manoeuvre. The same may not be true of, for instance, the large food producers like India or Indonesia. The mismatch of scale – between a highly national and sometimes local moral economy, on the one hand, and the global food

---

[8]This paper draws on initial literature review for that project, funded by DFID-ESRC (United Kingdom Department for International Development-Economic and Social Research Council Grant Reference ES/J018317/1) entitled 'Food riots and food rights: the moral and political economy of accountability for hunger'. It also draws on research undertaken as part of the ongoing *Life in a time of food price volatility* project, funded by UK Aid and Irish Aid.

economy within which local prices are set, on the other – is an instance of what Nancy Fraser calls 'misframing' (see Fraser 2011, 2013). The politics of provisions may look entirely different in a globalised food regime than in past periods of economic adjustment towards national food regimes, and will depend on a nation's position within the global food regime. It remains to be seen whether moral economy ideas can retain the same power to demand corrections to food markets when the market requiring intervention is global, and it is no longer clear where the authority to act lies.

References

Adams, A. 1993. Food Insecurity in Mali: Exploring the Role of the Moral Economy. *IDS Bulletin*, 24(4), 41–51.

Appadurai, A., *et al.* 1984. How moral is South Asia's economy? A review article. *The Journal of Asian Studies*, 43(3), 481–497.

Arezki, R. and M. Bruckner. 2011. Food prices and political instability. *IMF Working Paper*, no. 1162. Washington, DC: International Monetary Fund.

Arnold, T.C. 2001. Rethinking moral economy. *The American Political Science Review*, 95(1), 85–95.

Arora, A., J. Swinnen and M. Verpoorten. 2011. Food prices, social unrest and the facebook generation. Mimeo: University of Leuven, LICOS.

Bellemare, M. 2011. Rising food prices, food price volatility, and political unrest. *Munich Personal RePec Archive paper* no. 31888.

Bohstedt, J. 1992. The moral economy and the discipline of historical context. *Journal of Social History*, 26(2), 265–284.

Bohstedt, J. 2010. *The politics of provisions: Food riots, moral economy, and market transition in England, c. 1550–1850*. Farnham, Surrey/Burlington, VT: Ashgate.

Booth, W.J. 1994. On the idea of the moral economy. *American Political Science Review*, 88(3), 653–667.

Brinkman, H.-J. and C.S. Hendrix. 2011. Food insecurity and violent conflict: Causes, consequences, and addressing the challenges. WFP Occasional Paper. Rome: World Food Programme.

Bush, R. 2010. Food riots: Poverty, power and protest. *Journal of Agrarian Change*, 10(1), 119–129.

Clapp, J. 2009. Food price volatility and vulnerability in the global south: Considering the global economic context. *Third World Quarterly*, 30(6), 1183–1196.

Coles, A.J. 1978. The moral economy of the crowd: some twentieth-century food riots. *Journal of British Studies*, 18(1), 157–176.

Edelman, M. 2005. Bringing the moral economy back in … to the Study of 21st-century transnational peasant movements. *American Anthropologist*, 107(3), 331–345.

Fafchamps, M. 1992. Solidarity networks in preindustrial societies: Rational peasants with a moral economy. *Economic Development and Cultural Change*, 41(1), 147–174.

FAO. 2011. *The state of food insecurity in the world: How does international price volatility affect domestic economies and food security?* Rome: Food and Agricultural Organisation of the United Nations.

FAO. 2013. *Food and Agricultural Organisation Food Security Indicators*. http://www.fao.org/economic/ess/ess-fs/ess-fadata/en/ [Accessed 18 March 2013].

Fraser, N. 2011. Marketisation, social protection, emancipation: Towards a neo-polanyian conception of capitalist crisis. In: C. Calhoun and G. Derluguian, eds. *Business as usual: The roots of the global financial meltdown*, pp. 137–158. New York: New York University Press.

Fraser, N. 2013. A triple movement? Parsing the politics of crisis after Polanyi. *New Left Review*, 81: 119–132.

Ghosh, J. 2010. The unnatural coupling: Food and global finance. *Journal of Agrarian Change*, 10(1), 72–86.

Ghosh, J., J. Heintz, and R. Pollin. 2011. Considerations on the relationship between commodities, futures markets and food prices. Memorandum to the World Development Movement. New Delhi, India and Amherst, MA, US: Jawaharlal Nehru University / International Development Economics Associates (IDEAs) and Political Economy Research Institute (PERI), University of Massachusetts-Amherst.

Gilbert, C.L., and C.W. Morgan. 2010. Food price volatility. *Phil. Trans. R. Soc. B*, 365, 3023–3034.

Gore, C. 1993. Entitlement relations and 'unruly' social practices: A comment on the work of amartya sen. *Journal of Development Studies*, 29(3), 429–460.

Greenough, P.R. 1982. *Prosperity and misery in modern Bengal: The famine of 1943–1944*. Oxford: Oxford University Press.

Greenough, P.R. 1983. Indulgence and abundance as asian peasant values: A Bengali case in point. *The Journal of Asian Studies*, 42(4), 831–850.

Headey, D. 2013. The impact of the global food crisis on self-assessed food security. *The World Bank Economic Review*, 27(1), 1–27.

Heltberg, R., N. Hossain, A. Reva, and C. Turk. 2012. Coping and resilience during the food, fuel, and financial crises. *The Journal of Development Studies*, 49(5), 1–14.

Hossain, N. 2009. Reading political responses to food, fuel and financial crises: The return of the moral economy? *Development*, 52(3), 329–333.

Hossain, N. and D. Green. 2011. Living on a spike: How is the 2011 food price crisis affecting poor people? Oxfam GB Research Report. Oxford: Oxfam.

Hossain, N., R. Fillaili and G. Lubaale. 2010. Invisible impacts and lost opportunities: Evidence of the global recession in developing countries. *Journal of Poverty and Social Justice*, 18(3), 269–279.

Hossain, N. and J.A. McGregor. 2011. A 'lost generation'? Impacts of complex compound crises on children and young people. *Development Policy Review*, 29(5), 565–584.

IRIN (Integrated Regional Information Networks). n.d. 'Kenya's Unga Revolution'. UN Office for the Coordination of Humanitarian Affairs. Available from: http://www.irinnews.org/film/4882/ [Accessed 1 October 2013].

Islam, N. 2003. *Making of a nation, Bangladesh: An economist's tale*. DHaka: The University Press Ltd.

Ivanic, M. and W. Martin. 2008. Implications of higher global food prices for poverty in low-income countries. *Agricultural Economics*, 39(1), 405–416.

Jackson, P., N. Ward and P. Russell. 2009. Moral economies of food and geographies of responsibility. *Transactions of the Institute of British Geographers*, 34(1), 12–24.

Jones, G.A. and S. Corbridge. 2010. The continuing debate about urban bias: The thesis, its critics, its influence and its implications for poverty-reduction strategies. *Progress in Development Studies*, 10(1), 1–18.

Keyes, C. F. 1983. Peasant strategies in Asian societies: Moral and rational economic approaches: A symposium. *The Journal of Asian Studies*, 42(4), 753–773.

Lagi, M., K.Z. Bertrand and Y. Bar-Yam. 2011. The food crises and political instability in North Africa and the middle east. Mimeo: New England Complex Systems Institute. Available from: http://necsi.edu/research/social/food_crises.pdf [Accessed 1 October 2013].

Mason, N.M., T.S. Jayne and N. van de Walle. 2013. Fertilizer subsidies & Voting behavior: Political economy dimensions of input subsidy programs. Paper presented to the Agricultural & Applied Economics Association (AAEA & CAES Joint Annual Meeting), Washington, DC. Available from: http://ageconsearch.umn.edu/bitstream/149580/2/Fertilizer_subsidies_%26_election_out comes-AAEA_V2.pdf [Accessed 1 October 2013].

Minot, N. 2011. Transmission of world food price changes to markets in Sub-Saharan Africa. *IFPRI Discussion Paper* 01059. Washington, DC: International Food Policy Research Institute.

Moore, S.F. 1983. *Law as process: An anthropological approach*. London: Routledge and Kegan Paul.

Naylor, R.L. and W.P. Falcon. 2010. Food security in an era of economic volatility. *Population and Development Review*, 36(4), 693–723.

O'Brien, T. 2012. Food riots as representations of insecurity: Examining the relationship between contentious politics and human security. *Conflict, Security & Development*, 12(1), 31–49.

Orlove, B.S. 1997. Meat and strength: The moral economy of a chilean food riot. *Cultural Anthropology*, 12(2), 234–268.

Patel, R. and P. McMichael. 2009. A political economy of the food riot. *Review Fernand Braudel Center*, XXXII(1), 9–36.

Reuters. 2013. Kenya government makes partial U-turn on new tax after protests. July 03 2013. Available from: http://www.reuters.com/article/2013/07/03/kenya-tax-idUSL5N0F93LI20130703 [Accessed 20 August 2013].

Schneider, M. 2008. "We Are Hungry!" A Summary Report of Food Riots, Government Responses, and States of Democracy in 2008. Available from: http://www.academia.edu/238430/_We_are_

Hungry_A_Summary_Report_of_Food_Riots_Government_Responses_and_State_of_Democracy_in_2008 [Accessed 1 October 2013].

Scott, J.C. 1977. *The moral economy of the peasant: Rebellion and subsistence in Southeast Asia.* New Haven: Yale University Press.

Swinnen, J. 2011. 'The Right Price of Food'. *Development Policy Review*, 29(6), 667–688.

Thompson, E.P. 1971. The moral economy of the English crowd in the eighteenth century. *Past & Present*, 50, 76–136.

Thompson, E.P. 1991. *Customs in common.* London: Penguin.

Tilly, L.A. 1971. The food riot as a form of political conflict in France. *The Journal of Interdisciplinary History*, 2(1), 23–57.

Trentmann, F. 2007. Before 'fair trade': Empire, free trade, and the moral economies of food in the modern world. *Environment and Planning: Society and Space*, 25(6), 1079–1102.

Vanhaute, E. 2011. From famine to food crisis: What history can teach us about local and global subsistence crises. *The Journal of Peasant Studies*, 38(1), 47–65.

Von Braun, J. and G. Gebreyohanes. 2012. Global food price volatility and spikes: An overview of costs, causes, and solutions. *ZEF-Discussion Papers on Development Policy*, 161, 42.

World Bank. 2013. Bangladesh poverty assessment: Assessing a Decade of Progress in Reducing Poverty, 2000–2010. *Bangladesh Development Series Paper* 31. Dhaka: Bangladesh World Bank Country Office.

Websites

FAO GIEWS: Available from: http://www.fao.org/giews/pricetool/ [Accessed 16 August 2013].

FAO Food Security Indicators: Available from: http://www.fao.org/economic/ess/ess-fs/ess-fadata/en/ [Accessed 18 March 2013].

Life in a Time of Food Price Volatility research project: Available from: http://policy-practice.oxfam.org.uk/our-work/food-livelihoods/food-price-volatility-research [Accessed 16 August 2013].

Naomi Hossain is a political sociologist and a Research Fellow at the Institute of Development Studies at Sussex, currently based in Jakarta. She previously worked at BRAC Research and Evaluation Division in Dhaka. She researches the politics of poverty, including to date elite perceptions of poverty, informal or 'rude' forms of accountability in frontline service delivery, women's empowerment and unruly politics. She is presently leading two projects studying the social and political dimensions of food price volatility.

Devangana Kalita has a BA (Hon) degree in English Literature from Miranda House College, University of Delhi (India) and an MA in Gender and Development from the Institute of Development Studies, University of Sussex, UK. She is based in India and works as an independent activist. She also works as a part-time Research Assistant on the *Food riots and food rights: the moral and political economy of accountability for hunger* project.

The government of poverty and the arts of survival: mobile and recombinant strategies at the margins of the South African economy

Andries du Toit and David Neves

The paper is concerned with marginal populations affected by the 'truncated agrarian transitions' of the twentieth and twenty-first centuries: people displaced out of land-based employment without reasonable prospects for accumulation in the non-farm economy. It analyses the forms of economic agency of people living in the migrant routes and networks connecting the shantytowns of Cape Town and the rural Eastern Cape in South Africa. It describes the artful and hybrid nature of their livelihood strategies – strategies that involve the integration from 'below' of urban *and* rural spaces, formal *and* informal income, and which simultaneously take shape outside the regulatory spaces conferred by the state, *and* make use of the rights and opportunities created by law and formality. Far from being reduced to the 'outcast' condition of 'bare life', marginalized and poor people in South Africa pursue inventive strategies on uneven terrain, cutting across the dichotomies of official discourse and teleological analysis. This allows a more nuanced analysis of the nature and specificity of the agrarian transition in South Africa.

Introduction

In this paper, we consider responses from 'above' and 'below' to the 'truncated agrarian transitions' (Li 2011) by which people are displaced out of agricultural employment without significant prospects of employment in the non-farm economy. While the empirical facts relating to global un- and underemployment and working poverty are well established, the implications for policy and action are contested. One problem is that dominant competing interpretations are politically and ideologically overdetermined, shaped by teleological assumptions about the likely (or wished for) directions of change and characterized by sweeping and homogenizing generalizations. A more differentiated account is needed.

Sections of this paper have appeared in substantially different form, but under similar title, elsewhere: (du Toit 2011, du Toit 2012a). The research on which this paper draws was conducted over a period of ten years within a number of donor funded projects. We in particularly wish to acknowledge the support of the Chronic Poverty Research Centre (CPRC), the UK Department for International Development (DFID), the South African Treasury, USAID, the FinMark Trust and the Programme for Pro-Poor Policy Development (PSPPD) within the South African Presidency. We gratefully acknowledge our many intellectual debts to friends and colleagues, in particular Franco Barchiesi, Henry Bernstein, Ben Cousins, Colleen Crawford-Cousins, James Ferguson, Ruth Hall, Sam Hickey, Francie Lund, Hein Marais, Nicoli Nattrass, Jeremy Seekings and Barbara Tapela.

Our key concern is to ask what can be learned about the dynamics and implications of 'jobless de-agrarianization' (du Toit 2009) through an in-depth investigation of one geographically and historically delimited example: the livelihood profiles and strategies of people surviving within the migrant networks of the Eastern and Western Cape, South Africa. Though the parameters of unemployment and landlessness in South Africa are indeed 'extreme and exceptional' (Bernstein 1996) there are also important similarities with the course of de-agrarianization elsewhere. While we will resist the temptation to generalize our findings to the entirety of the 'global South', we hope that they may illustrate some of the limitations of dominant meta-narratives and suggest alternative avenues for exploration.

We begin with a summary of trends defining de-agrarianization in South Africa and elsewhere, and briefly discuss some of the dominant paradigms that have been deployed to understand them. We then turn to a consideration of the South African case: we describe the processes that drive persistent poverty and we consider the contributions and limitations of the bio-political strategies that have developed, if not to eradicate it, at least to manage it and render it governable. We then describe what we consider to be the most significant features of the strategies by which those straddling the urban-rural divide on the margins of the South African economy seek to survive, thrive or endure, and we reflect on what this means for our understanding of economic marginality in South Africa. In the final section, we explore what this geographically and historically delimited investigation might imply for the task of making sense of 'truncated agrarian transitions' more broadly.

Urbanization, underemployment and marginalization in the global South

Almost half of all the households in South Africa survive on the margins of the formal economy: farm workers, domestic workers, landless or land-poor people in the rural areas, and unemployed or underemployed slum dwellers eking out an existence in the informal sector. No longer able to provide for themselves on the basis of agricultural production, they also have proved unable to find gainful opportunities in the formal sector: they constitute a 'marginal working class' (Seekings and Nattrass 2005) with few opportunities for social mobility or advancement. More than 36 per cent of South Africans are unemployed according to Statistics South Africa's 'expanded' definition (Statistics South Africa 2013), while one third of the employed population count among the working poor (Statistics South Africa 2012). Since the end of Apartheid, the South African economy has been growing steadily, but for this population, *contra* Dollar and Kraay (2001), growth has not been 'good for the poor': without land, poorly educated and mostly unskilled, without access to jobs and without significant prospects for prosperity in the informal sector they appear increasingly to be 'left behind' – superfluous to the labour needs of the economy.

South Africa's unemployment rate and numbers of working poor are high by international standards. But it is not the only place where large populations face un- or underemployment. The International Labour Organization estimates that more than a billion people globally are either unemployed or working in poverty (ILO 2010, 2013). Many of them are trapped in land-based livelihoods with little chance of prospering, or face being moved out of agricultural employment with scant prospects for gainful employment in other sectors. These patterns are especially marked in South East Asia and sub-Saharan Africa: in India, economic growth has led to the creation of a dualistic economy with wealth concentrated in a tiny service sector while 80 per cent of the population eke out a marginal existence in the agrarian and informal sectors (Harriss-White 2012). In Indonesia, the transformation of agriculture by large-scale plantation agriculture and outgrower schemes looks likely to push millions off the land without much chance of being absorbed in the industrial proletariat (Li 2011, 2009a). In sub-Saharan Africa (with South Africa a notable exception), few significant industrial sectors have emerged even in periods of sustained economic growth. More than 60 per cent of the population still depends on agricultural empoyment, but pressures for urbanization and migration are intensifying and the prospects for 'job real-location' out of agriculture into industry and services are not positive (Havnevik *et al.* 2007, ILO 2010). In these contexts, de-agrarianization and rural-urban migration seem likely to be not the precursors for Rostowian 'take off into self-sustained growth'; rather, they may well be drivers of 'urban involution', relegating hundreds of millions of displaced un- or under-employed people to the 'planet of slums' (Davis 2004).

The similarities and dissimilarities between the South African dynamics of rural-urban migration, (un)employment, poverty and inequality and those experienced elsewhere raise interesting questions about the likely directions of change in the future and the scope for responses in the present. Is the South African case a kind of perverse singularity where uniquely irrational and racist policies have led to outcomes that, dire as they are, are at most of admonitory relevance elsewhere? Or, on the contrary, does it constitute a kind of mirror image of the future awaiting other countries in the global South that have not yet made a successful transition to an industrial economy? If the truth lies somewhere between these extremes, should we imagine a smooth continuum? Or are there a limited number of clearly distinguishable path-dependent scenarios, each with distinct character-istics? And how are we to evaluate the policies for 'poverty alleviation', 'development' and 'social protection' that governments and development agencies advance in response?

Answering such questions requires a clearer sense of precisely in which respects the South African case is similar to those of other places in the 'global South' and where it is different. In this paper, we argue that light can be cast on this by a more careful inves-tigation of the terms upon which such populations are incorporated into the political and economic formations of present-day capitalism – and an understanding of what this means for the differential impacts of 'growth', 'development' and employment. This is a complex subject. The relationships between growth, employment and human development are conceptualized in divergent and competing ways in scholarly and policy literature. We cannot deal with this literature in any great depth. We do, however, wish to highlight the extent to which competing interpretations are grounded in stark, often overly-simplifying meta-narratives about the likely and desired directions of economic and political change.

One of the dominant ways of understanding the dynamics of employment, marginaliza-tion and economic change is provided by those who are optimistic about the ability of capi-talist economic growth to deliver prosperity and wellbeing to those who participate in it. This approach is well represented, for example, in the most recent World Development Report (WDR) by the International Bank for Reconstruction and Development, simply entitled *Jobs* (World Bank 2012). Obviously, much can be said about the merits and limits of the Bank's newly discovered concern with employment and its revised approach to labour market policy. Our remarks here, however, relate not to the Bank's policy pre-scriptions but to the ways in which it conceives the relationships between employment, development and economic change. Most important is the great centrality accorded to the notion of 'transformational jobs' that can serve at one and the same time as the basis of social entitlements, identity and social cohesion, and productivity gains (8). In this respect, perhaps the most arresting aspect of the Bank's discourse is its optimism. It is not that the Bank denies the existence of working poverty, unemployment, forced labour and the like. Rather, it sees them as *residual* in nature: the result of insufficient or lagging development, to be erased by the application of sensible policies for inclusive growth. This optimism is based on a teleological understanding of economic change. What are these 'transformational' jobs and what policies are needed to create them? The answer, for the Bank, depends primarily on the place of a given country on what it calls its 'development path': the trajectory by which that country travels from being 'agrarian' to being 'urbanizing' or 'formalizing' (e.g. 18–21). As in the 2008 'Agriculture' WDR, development is understood to be essentially a process of agrarian transition by which rural peasants travel to become, in good time and within a generation or two, urban workers (Li 2009a). Although countries may encounter different special circumstances along the way (demographic issues, resource conflicts), the basic assumption is that all of these problems are optimally solvable by getting the 'policy mix' right, so that the potential positive feedbacks between economic growth, technological innovation, productivity growth and increasing prosperity can be maximised. The notion that de-agrarianization may lead to industrialization for some countries but not for others, or that economic processes may themselves stall the journey from rural peasant to urban worker, is nowhere entertained. Neither is the possibility that global resource scarcity, fragile ecosystems, climate change and similar inconveniences pose any real threat to the prospect of inexhaustible and limitless growth.

Other scholars have taken a different view of the potential of capitalist economic development to bring prosperity to the global South. Rather than seeing the persistence of un- and underemployment as being merely a 'residue' or as an aberration, they have argued that it is better understood as structural – an outcome of the routine functioning of the core institutions of the global market economy and of the path of economic growth. The most familiar of these views is, of course, the Marxist idea that the creation of a 'reserve army of labour' is one of the unavoidable outcomes of capitalist development (Engels 1943, Bernstein 2002). Similar interpretations also have been developed outside the Marxist canon: Zygmunt Bauman has argued that the production of 'human waste' – migrants, refugees and the unemployed – is an inevitable consequence of the social changes wrought by modernity (Bauman 2004). For the followers of Giorgio Agamben, the differential inclusion of such marginalized populations within legal and political frameworks reduces them to the status of 'bare life' – subject to law but not recognized as civil or political beings (Agamben 1998, Mbembe 2003, Downey 2009). For Mbembe, Bauman and Agamben, these 'waste lives' – migrants working in sweat shops, displaced populations, people in concentration camps or detention centres – are *necessary* products of modernity and modern political sovereignty. Interestingly, these writers seem unconcerned with actually investigating the quotidian realities of life in the 'wastelands' or 'necro-political zones': what it takes to feed a child, to go to school, to care for the sick or to plan for the future. Rather, these phenomena seem to function merely as symbols in a symptomatic reading intended to lay bare the empty promises and dark sides of capitalist progress and enlightened modernity.

At the time of writing, however, perhaps the most influential concept that has evolved in response to the prospect of a global population of un- or underemployed people is the notion of precarity. The precarity literature initially focused on the growing prominence of casual, temporary and 'flexible' work in post-industrial societies (Neilson and Rossiter 2008). In recent years, the term has been extended to propose the existence of a global precarious class or 'precariat' comprised of all manner of marginalized and insecure workers: under-employed 'freeters' (freelance workers) in Japan, migrant factory workers in China, educated and unemployed youth in the Middle East and unorganized migrant, domestic, casual and informal workers in the global South (Standing 2011, Mosoetsa and Williams 2012).

These counter-narratives are important, but they are often articulated in ways that limit their utility. There is a risk of being exactly as teleological as the Bank. Instead of seeing economic growth, progress and development as forces that will liberate and empower the world's working populations, critics all too easily invoke the spectres of 'neoliberalism' and 'globalization' as 'big Leviathan' or catchall explanations for everything and anything that is going wrong (Collier 2012). The most problematic examples can be found in the rhetoric of modern-day European critical theory, which is too often characterized by an apparent disregard for empirical complexity or specificity. Sweeping and epochal generalizations are made about 'the' state, about 'the multitude' and 'the modern day office' with only the most sketchy attempt at an encounter with the messiness of empirical phenomena (Hardt and Negri 2004, Gregory 2011). Similarly, there are clearly problems with the notion that we can use the concept of the 'precariat' to lump together the enormously divergent and variegated groups of vulnerable, precarious or flexible workers.

A more differentiated account is necessary. The WDR's teleological and normative focus on the virtuous circle of 'transformative jobs', 'development' and growth does not help, for it cannot deal effectively with exceptions and counter-examples. Neither is it enough to lump together those marginalized and impoverished by capitalist growth simply by way of reference to non-inclusion, precarity or externality. Part of the problem is the nature of the dichotomizing assumptions that underpin both optimistic and pessimistic teleologies – assumptions that simplistically contrast 'precarity' with 'security', 'bare' life with 'political life', 'inclusion' with 'exclusion' – and, for that matter, 'adverse' with 'beneficial' incorporation. Such modes of analysis flatten and oversimplify the uneven and complex terrain of what Mezzadra and Neilson have called 'differential inclusion' (Mezzadra and Neilson 2012).

In this paper, we explore the dynamics of employment and marginality in one geographically and temporally delimited context. Our argument draws principally on two traditions. On the one hand, we draw on the insights of Marxist political economy – particularly the work of those who emphasize the variegated interests and natures of groupings produced by the 'structural fragmentation of labour' (Bernstein 2002, Lerche 2010). Secondly, we interpret the phenomena we find there through the lens of a Foucauldian analysis of biopolitics (Li 2009b, Foucault 2010). At the same time we hope to avoid the totalizing and reductive tendencies that characterize some variants of both Marxist and Foucauldian analysis. Our aim is to produce what Collier (2009) has called a 'topological' account of economic agency as it takes shape at a particular location and at a given moment in history. Rather than pitching our analysis at what is not there (stability, decent work or even 'transformative jobs'), our aim is to look rather more dispassionately at what *is* there, to trace the processes and dynamics that shape the current terrain, and at the strategies and devices employed by those who seek to survive and thrive – or just to endure (Povinelli 2011) – at the margins of the global economy. Our purpose is not to come up with an authoritative counter-narrative about 'the logic' of 'neoliberalism' or 'capitalism' *as such*, but rather to illustrate the ambiguous and contradictory nature of the struggles that are taking place, and to allow more nuanced judgements about similarities and differences between late capitalist agrarian landscapes in South Africa and elsewhere.

Adverse incorporation in post-Apartheid South Africa

We now turn to a brief description of the specific processes that produce marginality and persistent poverty in South Africa. These are strongly related to the nature and direction of de-agrarianization (Bryceson and Jamal 1997) and its local trajectories. Not only has the contribution of agriculture to household income and food security steadily diminished, but millions of households also face little prospect of finding employment in the non-farm sector. This is in part a result of the impacts of colonial dispossession, segregation and Apartheid policies and the resultant deeply unequal distribution of land (May 2000, Andrew *et al.* 2003, Lahiff and Cousins 2005, Walker 2008). Apartheid created an impoverishing spatial economy which at the national level confined a third of the population to distant and overcrowded Bantustans; in the urban areas it relegated black South Africans to dormitory townships and informal settlements situated far from markets and economic opportunity (Smith 1992, Christopher 1994, Harrison *et al.* 2008). Another important Apartheid legacy is the deeply unequal distribution of human capital stemming from Verwoerdian policies of education. And while racial identity is no longer the *formal* basis of socially sanctioned inequality, South African social and political life is still marked by the continuing salience of racial and national identities in sometimes violently enforced processes of inclusion and exclusion in a wide range of formal and informal contexts.

But Apartheid policies and the difficulty of erasing their legacy are only part of the story. It is also important to understand the importance of other features of South African society, many of which preceded National Party rule and which persist after its end. Here, three phenomena are particularly important. The first is the coming into being of a highly developed, concentrated and vertically integrated core economy characterized by capital intensity and spatially extensive distribution systems. The second is the continued commitment to authoritarian and modernist conceptions of 'development'. The third is the evolution of a distributional regime that institutionalizes deep distinctions between insiders and outsiders and which is premised on the centrality of the worker-citizen nexus.

The first feature is fairly well understood. The South African economy is dominated by an industrial complex centred around mineral beneficiation and cheap energy (Fine and Rustomjee 1996). Growth has tended to be capital intensive, and the gross output elasticity for employment has tended to be low (Black 2010). Alongside this industrial complex has developed a highly concentrated and vertically integrated food processing and retail sector characterized by efficient and spatially extensive distribution systems that reach deep into rural areas. The agro-food sector is characterized not only by concentration and agribusiness penetration on the production side, but also by the downstream development of buyer-driven value chains governed by supermarkets. By 2010, nearly 70 per cent of food marketing was in formal retail, of which 94 per cent was controlled by six supermarket retailers, while 80 per cent of processed food staples are in the hands of just four corporations (Greenberg 2010, Igumbor *et al.* 2012, Bernstein 2013). While supermarkets doubtlessly help ensure cheap food for those workers who can pay, their supply lines have tended to marginalise small producers and small and medium enterprises (SMMEs), while their retail activities have tended to depress the prospects for the local informal sector (D'Haese and Van Huylenbroeck 2005, Mather 2005). South Africa thus has a problem unusual among developing or middle income countries: the 'commercial space' available at the 'base of the pyramid' for 'upgrading' and the development of local industries and informal markets is absent or already dominated by big capital (Philip 2010).

Secondly, an important role seems also to have been played by the existence of deeply held commitments, on the right and the left of the political spectrum, to ideological narratives of modernising progress, Eurocentrically conceived global integration and top-down control. These features were central aspects of the authoritarian and racist state that took shape under colonial rule and which was perfected by the Apartheid government, but in many respects the post-Apartheid government has inherited this tradition (Louw 1997). Debates in a wide range of areas (agricultural policy, rural development, urban planning and industrial relations) are influenced by strongly-held normative preferences for the institutional and organizational forms associated with Fordism, Taylorism, bureaucratic centralism and formality. The ambivalent and sometimes even antagonistic stance of metropolitan governments and national settlement policy to informal traders and slum settlements (Lund and Skinner 2004), the predilection of the ruling party for large-scale as opposed to small-scale agriculture (Aliber and Hall 2012) and the adoption of forms of industrial relations that suit large-scale, bureaucratic organizations rather than small enterprises and SMMEs all seem to be connected not simply to the embrace of 'neoliberal policies' but also to salience – on the right as well as the left – of authoritarian and Fordist approaches to modernization, scale and progress.

Thirdly, one of the more significant legacies of this modernism lies in its impact on the South African distributional regime – the ways in which industrial policy, labour market regulation, industrial relations and social policy shape the distribution of the benefits and costs of growth in the economy. As Seekings and Nattrass have pointed out, while the South African distributional regime has undergone an important process of de-racialization in recent years, some of its key features have persisted in significantly similar forms since 1920s. The struggles between capital and labour in the early part of the twentieth century were resolved by a social pact pivoting on the citizen-worker nexus: the notion that the platform for the attainment and allocation of social claims and entitlements should be provided by the social identities and institutions developing around formal sector employment (Barchiesi 2011). This created benefits and social entitlements for the middle classes and for the white, urban, industrial working class. Rural black South Africans were excluded, while the place of the urban black working class was intensely contested (Seekings and Nattrass 2005). An essential part of this arrangement was the institutionalization in the rural areas of the racially bifurcated power of colonial governance, which distinguished urban 'citizens' from the black 'subjects' of customary rule (Mamdani 1996).

By the end of Apartheid, the economic formations of capitalism in South Africa were characterized by institutional arrangements and socio-economic processes that undermined the economic agency of a large part of the population. Note that we do not use the word *exclusion*: as we have argued elsewhere, the situation of South Africa's poor is not simply that they are 'outside' the 'first economy', disconnected from the opportunities of the market. It is better characterised as 'adverse incorporation' (Bracking 2003, du Toit and Neves 2007): millions of poor, black South Africans find themselves excluded from the economy as farmers, growers, producers, workers and traders but included as citizens and as the consumers of the manufactured goods and services created by the South African core economy (du Toit 2011). Even at the 'core' of the manufacturing and mining economy, flexibilization and outsourcing reduce the worse-off workers to a situation not much better than informal workers. These problems are not the result of a development deficit and they cannot be addressed simply by removing obstacles to growth. Rather, they are the path-dependent outcomes of the character and direction of capitalist development itself.

While the transformation of South African society and the reduction of poverty have been a central commitment of the post-Apartheid government, policy frameworks for at least the first 10 years after 1994 tended to ignore or avoid these issues. Economic policies were premised on the notion that the dynamism of markets in a growing economy would in themselves ensure the kind of jobs that would provide pathways out of poverty. This required a focus on rapid improvements in the efficiency, competitiveness and productivity of labour, and on providing access to markets for South Africa's black entrepreneurs and farmers (Nattrass 2001, McCord 2005, Jacobs 2008), without paying much attention to the ways in which the key institutions and the distribution of power in these markets themselves routinely produced marginalization and disempowerment.

As a result, growth in the South African economy in 1995–2000 was biased against the poor. Real per capita household expenditures declined for poorer people, and the absolute numbers of poor people increased (Özler and Hoogeveen 2005). The period between 2000 and 2005 saw a small reduction in the total poverty headcount and a small increase in the incomes of the poor, arguably as a result of the increase in social grants (Van der Berg et al. 2006). But poverty for millions of black South Africans remained acute, and the distribution of income became more unequal (Leibbrandt et al. 2010). In 2005, Seekings and Nattrass calculated that 45 per cent of income and assets accrued to a tiny urban elite, constituting 12 per cent of households, while poverty was concentrated in a heterogeneous underclass of poor and landless urban and rural un and under-employed South Africans, who comprised 41 per cent of households but who received only 10 per cent of income. Importantly, this marginal class includes both the unemployed and the working poor: it comprises not only domestic workers and farm workers but also flexibilized and casualized workers eking out an existence in the bottom tiers of the formal sector. Crucially, Seekings and Nattrass argue that this group was likely to remain poor: lack of access to social networks, lack of skills and the costs of seeking employment constituted structural barriers to entry into stable employment and consequently to social mobility (Seekings and Nattrass 2005, 252–8).

Since the mid-2000s, policy discourse has shifted to allow for greater recognition of the existence of structural inequality. This began with the articulation by Thabo Mbeki of his 'two economies' theory, and continued with policy frameworks such as the Initiative for Accelerating Shared Growth in South Africa (ASGISA), the second Industrial Policy Action Plan, the New Growth Path and the National Development Plan (du Toit and Neves 2007, du Toit 2012b). In these documents, there seems to be at least partial recognition of the structural nature of inequality and of the need to explicitly challenge marginalizing processes (Philip 2010). But, at present, there is little probability of this being turned into implementable policy. Rather than a clear change of direction, these policy documents (emerging with the end of Mbeki's presidency and the start of Zuma's) seem to herald a period of ever-widening contestation and struggle within and around the state, in which disparate, ill-matched and sometimes contradictory policies are pursued by rival political factions and interest groupings (Marais 2011).

The government of poverty

While post-Apartheid economic and development policies have avoided squarely confronting the structural inequalities perpetuated by the nature of the core economy, neither have they simply abandoned poor South Africans to the logic of the market. Here, it is useful to consider carefully the calculations involved in what Tania Li has called the biopolitics of 'making live and letting die': the way governments and administrations decide whether and how to invest in the health, productivity and longevity of particular populations, differentially positioned within society (Li 2009b). Particularly useful has been the ability of these forms of analysis to explore the *differential* governance of marginal and labouring populations (Ong 2006).

In South Africa, these calculations have been decisively shaped by the fact that the poor, while economically disenfranchised, are politically central: citizens with votes in a polity where the legitimacy of the government depends on its claim to represent their needs and interests. At the same time as post-apartheid economic growth contributed to the perpetuation and exacerbation of economic marginalization, it has also been characterized by the bio-political incorporation of poor and black people within the distributional regime. An important aspect of post-Apartheid politics and policy relates to the 'government of poverty' – the ways by which the state has sought to construct poverty and poor populations as objects of knowledge and management (du Toit 2012b). One consequence was the rapid increase in the production of quantitative knowledge about poverty and poor people required for the juridico-legal and technical decision-making needs of the new administration (Seekings 2001). Another was a massive investment in 'make live' policies aimed at improving the well-being and productivity of the population. A rapid expansion of electrification and water reticulation was one aspect of this, and so was a massive investment in health and education. Most significant, however, has been the formal de-racialization of social protection policy and the roll-out of cash transfers by the Department of Social Development. Since the 1990s, expenditure on social grants has increased dramatically, growing at almost 20 per cent per year between 2001 and 2007. In 2009/10, expenditure on social security was R 73 billion, totalling more than 3 per cent of gross domestic product (GDP), unusually high for a middle-income country (Neves *et al.* 2009).

The co-existence of 'social-democratic' and 'neoliberal' elements in post-Apartheid policy is far from a contradiction. Both aspects are premised upon the extension of a distributional regime founded on the citizen-worker nexus. The continued normative and ideological centrality of the notion of employment as the basis of social integration and entitlement has been a key element both of the rapid roll-out of cash transfers and of the emphasis on market-led economic transformation. This has shaped both the scale of post-Apartheid 'make live' policies and their limitations. While spending has been relatively high for a middle-income country, social policy has been based on the assumption of full employment and the 'productivist' notion that access to social entitlements needs to be founded first and foremost on the attainments of the independent, self-activating, employed able-bodied citizen-worker (Ferguson 2007a, Barchiesi 2011).

This has also meant that the ability of the South African state to incorporate the population into the distributional regime is limited. While policies aimed at extending a bio-politics founded on the citizen-worker nexus may be logically coherent, they run into serious economic and political impasses. The political centrality of the 'marginal working class' means that the post-Apartheid regime has to put the economic and social demands of the vast mass of poor and marginalized citizens centre stage but it is unable to give sustainable effect to these demands. For the reality that millions of those able-bodied workers are unable to find employment, post-Apartheid social policy has not been able to formulate a response: at best it can only deal with the fallout of marginalising growth.

Attempts to find alternatives to a biopolitics founded on the citizen-worker nexus have found little traction. Calls for a universal citizen's grant, for instance, attempt to push the politics of 'equitable share' beyond the limitations of a productivist approach and to delink social protection from assumptions about full employment (Ferguson 2007b). Yet the South African campaign for a 'Basic Income Grant' has run aground. This is partly due to fiscal conservatism and political caution on the part of the African National Congress (ANC) (Marais 2011). But it should also be noted that besides lukewarm support from Congress of South African Trade Unions (COSATU), its proposals have found little popular support. Calls for a citizen's grant are noticeably absent, for instance, among the demands articulated in ongoing service delivery protests: it seems that poor people themselves are as wedded as neoliberal policymakers to the dream of finding independence and self-respect through 'proper jobs'.

Another alternative is linked to the hope that if formal employment has not succeeded as an avenue by which poor people in South Africa can benefit from capitalist growth, the informal sector could achieve this. Here, the focus on 'worker-citizens' is to some extent replaced by the imperative to create what Aihwa Ong calls 'enterprising citizens' (Ong 2006). This has led to the argument in some policy circles that the informal sector is not a transitional terrain destined to fade away as the formal sector grows, but could be seen as a source of employment in its own right, or even a 'buffer between employment and unemployment'. This optimistic assessment of the informal economy has been most enthusiastically articulated by employment agency Adcorp's labour economist Loan Sharp, who has all but denied the existence of unemployment in South Africa and who has touted the informal sector as an engine for growth (Business Day Live 2013, Van der Berg 2013). The reality, however is that the informal sector in South Africa is by developing country standards comparatively small, and disproportionally concentrated in the retail sector (Chichello *et al.* 2005). However compelling the notion of the informal sector as an engine for growth may be, the structural context described in our previous section seriously limits the extent to which the informal sector can create jobs.

A third set of possibilities relate to a resuscitated politics of rural containment: the notion that rural and agrarian development can reduce or hold back urbanization, create alternative avenues for food security, relieve pressures on the job market and even preserve rural institutions and forms of life. Although these ideas do not constitute a formal or coherent programme, such notions certainly animate the Comprehensive Rural Development Plan (CRDP) of the South African Department of Rural Development and Land Reform, which has recently announced its intention to 'rekindle the class of black commercial farmers which was destroyed by the Natives Land Act of 1913' (Department of Rural Development and Land Reform 2013), and the policies of the Department of Traditional Affairs, whose Minister has lauded the positive role played by the rule of 'Kings' in South Africa (Department of Traditional Affairs 2011). The notion that rural populations should be governed differently has been given explicit articulation in the provisions of the Communal Land Rights Act (now struck down), which attempted to give traditional councils wide ranging powers over communal land; the Traditional Leadership and Governance Framework Act of 2003, which entrusted black rural South Africans to the authority of 'traditional' councils; and the controversial Traditional Courts Bill, which proposes that such courts be given extensive new powers and that their judgements will not be subject to appeal (Cousins and Claassens 2008).

This of course, is a political project with a long history. Apartheid's differential incorporation of rural subjects by way of notionally independent agrarian Bantustans was a central component in the exclusion of rural people from the entitlements afforded to citizen-workers. But while the idea of a bucolic rural countryside governed by tribal chiefs may have nostalgic appeal for some, it is neither politically nor economically sustainable. The decline of Bantustan agriculture and the incorporation of rural South Africa into the distribution and production systems of its corporate agri-food system undermines the basis of such a project, and the notion that the 'bifurcated power' of colonial governance should be reproduced in post-Apartheid times has met with significant opposition. Significant opportunities for personal aggrandizement and political patronage are clearly offered by these proposals but they hold little prospect of resolving the difficulties created by the triumph of capital in the countryside.

The arts of survival

Thus far we have focused on how poor, black, marginalized households are structurally located within the formations of South African capitalism, and how they are incorporated within the institutional systems and frameworks of post-Apartheid bio-politics. This, however, is only a part of the story. The 'global assemblages' that shape the life chances of individuals and populations in present-day capitalism do not consist only of the formal and official formations of 'territory, authority and rights' (Sassen 2008) but also on the social institutions and responses of the members of those populations.

Here, we draw on almost a decade of research into the livelihood profiles and strategies of poor and marginalized 'African' households existing within the Eastern and Western Cape provinces (du Toit 2004, De Swardt *et al.* 2005, Neves and du Toit 2013) conducted under the auspices of the Chronic Poverty Research Centre (CPRC), USAID and the Programme to support Pro-poor Policy Development (PSPPD). This research relied greatly on in-depth quantitative and qualitative work (livelihood profiles, life histories and enterprise studies) to explore the structural limitations faced by impoverished households located in the Alfred Nzo District of the rural Eastern Cape and in the shantytowns of Khayelitsha within metropolitan Cape Town. These are, of course, only two of the contexts that make up the variegated landscapes created by migrancy in South African. Furthermore, recently urbanised township dwellers and residents of the former homeland areas do not constitute the entirety of its poor population. At the same time, much can be understood about the nature of South African poverty more generally by looking at the situation and responses of those affected by marginalizing forms of de-agrarianization. These phenomena link town and countryside, urban and rural, metropolitan modernity and the enclaves of townships and communal areas. From this research, and from the wider literature on migrancy, livelihoods and informal sector self-employment in South Africa, a fairly detailed picture emerges.

Firstly, migration and the ability to communicate, travel and effect resource transfers across space remain central to the survival of poor people. These practices link urban and rural locales, and connect households to markets in vital ways. The evolution of these strategies is captured neither by narratives of inexorable, unidirectional urbanization, nor of simply circular migration. Contrary to earlier expectations, post-Apartheid South Africa did not see the rural poor moving *en masse* to the cities from which they were previously excluded by legal fiat (Mabin 1990). The absence of a decisive 'urban transition' is due to the continuing precariousness of urban livelihoods, and the affordances offered by rural areas as zones of retirement and retreat from urban shocks and labour markets. Survival strategies depend on maintaining an extended system in which rural *and* urban outposts, agrarian *and* urban livelihoods all play an important role, and in which resources and people flow in both directions (du Toit and Neves 2008a).

Secondly, one of the lasting consequences of decades of migrancy has been the reconstitution of African households. Migration meant that African households in the twentieth century became increasingly fluid, stretched and porous: characterized by changing membership, intimately connected to distant places and open to competing claims from members and dependants (Ngwane 2003, Ross 2003, Russel 2004). While South African society is predominantly urbanised, what has emerged in the last two decades is the development of many-rooted, 'rhizomic' networks in which widely dispersed households are connected by overlapping and transient membership (du Toit *et al.* 2007).

Thirdly, livelihood strategies are reliant on the existence of elaborate, carefully negotiated and culturally encoded practices of reciprocal social exchange and 'distributive labour' (Ferguson 2007a). These are embedded within – and work to nurture and sustain – tightly knitted, closely contested, sometimes constrictive but almost always essential social networks founded on kinship, contiguity, patronage, politics and friendship. Spatially extensive and highly complex networks of reciprocal exchange arise that involve the maintenance of intimate and economically vital connections between distant locales – and the transfer across space of economic resources, information, social obligations, credit and debts, productive activities, opportunities and persons. These networks play a vital role in ensuring survival, and enable the mitigation of shocks and the distribution of economic resources and opportunities – but they can also function to transmit shocks and impose costs. Though often disregarded and analytically elusive for quantitative measurement and specification, they function, alongside the formal systems of corporate networks and political institutions, as a key vector of connectivity within the South African economy: a matrix that works to integrate (albeit from below) the sundered and segmented spaces of South African society (du Toit and Neves 2008b).

The notion of 'social capital' on its own does not adequately illuminate the significance or implication of these practices. Quite unlike 'generalized social trust', they are a form of social practice the benefits and costs of which accrue to individuals. Their impact is mediated by gender and cultural identity, by the nature of the social relationships involved, by individuals' access (or not) to resources, by personal histories, and by highly specific local prescriptive codes and expectations. This also means that they should not simply be understood as a universal social good. The negotiation of obligations and claims within these social networks is not a simple expression of solidarity: instead it constitutes an intimate politics the stakes of which can be dire and desperate. Divorce, expulsion from households, violence, gossip and accusations of witchcraft are as much part of these politics as the creation of savings clubs and mutual support. The bargains exacted are not guaranteed to be fair or equal. Gender is central to these struggles and contestations. Those who are less powerful, who are not accorded patriarchal privilege, or who have fewer resources to begin with – elderly women, single mothers and unemployed youth who are not able to constitute their own households – can end up isolated, abandoned or exploited.

Fourthly, within this context, individuals and households pursue complex strategies based on knitting together a heterogenous mix of activities and income streams that includes agrarian production, reciprocal exchange, formal and informal employment, and receipt of state cash transfers. These strategies deploy a kind of improvisatory, recombinant *bricolage* in which the aim seems to be to bring together a wide variety of activities so that they supplement and complement each other to constitute a whole which, with luck and skill, can be more than the sum of its parts. The ability to co-ordinate action, win support, and manufacture consent within and between households is crucial here. The reductive, hand-to-mouth connotations of 'survivalism' do not do justice to what is involved: survival requires not only a willingness for super-exploitation and self-exploitation, but also knowledge, know-how, connections, experience, co-operative ability and an ability to negotiate conflictual terrain, eke out meager resources, spot transient opportunities and bend them to one's purpose (du Toit and Neves 2006).

Within these recombinant strategies, access to state cash social transfers is probably the single most important component. Cash transfers are predictable and accrue to people universalistically as citizens. They play a vital role in anchoring distributive strategies, lubricating the gears and supplementing the benefits from processes of reciprocal exchange. Despite policymakers' perennial concern with 'targeting' social protection, informal systems of reciprocal exchange allow a significant amount of 'retargeting' from below, ensuring that many who are not formally entitled to social grants benefit directly or indirectly. The functioning of formal social protection policies is thus greatly influenced by the manner in which these transfers articulate with lay practices of mutuality and 'informal social protection' (Bracking and Sachikonye 2008). This may involve contested and even exploitative pressures for benefit sharing (Sagner and Mtati 1999). But cash transfers – particularly the higher-value pensions and disability grants – also subsidize job-seeking behaviour, informal businesses and agricultural production and investments in assets such as homesteads, and so sustain the vanishingly small ledges on which survival depends (du Toit and Neves 2008b, Neves *et al.* 2009, Ferguson 2013).

Agricultural production is a second important component. Land-based activities, despite their marginality, continue to be significant in augmenting the livelihood activities of many vulnerable households: even small-scale and marginal activities can, in aggregate terms, be significant (Andrew *et al.* 2003). Additionaly, agriculture plays a key role in grounding and anchoring the redistributive economy of practices of reciprocal exchange (Ferguson 2013). Important as agriculture is, it does, however, play largely a supplementary role (Aliber and Hart 2009). Hopes that agricultural activity can support a form of 'accumulation from below' for 'petty commodity producers' seem misplaced in the context of land scarcity and unchallenged supermarket and agribusiness power (Cousins 2013).

Informal self-employment in the non-farm sector is the third significant contributor to the livelihoods of the impoverished within migrant networks. Our research on the informal sector (Neves and du Toit 2013) seems to indicate that poor black South Africans can indeed be resourceful entrepreneurs – but often not in the way envisaged in business school. Informal enterprises are not self-contained firms: rather, they are components of a larger household economy, sustaining (and in turn sustained by) intra-household transfers of resources, goods, obligations, labour and income. Neither can they be understood simply in terms of the rational pursuit of profit or maximising income: rather, they serve a range of different ends including food security, the deflection of redistributive claims, relocation within social networks or establishing footholds in urban or rural space (Neves and du Toit 2012a, 2012b).

Above all, informal economic activity is characterized by its complex relationship with the formal sector. Not only are the informal and formal sectors connected through various capital, productive and employment linkages; hybrid household strategies frequently comprise arrangements whereby formal and informal economic activities subsidize, supplement and complement one another. Informal sector activity is often dependent on formal sector income. Conversely, formal sector employment can in its turn depend on rural and informal work. The unpaid care-work of rural women, for instance, provides a massive but at present unquantified and unseen subsidy to urban wages and household reproduction.

Most crucially, however, informal economic activity is beleagured by the formal economy, subsisting in its shadows and crowded out by competition from large and well-resourced corporates. Informal entrepreneurs survive by trading in tiny economic niches defined by locational advantage or by culturally specific desires and markets not yet occupied by 'big retail' (Neves and du Toit 2013). Here too they rely on social technologies to manage distance, as when social networks are used to access markets or co-ordinate production (du Toit and Neves 2006), or in family enterprises in which, for example, livestock agriculture, livestock trade, meat processing, retail and food preparation are linked in a single tightly-knit, farm-to-fork, vertically-integrated intra-family production network (Neves and du Toit 2012a, 2012b).

Mobile and recombinant strategies on an uneven terrain

The terrain we have described here and the strategies that are deployed on it do not fit neatly with teleological and sweeping narratives about the relationship between development, economic transformation and employment. On the one hand, optimism about the capacity of capitalist growth to support 'inclusive' economic transformation seems misplaced: policy prescriptions that assume a natural path from an agrarian to a post-industrial service economy, available to all who apply the right policy mix, offer little guidance in a context where there is such a deep structural mismatch between the employment that can be generated and the jobs that are needed. But if the optimistic development vision of the Bank seems misplaced, so is the dystopian pessimism of those who see capitalist development, modernity and 'neoliberalism' simply as processes that inevitably produce 'waste lives' and political exclusion. While the terrain of adverse incorporation is unfavourable, it is still a terrain upon which forms of agency can take shape and upon which it is possible to deploy strategies for survival, dignity and betterment – however constrained and limited. Desperate as the situation of those sunk in deep rural poverty in the Eastern Cape is, that situation is not well understood through narratives of 'bare life' or the denial of rights (Westaway 2012). Far from being abandoned by the state or excluded from its laws, South Africa's marginalized poor live assertively within them, assembling survival strategies out of fragmented resources, evading the the law when they need to but insisting, when they can, on the rights and entitlements due to them as citizens.

If the livelihood journeys of marginal and poor South Africans are not to be understood in terms of either simple urbanization or rural containment, how should they be described? If the marginal rural areas and shantytowns of South Africa are neither places of beneficial progress nor 'zones of exclusion', what are they? Rather than these reductive, unidirectional metaphors, a better approach would be to look in more detail at the uneven ways in which the networks, systems and institutions of corporate capitalism, of bureaucratic governance, of democratic citizenship and of informal sociality are *everywhere* present; the ways these networks *differentially* incorporate regions, groups, households and individuals; the ambiguous, provisional, constrained and double-edged strategies that those caught in these networks evolve in response, and the unequal distribution of benefits and costs that results. In other words, what is needed is a careful analysis of the ways in which the structuring of economic and social domains and the responses to them involve incorporation *and* externality, the extent to which they can be both advantageous and disadvantageous, and the way this shapes the resilience *and* the sensitivity of livelihood ecologies and survival strategies.

We are not here simply making a case for pure ambiguity. The critique of reductionist and teleological narratives leads to a recognition of complexity, but there is little point in making a theoretical fetish of complexity and indeterminacy in themselves. Rather, we would like to point out some of the key sites of contestation and struggle that emerge on the socio-political terrain of post-Apartheid South Africa. What follows is a provisional attempt to define some of the characteristics of this terrain and the responses that evolve on it.

In our argument so far we have sought to describe more concretely the liminal conditions usually invoked by references to 'precarity', 'exclusion' and 'fragmentation'. This has involved a careful investigation of the *terms* of inclusion and the determinants of both vulnerability and resilience. This general orientation has shaped our understanding of migrant networks. We see migration as one component of a set of mobile and recombinant strategies deployed in political and economic space. These are strategies by which people manage to create a measure of provisional coherency and functionality in spite of and across geographic divides and governmental exceptions: strategies that allow them to connect rural and urban contexts, link formal and informal livelihoods, move between legality and illegality, and evade or invite governmental attention. These strategies depend on a wide repertoire of technologies, moves, arrangements and practices by which those subject to 'differential incorporation' can locate themselves within (or remove themselves from) geographical spaces, legal frameworks, customary practices and markets. Moving house, seeking jobs, visiting or taking care of relatives, attachment to (or detachment from) households, building rural homesteads and burying the dead all form part of this repertoire; so do the use of information and communication technologies (ICTs) and bank accounts (Skuse and Cousins 2007, 2008). Rather than focus on the movement of persons from rural homestead to town and back again), analysis should be focused on the many-rooted, rhizomic networks themselves: how they stitch different zones and spaces together, how people act within them to exploit or cushion chance and fate, and how this shapes the course of social differentiation.

This, however, is not all that can be said. Perhaps one of the most central features of the differential incorporation of poor and marginalized people into the South African social formation is the the fact that all of these interactions take place in a context defined by a new politics of citizenship. One of the most significant and distinctive aspects of the South African case is the highly ambiguous and unstable situation in which an impoverished and vulnerable population is at one and the same time economically redundant and politically central. The government of poverty in South Africa is characterized at one and the same time by the economic 'abandonment' of the population (Povinelli 2011) *and* its embrace.

Here, our argument takes us beyond the specific situation of de-agrarianized' populations, and touches on the nature and dynamics of political contestation and struggle in post-Apartheid South Africa more broadly. This is of course a vast and complicated issue lying for the most part outside the purview of this contribution. It is, however, necessary to make some remarks that help to locate the politics of citizenship within our reading of the gaps and points of instability of the post-Apartheid distributive regime: for just as poor people in South Africa are differentially incorporated in the formations of South African capitalism, so is their political incorporation as citizens partial, uneven and crisis ridden.

Firstly, consider the problems besetting the project of a South African distributive politics centred on the citizen-worker nexus. As T.H. Marshall has pointed out, a central component of the ability of welfare states to deliver and institutionalize broad social compacts in the twentieth century was related to the way workers' status as citizens created the conditions of possibility for forms of workplace struggle by which they could progressively increase their social and economic status in society. Civil rights in the workplace became the vehicles for the attainment of social and economic rights (Marshall 2009). In South Africa, where full-time employment and trade union membership are available only for a minority of workers, this 'royal road' for the realization of working people's social and economic demands is obstructed. Instead of trade union struggles being a vehicle whereby the social standing of the broad mass of the 'labouring classes' can be improved, they are continually at risk of appearing as institutions that represent sectional interests, while the social and political struggles of marginal workers take place outside the key institutions of the industrial relations system.

Secondly, as important as the crisis of the ideal of the 'worker citizen' are the consequences of the institutional forms taken by parliamentary democracy and proportional representation on South Africa. The strong form of proportional representation that governs the functioning of the South African parliament, coupled with the unchallenged political legitimacy of the African National Congress, means that the links between local politics and parliamentary representation are severed. In a context where only one party can govern, and where people vote for a party list, political competition and contestation turns inward and is governed more by intra-party power relations than by the risk of electoral dissent (Mattes 2002). The institutions whereby claims and entitlements in South Africa can be formally and politically contested are thus only partially available to its poor and marginalized citizens.

This has important and far-reaching consequences. Above all, it means that despite the institutionalization of parliamentary democracy, the most important means for marginalized and politically restive populations to register their demands is through the organizational forms and vocabularies of extra-parliamentary protest. This provides a useful vantage point from which to understand the wide range of forms of popular mobilization – 'service delivery protests', 'xenophobic violence' and unprotected strikes – that characterize the South African political scene at the time of writing. The dynamics of these struggles are of course complex, varied and locally specific, and they need to be analysed in depth in their own right. We will confine ourselves to pointing out that in part, these struggles can be seen to embody an 'insurgent citizenship' whereby the traditions of popular struggle are invoked as means to make claims for resources within the post-Apartheid distributive regime (Von Holdt et al. 2011) – but that at the same time they can also be read as counter-hegemonic struggles contesting the legitimacy of the new social order as such (Hart 2013).

Conclusion

While the phenomena we describe here are distinctively South African, we believe they are of relevance elsewhere. The depopulated rural landscapes and impoverished shantytowns of South Africa may not in any simple sense prefigure the future awaiting peasants in South and South East Asia or East and Central Africa – but the dynamics of marginalization and displacement experienced in South Africa clearly do find echoes elsewhere, and the forms of governmentality and corporate strategies that arise in response to them may be adopted by others.

One important task is thus to develop a more differentiated approach to understanding the terrains created by these transitions. If concepts such as 'the precariat', 'fragmentation', 'bare life' and so on are to be criticized for a lack of differentiation, the same can be said for notions such as 'truncated agrarian transition'. It may be true that de-agrarianization, urbanization and proletarianization in South Africa, India and Indonesia do not follow the paths they did in Northern Europe and the US, but to say this is only to describe their trajectories negatively, in terms of what they are not. What is needed is a better sense of their *positive* features, of the differentiation in the kinds of social and economic formations that typify these 'truncated transitions', and some kind of schema or typology that allows an understanding of these differences.

By this we do not mean the Weberian comparison of ideal types. Rather, it requires the careful comparison of the dynamism of social and economic formations in different contexts. In this paper, we have concentrated our attention particularly on what we have called the terms of incorporation: the nature and implications of the institutions, practices and arrangements that shape the ways in which people are located within the larger social and economic formations of present-day South African capitalism. Our research started in the early 2000s as an exploration of what we have elsewhere called the 'vertical' (i.e. value chain/filiére) and 'horizontal' (social, political and institutional) linkages that locate people and activities within such formations (Bolwig et al. 2010). In this paper, we have abandoned this rather schematic mapping in favour of a focused analysis of three kinds of integrative social and economic formation: the organization of capitalist production and exchange in value chains and production networks, the differential inclusion of populations under modern biopolitical forms of governmentality, and the networks of sociality, reciprocal exchange and distributive labour formed by vulnerable and marginalized populations themselves. Though these do not constitute an exhaustive list of issues for investigation, they seem to us to central to an understanding of the terrain on which social and economic agency is formed.

They also offer a basis on which to specify both aspects of similarity and of distinctiveness and thus allow the beginnings of a differentiation among post-agrarian landscapes. For one thing, the high levels of concentration and vertical integration in the formal economy, coupled with the efficiency and spatial extensiveness of South African corporate and retail networks, set South Africa apart from what can be found in other parts of sub-Saharan Africa as well as South and Southeast Asia. Also distinctive is the centrality of the politics of democratic citizenship, the significance of political discourses that legitimize the aspirations and demands of the impoverished majority, the unchallenged political hegemony of the populist traditions invoked by the ruling party, and the enduring salience of racial identity as a marker of both inclusion and exclusion. These factors all mean that the politics of the South African distributional regime are particularly urgent, central and contested. In other contexts, where the informal economy is less penetrated by corporate capital, or where biopolitical regimes are differently constituted, the terms of incorporation will look very different – and so will the possibilities for socio-economic and political agency.

This creates an opportunity for interesting comparative work. There is clearly profit in setting the picture we have drawn here alongside some of the careful and detailed work that is already being done on the differential incorporation of marginalized people in biopolitical regimes in India (Gupta 2012) South East Asia (Li 2007, Ong 2007) and China (Ngai 2005), to mention just a few examples. In addition, the ongoing process of commercialization and corporate penetration that is already happening elsewhere in sub-Saharan Africa indicates that some of the dynamics we have described here may soon become features of other African economies (Hall 2012). A closer investigation of these issues is necessary if we are to take the analysis of post-agrarian societies beyond narratives that simply prophesy either triumph or disaster.

References

Agamben, G. 1998. *Homo Sacer: Sovereign power and bare life*. Palo Alto: Stanford University Press.

Aliber, M. and R. Hall. 2012. Support for smallholder farmers in South Africa: challenges of scale and strategy. *Development Southern Africa*, 29(4), 548–562.

Aliber, Michael, and Tim G. B. Hart. 2009. Should Subsistence Agriculture Be Supported as a Strategy to Address Rural Food Insecurity? *Agrekon* 48(4).

Andrew, M., A. Ainslie, and C. Shackleton. 2003. *Land use and livelihoods*. Evaluating Land and Agrarian Reform in South Africa, No. 8. Bellville: University of the Western Cape.

Barchiesi, F. 2011. *Precarious liberation: workers, the state, and contested social citizenship in post-Apartheid South Africa*. Albany: SUNY Press.

Bauman, Z. 2004. *Wasted lives: modernity and its outcasts*. Cambridge: Polity.

Bernstein, H. 1996. South Africa's agrarian question: extreme and exceptional? *Journal of Peasant Studies*, 23(2–3), 1–52.

Bernstein, H. 2002. Land reform: taking a long(er) view. *Journal of Agrarian Change*, 2(4), 433–463.

Bernstein, H. 2013. Commercial agriculture in South Africa since 1994: 'natural, simply capitalism'. *Journal of Agrarian Change*, 13(1), 23–46.

Black, A. 2010. *Tilting the playing field: labour absorbing growth and the role of industrial policy*. (No. 279). Cape Town: Centre for Social Science Research, University of Cape Town.

Bolwig, S., *et al.* 2010. Integrating poverty and environmental concerns into value chain analysis: a conceptual framework. *Development Policy Review*, 28(2), 173–194.

Bracking, S. 2003. The political economy of chronic poverty. IDPM: Working Paper 23. *SSRN eLibrary*. Available from: http://papers.ssrn.com/sol3/papers.cfm?abstract_id=1754446 [Accessed 30 October 2013].

Bracking, S. and L. Sachikonye. 2008. *Remittances, poverty reduction and informalisation in Zimbabwe, 2005-6: a political economy of dispossession*. Brookes World Poverty Institute Working Paper. Manchester: BWPI.

Bryceson, D.F. and V. Jamal. 1997. Farewell to farms : de-agrarianisation and employment in Africa. Research Series ; 10. Aldershot: Ashgate.

Business Day Live. 2013. South Africa's informal sector still growing. *Business Day Live*. Available from: http://www.bdlive.co.za/economy/2013/02/12/south-africa-s-informal-sector-still-growing [12 June 2013, Accessed 30 October 2013].

Cichello, P. 2005. Hindrances to Self Employment Activity: Evidence from the 2000 Khayelitsha/ Mitchell's Plain Survey. *CSSR Working Paper*, no. 131. Cape Town: Centre for Social Science Research.

Christopher, A.J. 1994. *The Atlas of Apartheid*. Routledge.

Collier, S.J. 2009. Topologies of power: foucault's analysis of political government beyond 'governmentality'. *Theory, Culture & Society*, 26(6), 78–108.

Collier, S.J. 2012. Neoliberalism as big Leviathan, or … ? A response to Wacquant and Hilgers. *Social Anthropology*, 20(2), 186–195.

Cousins, B. 2013. Smallholder irrigation schemes, agrarian reform and 'accumulation from above and from below' in South Africa. *Journal of Agrarian Change*, 13(1), 116–139.

Cousins, B. and A. Claassens, eds. 2008. *Land, power & custom: controversies generated by South Africa's communal land rights act*. Juta and Company Ltd.

Davis, M. 2004. Planet of slums: urban involution and the informal proletariat. *New Left Review*, 26, 5–34.

Department of Rural Development and Land Reform. 2013. *Speech by the minister of rural development and land reform, Nkwinti, G E (mp): 2013 Budget – policy speech 'building vibrant, equitable, and sustainable rural communities' 31 May 2013*. Cape Town: Department of Rural Development and Land Reform.

Department of Traditional Affairs. 2011. *Department of traditional affairs strategic direction: delivered by Prof. MC Nwaila*. Pretoria: Department of Traditional Affairs. Available from: http://www.dta.gov.za/index.php/speeches/deputy-general/75-department-of-traditional-affairs-strategic-direction-delivered-by-prof-mc-nwaila.html [Accessed 1 September 2013]

De Swardt, C., T. Puoane, M. Chopra, and A. du Toit. 2005. Urban poverty in Cape Town. *Environment and Urbanization*, 17(2), 101–111.

D'Haese, M. and G. Van Huylenbroeck. 2005. The rise of supermarkets and changing expenditure patterns of the poor rural households case study in the Transkei area, South Africa. *Food Policy*, 30, 97–113.

Dollar, D. and A. Kraay. 2001. *Growth is good for the poor*. Policy Research Working Paper Series 2587. Washington, DC: The World Bank. Available from: http://elibrary.worldbank.org/doi/book/10.1596/1813-9450-2587 [Accessed 30 October 2013].

Downey, A. 2009. Zones of indistinction: Giorgio Agamben's 'bare life' and the politics of aesthetics. *Third Text*, 23(2), 109–125.

Du Toit, A. 2004. 'Social exclusion' discourse and chronic poverty: a South African case study. *Development and Change*, 35(5), 987–1010.

Du Toit, A. 2009. Adverse incorporation and agrarian policy in South Africa or, how not to connect the rural poor to growth. Available from: http://repository.uwc.ac.za/xmlui/handle/10566/65 [Accessed 30 October 2013].

Du Toit, A. 2011. *The government of poverty and the arts of survival: de-agrarianization, 'surplus population' and social policy in South Africa*. The Hague: Institute for Social Studies.

Du Toit, A. 2012a. The government of poverty and the arts of survival: responses to structural poverty and marginality in South Africa. *Paper prepared for Towards Carnegie 3: Conference on Strategies to Overcome Poverty & Inequality University of Cape Town, 3–7 September*.

Du Toit, A. 2012b. *The trouble with poverty: reflections on South Africa's post-Apartheid anti-poverty consensus*. (PLAAS Working Paper No. 22). Bellville: Institute for Poverty, Land and Agrarian Studies, University of the Western Cape.

Du Toit, A. and D. Neves. 2006. *Vulnerability and social protection at the margins of the formal economy case studies from Khayelitsha and the Eastern Cape*. Bellville: Institute for Poverty, Land and Agrarian Studies, University of the Western Cape.

Du Toit, A. and D. Neves. 2007. In search of South Africa's second economy. *Africanus*, 37(2), 145–174.

Du Toit, A. and D. Neves. 2008a. *Informal social protection in post-Apartheid Migrant networks: vulnerability, social networks and reciprocal exchange in the Eastern and Western Cape, South Africa*. (No. 74). Manchester: Brooks World Poverty Institute. Available from: http://ideas.repec.org/p/bwp/bwppap/7409.html [Accessed 30 October 2013]

Du Toit, A. and D. Neves. 2008b. *Trading on a grant: integrating formal and informal social protection in post-Apartheid Migrant networks*. (No. 75). Manchester: Brooks World Poverty Institute.

Du Toit, A., A. Skuse, and T. Cousins. 2007. The political economy of social capital: chronic poverty, remoteness and gender in the Rural Eastern Cape. *Social Identities*, 13(4), 521–540.

Engels, F. 1943. *The condition of the working-class in England in 1844*. London: G. Allen & Unwin, Limited.

Ferguson, J. 2007a. Distributive labor and survivalist improvisation: productionist thinking and the misrecognition of the urban poor. SANPAD Conference, The Poverty Challenge 2007: Poverty and Poverty Reduction in South Africa, India, and Brazil. 26-29 June, Durban.

Ferguson, J. 2007b. Formalities of poverty: thinking about social assistance in neoliberal South Africa. *Africa Studies Review*, 50(2), 1–32.

Ferguson, J. 2013. How to do things with land: a distributive perspective on rural livelihoods in Southern Africa. *Journal of Agrarian Change*, 13(1), 166–174.

Fine, B. and Z. Rustomjee. 1996. *The political economy of South Africa: from minerals-energy complex to industrialisation*. London: C. Hurst & Co. Publishers.

Foucault, M. 2010. *The birth of biopolitics: lectures at the Collège de France, 1978-1979*. New York: Picador.

Greenberg, S. 2010. *Contesting the food system in South Africa: issues and opportunities*. (No. 42). Bellville: Institute for Poverty, Land and Agrarian Studies, University of the Western Cape.

Gregory, T. 2011. The rise of the productive non-place the contemporary office as a state of exception. *Space and Culture*, 14(3), 244–258.

Gupta, A. 2012. *Red tape: bureaucracy, structural violence, and poverty in India*. Durham: Duke University Press.

Hall, R. 2012. The next great trek? South African commercial farmers move north. *Journal of Peasant Studies*, 39(3–4), 823–843.

Hardt, M. and A. Negri. 2004. *Multitude: war and democracy in the age of empire*. New York: Penguin.

Harrison, P., A. Todes, and V. Watson. 2008. *Planning and transformation: learning from the post-Apartheid experience*. New York: Routledge.

Harriss-White, B. 2012. Capitalism and the common man: peasants and petty production in Africa and South Asia. *Agrarian South: Journal of Political Economy*, 1(2), 109–160.

Hart, G. 2013. *Rethinking the South African crisis: nationalism, populism, hegemony*. Scottsville: University of Kwazulu Natal Press.

Havnevik, K., D. Bryceson, P. Matondi, and A. Beyene. 2007. *African agriculture and the world bank: development or impoverishmen ?* (p. 1/75). Nordiska Afrikainstitutet, Uppsala 2007 1.

Igumbor, E.U., *et al.* 2012. 'big food,' the consumer food environment, health, and the policy response in South Africa. *PLoS Med*, 9(7), 1–7.

ILO. 2010. *Global employment trends january 2010*. Geneva: International Labour Organization.

ILO. 2013. *Global employment trends 2013 - recovering from a second jobs dip*. Geneva: International Labour Organization.

Jacobs, P. 2008. Market development and smallholder farmers: a selective literature survey (Background paper for the 'second economy project') October 2008.

Lahiff, E. and B. Cousins. 2005. Smallholder agriculture and land reform in South Africa. *IDS Bulletin*, 36(2), 127–131.

Leibbrandt, M., *et al.* 2010. *Trends in South African income distribution and poverty since the fall of Apartheid*. (No. 101). Paris: Organization for Economic Co-operation and Development.

Lerche, J. 2010. From 'rural labour' to 'classes of labour': class fragmentation, caste and class struggle at the bottom of the Indian labour hierarchy. In: B. Harriss-White and J. Heyer, eds.

The comparative political economy of development. Africa and South Asia. London: Routledge, pp. 66–87.

Li, T.M. 2007. *The will to improve: governmentality, development, and the practice of politics.* Durham: Duke University Press Books.

Li, T.M. 2009a. Exit from agriculture: a step forward or a step backward for the rural poor?. *Journal of Peasant Studies*, 36(3), 629–636.

Li, T.M. 2009b. To make live or let die? Rural dispossession and the protection of surplus populations. *Antipode*, 41, 66–93.

Li, T.M. 2011. Centering labor in the land grab debate. *Journal of Peasant Studies*, 38(2), 281–298.

Louw, E. 1997. Nationalism, modernity and postmodernity: comparing the South African and Australian experiences. *Politikon*, 24(1), 76–105.

Lund, F. and C. Skinner. 2004. Integrating the informal economy in urban planning and governance: A case study of the process of policy development in Durban, South Africa. *International Development Planning Review*, 26(4), 431–456.

Mabin, A. 1990. Limits of urban transition models in understanding South African urbanisation. *Development Southern Africa : Quarterly Journal*, 7(3), 311–322.

Mamdani, M. 1996. *Citizen and subject: contemporary Africa and the legacy of late colonialism.* Princeton: Princeton University Press.

Marais, H. 2011. *South Africa pushed to the limit: the political economy of change.* Claremont: UCT Press.

Marshall, T.H. 2009. Citizenship and social class. In: J. Manza and M. Sauder, eds. *Inequality and society.* New York: W. W. Norton and Co, pp. 148–154.

Mather, C. 2005. The growth challenges of small and medium enterprises (SMEs) in SouthAfrica's food processing complex. *Development Southern Africa*, 22(5), 607–622.

Mattes, R.B. 2002. South Africa: democracy without the people? *Journal of Democracy*, 13(1), 22–36.

May, J. 2000. The structure and composition of rural poverty and livelihoods in South Africa. In: B. Cousins, ed. *At the crossroads: land and agrarian reform in South Africa into the 21st Century.* Bellville: Programme for Land and Agrarian Studies, pp. 21–34.

Mbembe, A. 2003. Necropolitics. *Public Culture*, 15(1), 11–40.

McCord, A. 2005. A critical evaluation of training within the South African national public works programme. *Journal of Vocational Education and Training*, 57(4), 563–588.

Mezzadra, S. and B. Neilson. 2012. Between inclusion and exclusion: on the topology of global space and borders. *Theory, Culture & Society*, 29(4–5), 58–75.

Mosoetsa, S. and M. Williams. 2012. *Labour in the global South: challenges and alternatives for workers.* Geneva: International Labour Organization.

Nattrass, N. 2001. High productivity now: a critical review of South Africa's growth strategy. *Transformation: Critical Perspectives on Southern Africa*, 45, 1–24.

Neilson, B. and N. Rossiter. 2008. Precarity as a political concept, or, fordism as exception. *Theory, Culture & Society*, 25(7–8), 51–72.

Neves, D., *et al.* 2009. *The use and effectiveness of social grants in South Africa.* Cape Town: Finmark Trust.

Neves, D. and A. du Toit. 2012a. Money and sociality in South Africa's informal economy. *Africa*, 82 (Special Issue 01), 131–149.

Neves, D. and A. du Toit. 2012b. Informal agro-food chains: food on the margins of South Africa's formal economy. Presented at the Towards Carnegie 3: Conference on Strategies to Overcome Poverty & Inequality, Cape Town.

Neves, D. and A. du Toit. 2013. Rural livelihoods in South Africa: complexity, vulnerability and differentiation. *Journal of Agrarian Change*, 13(1), 93–115.

Ngai, P. 2005. *Made in china: women factory workers in a global workplace.* London: Duke University Press.

Ngwane, Z. 2003. 'Christmas time' and the struggles for the household in the countryside: rethinking the cultural geography of Migrant labour in South Africa. *Journal of Southern African Studies*, 29 (3), 681–699.

Ong, A. 2006. *Neoliberalism as exception: mutations in citizenship and sovereignty.* Durham: Duke University Press.

Ong, A. 2007. Neoliberalism as a mobile technology. *Transactions of the Institute of British Geographers*, 32(1), 3–8.

Özler, B. and J.G. Hoogeveen. 2005. Poverty and inequality in post-Apartheid South Africa not separate, not equal : poverty and inequality in post-Apartheid South Africa, (739).

Philip, K. 2010. Inequality and economic marginalisation : how the structure of the economy impacts on opportunities on the margins. *Law, Democracy and Development*, 14, 1–28.

Povinelli, E.A. 2011. *Economies of abandonment: social belonging and endurance in late liberalism.* London: Duke University Press Books.

Ross, F. 2003. Dependents and dependence: a case study of housing and heuristics in an informal settlement in the Western Cape. *Social Dynamics*, 29(2), 132–152.

Russel, M. 2004. *Understanding black households in southern Africa: the African kinship and western nuclear family systems.* (No. 67). Cape Town: Centre for Social Science Research, University of Cape Town.

Sagner, A. and R.Z. Mtati. 1999. Politics of pension sharing in urban South Africa. *Ageing & Society*, 19(4), 393–416.

Sassen, S. 2008. *Territory, authority, rights: from medieval to global assemblages.* Princeton: Princeton University Press.

Seekings, J. 2001. The uneven development of quantitative social science in South Africa. *Social Dynamics*, 27(1), 1–36.

Seekings, J. and N. Nattrass. 2005. *Class, race, and inequality in South Africa.* Yale University Press.

Skuse, A. and T. Cousins. 2007. Managing distance: rural poverty and the promise of communication in post-Apartheid South Africa. *Journal of Asian and African Studies*, 42(2), 185–207.

Skuse, A. and T. Cousins. 2008. Getting connected: the social dynamics of urban telecommunications access and use in Khayelitsha, Cape Town. *New Media & Society*, 10(1), 9–26.

Smith, D.M. 1992. *The Apartheid city and beyond: urbanization and social change in South Africa.* Johannesburg: Routledge.

Standing, G. 2011. *The precariat: the new dangerous class.* London: Bloomsbury Academic.

Statistics South Africa. 2012. *Poverty profile of South Africa: application of the poverty lines on the LCS 2008/2009 /.* Pretoria.

Statistics South Africa. 2013. *Quarterly labour force survey: quarter 1 (January to March), 2013 - press statement (press statement).* Pretoria: Statistics South Africa.

Van der Berg, S. 2013. Adcorp's employment and unemployment figures are not taken seriously by researchers – yet they can do much harm. *Econ3×3*. Available from: http://www.econ3x3.org/article/adcorp%E2%80%99s-employment-and-unemployment-figures-are-not-taken-seriously-researchers-%E2%80%93-yet-they [Accessed 30 October 2013].

Van der Berg, S., et al. 2006. *Trends in poverty and inequality since the political transition.* (No. 06/104). Cape Town: Development Policy Research Unit.

Von Holdt, K., M. Langa, S. Molapo, N. Mogapi, K. Ngubeni, J. Dlamini, and A. Kirsten. 2011. *The smoke that calls: insurgent citizenship, collective violence and the struggle for a place in the new South Africa.* Johannesburg: Centre for the Study of Violence and Reconciliation.

Walker, C. 2008. *Landmarked: land claims and land restitution in South Africa.* Cape Town: Jacana Media.

Westaway, A. 2010. Rural poverty in the Eastern Cape Province : legacy of Apartheid or consequence of contemporary segregationism ? Paper presented at the conference on Inequality and structural poverty in South Africa: Towards inclusive growth and development', Johannesburg, 20 September 2010.

World Bank. 2012. *World Development Report 2013: jobs.* Washington, DC: The World Bank.

Andries du Toit is the Director of the Institute for Poverty, Land and Agrarian Studies (PLAAS). He has a PhD in Comparative Studies from the University of Essex. His work focuses on developing a critical understanding of the politics of knowledge production in the government of poverty and marginal livelihoods. University of the Western Cape, Institute for Poverty, Land and Agrarian Studies, Private Bag X 17, Bellville, 7535 South Africa.

David Neves joined PLAAS in 2006. His research is concerned with the strategies of people coping with the consequences of structural poverty and unemployment in urban and rural contexts.

Rural-urban migration in Vietnam and China: gendered householding, production of space and the state

Minh T.N. Nguyen and Catherine Locke

The transition from state socialism to market socialism in Vietnam and China has been characterized by unprecedented rural-urban migration. We argue that this migration is integral rather than incidental to the gendered reproduction of state and society. A review of the emerging literature on trans-local householding explores the process whereby the reflexive engagement of the state and the household remakes rural-urban differentiation in ways that are deeply gendered and classed. As such, state regulation and control of migrants are part of a process of reconfiguring state-society relations in which the production of space and the symbolic valuation of ruralness and urbanness have become a central trope.

Introduction

In the latter half of the last century, both Vietnam and China underwent two major social transformations that have fundamentally reconfigured the relationship between rural households and the state (Kerkvliet and Selden 1998; Perry and Selden 2010; Whyte 2010). The first was the socialist revolution followed by successive policies of land reform and collectivization between 1945 and the 1970s. The state initially redistributed agricultural land to peasant households and later reorganized land and labor for large-scale production through agricultural collectives. A socialist welfare system was established, which went hand in hand with state interventions in family life. Equity was promoted in both productive and reproductive spheres through policies targeting gender discrimination, gerontocratic control and elite privilege (Phạm 1998; Hershatter 2007). The second transformation, which started in the late 1970s in China and the late 1980s in Vietnam, saw a redeployment of land to peasant households together with reduced state control of labor, pricing and accumulation, creating a market economy managed and directed by the state, namely market socialism. In both countries, the state has since exercised less intrusion into the daily life of ordinary people and their mobility, while reinstating the household as the basic unit of welfare provision. While the household economy existed only at the margin of the previous economy, the household now regained its primary role in production and reproduction.

Liberalization and market integration have brought about enormous economic and social changes in both countries. Gender differentiation has sharpened as welfare provisions designed to support equity have been dismantled and responsibility for social

---

*We would like to thank the three anonymous reviewers for their extremely helpful comments.

provisioning has been reallocated to households. There has been a dramatic decline in absolute poverty while the high growth rate has been sustained over a long period, despite fluctuations and environmental problems. Wealth accumulation and urbanization have taken place at a faster rate than ever before in history; in China, the number of million-aires has reached hundreds of thousands and China's mega urban centers are emerging as global cities (Hsing 2010; Jacka and Sargeson 2011). The second transformation has been characterized by much greater population mobility, internally and internationally. In the last three decades, hundreds of millions of Chinese and millions of Vietnamese have been migrating from rural areas to urban centers as hundreds of thousands others have gone overseas (World Bank 2008a)[1] as contract and illegal workers (Chu 2010; Dang et al. 2010; Nyíri 2010). A large number of others have emigrated to study and for leisure (Dang et al. 2010; Nyíri 2010; Fong 2011), becoming part of global circuits of labor, ideas and lifestyle.

Our focus is on rural-urban migration, which we argue is integral rather than incidental to market socialism, and its role in the ongoing reconfiguration of state-citizen relationships in Vietnam and China. Until recently, studies of rural-urban migration have focused mostly on rural-urban inequalities, gendered experiences of migrants in the city, and the life of 'left-behind' family members. These phenomena were often treated in isolation from each other rather than as components of a long-term and dynamic process of social change in which the nature of rural households is redefined. A false dichotomy between migrants and the left-behind obscures the processes in which households are iteratively organizing and deploying labor for multiple purposes in response to shifting systemic con-ditions while recreating gender and generational relations.

In contrast, an emerging literature on Vietnam and China examines the ways in which households straddle rural and urban areas, with generations of men and women simul-taneously undertaking care and livelihood activities from different locations. We review this literature in order to link rural-urban migration with changing state-citizen relations under market socialism. We argue that rural households' multi-locational functioning interacts with state visions of development, control mechanisms and discursive categories in ways that marginalize migrants but simultaneously render migration increasingly imperative for market socialism and for household reproduction. Our analysis contributes to the existing literature on the ways in which migration is integral to the restructuring of state-society relations. Whilst one strand of this work focuses on international migration (Ong 1999; Silvey 2004; Porio 2007; Chan and Tran 2011; Pearson 2012), we develop insights about rural-urban migration that complement those for other places, such as Latin America and Africa (Hojman 1989; Lawson 1998; Francis 2002). Vietnam and China represent particular cases as transitional economies in which the state, despite liberal-inspired reforms, has been seeking to shape economic development, rural-urban migration and household relations to a much greater degree than other politi-cal regimes.

In the next section, we introduce the theoretical lens for the review which combines two concepts, namely Douglass's (2006) concept of *householding* and Lefebvre's (1991) theory of the *production of space*. We then proceed to analyze the changing forms of rural-urban

---

[1]According to the World Bank online database, International migrants from China numbered 685,775, and from Vietnam 69,307, as of 2008. The data were based on national population censuses and the number is likely to be much higher when taking into account illegal and cross-border migrants.

inequalities rooted in the rural-urban relationship of the previous eras. While the forms have altered, the relationship remains unequal, with significant bearings on the social status of rural migrants. We argue that it is from rural migrants' reflexive engagements with this institutional context that translocal householding emerges, and we will examine how this produces a life 'in between' that is deeply gendered. Finally, we explore how this life 'at the margin' is producing ambivalent spaces for householding and for negotiating citizenship across different locations. This differentiated process invests rural(ness) and urban(ness) with meanings about gender and class.[2]

Householding and the production of space

Douglass (2006; 2012) develops the concept 'global householding' to analyze the social reproductive dynamics that are emerging in Pacific Asia in response to rapid changes in the world order. It refers to the 'ways in which the processes of forming and maintaining household through time' (Douglass 2012, 4) evolve in the face of mobility, population ageing and neo-liberalism. According to Douglass, 'all the key dimensions of the life of the household, including marriage/partnership, bearing, raising and educating children, managing daily life, earning income and caring for elders and non-working members' (2012, 4) have to be understood as projects sustained across multiple locations and through varied social formations. He argues that neo-liberal policy regimes promote ideo-logical functions of households that sustain gender, class and racial differences for the sake of capitalism. Meanwhile, '"transitional economies" such as Vietnam and China [...] maintain the authority of socialism but are dismantling public support for household repro-duction while dispossessing rural and lower income households of land and other assets' (2012, 10). His conceptualization thus also enables analytical attention to the householding processes that are occurring through rural-urban migration.

In post-reform Vietnam and China, the rural household has been reproducing its labor and gendered social relations across and between places and spaces (Oakes and Schein 2005; Jacka 2012; Nguyen forthcoming). Unlike the 'global household' that Douglass ana-lyses, their spatial practices are heavily circumscribed by the sovereign state and firmly embedded within specific historical and cultural contexts. Lefebvre's notion of 'production of space' (1991) is therefore apt to capture this interaction. According to Lefebvre, space is produced through a dialectical interplay between 'the perceived, the conceived, and the directly experienced' (1991, 246). These constitute *spatial practices* by human beings, lived spaces (*spaces of representation*), and the ways in which they are conceived by state actors and the spatial imaginary of the time (*the representation of space*). Space is 'politically instrumental in that it facilitates the control of society, while at the same time being a means of production by virtue of the way it is developed' (349). It is central to the 'reproduction of the social relations of production', including the hierarchical ordering of locations and class structures (349). The urban-rural trope therefore must be analyzed as a complex social construction that is fundamental to capitalism's relations of domination.

In Vietnam and China, the shifting social construction of spatial categories has been intrinsic to the changeover from state socialism to market socialism. This process, we

---

[2]Following Prota and Beresford (2012), we see class structure as a historically contingent set of relationships amongst producers to the means of production and with the ruling classes that sustains the unequal extraction of surplus from producers.

show, is political in the ways in which it reconfigures gender and class relations via means of reproducing 'the rural' and 'the urban' for market socialism. Combining Lefebvre's conception of space with Douglass's notion of *householding* enables us to analyze the reflexive processes in which rural-urban migrants' strategies for sustaining the household interplay with state governance of market socialism. In particular, it allows us to explore how they are inscribed in the production of space and the ongoing restructuring of rural and urban identities.

Changing social relations of production

Although state socialism was aimed at creating a classless society, we argue that it in fact elaborated pre-existing differences between urban and rural areas by privileging the party membership, its leadership and urban workers over the rural peasantry. Rather than creating entirely new class relations, market socialism in turn remade the social relations of state socialism in ways that reify rural-urban differentiation. This section elucidates this argument by tracing the changing social relations of production in China and Vietnam over time.

The size of the economy aside, there are a number of differences between China's and Vietnam's experiences with state socialism. First of all, collectivization was more far-reaching in China in terms of scale, duration and the actual outcomes for productivity and rural social organization. Despite the Great Leap Forward policy with which collective production was brought to the extremes, Chinese cooperatives, which were larger and more industrially oriented, provided the conditions for rural entrepreneurship to evolve (Kerkvliet and Selden 1998; Ye *et al.* 2010). The cooperatives in northern Vietnam were, in contrast, more fragmented and had to be maintained alongside fighting the American war in a divided country. The household in Vietnam had somewhat greater autonomy in agricultural production than its counterpart in China, where household labor was managed and controlled to a greater degree.[3] Secondly, state interference in social reproduction was stronger in China than in Vietnam, where the Vietnamese state, especially at the local level, was less able and less willing to do so (Kerkvliet *et al.* 1998; Kerkvliet and Selden 1998; Kerkvliet and Marr 2004; Koh 2004; Kerkvliet 2005). State socialism sought to reshape social reproduction through propaganda and incentives, social surveillance, social criticism, re-education, violence and imprisonment, but the will and reach of the Chinese state were far greater than those of the Vietnamese state. Key elements of social reproduction such as marriage and child bearing, child rearing and socialization, and peasant/worker entitlements were more strictly regulated in China than in Vietnam. An example is the draconian application of the one-child policy in China, compared with the two-child policy in Vietnam that was pursued mainly through discriminatory entitlements and social pressure.

These differences influenced the ways in which regional development took place following the reform, which began a decade earlier in China than in Vietnam, where actual restructuring only started in the late 1980s. The rural entrepreneurship emerging from the former Chinese cooperatives made it possible for the government to implement a policy to urbanize rural areas, the Township and Village Enterprises program or TVEs (O'Connor 1998; Perry and Selden 2010), together with developing core urban

---

[3]While, in both countries, households were officially allowed to farm five percent of the local agricultural land independently of the cooperative, the proportion of land privately worked by Vietnamese households ranged between 7–20 percent (Kerkvliet and Selden 1998).

Table 1. Urban population of Vietnam and China (percent of total population) (World Bank 2010).

Year	1980	1990	2000	2010	2011
China	19	26	36	49	51
Vietnam	19	20	24	30	31

The data are based on national statistics, which do not capture floating migrants. The populations of Vietnam and China, respectively, stood at about 78 million and 1.263 billion in 2000, and 88 million and 1.338 billion in 2010.

metropolitan areas and opening up its coast to global capital. This dual-track urbanization, alongside policies to control migration, was aimed at limiting urban growth in existing urban areas by simultaneously attracting the rural populations to smaller towns and diverting migrants to the periphery of major cities where they could be accommodated by local farmers (McGee 2009; Hsing 2010).

In contrast, Vietnam's urban development in the early reform period was less planned by the state and was instead more driven by the 'popular sector' comprising individuals and households taking advantage of the ambiguity in urban property rights and support by local governments (Leaf 1999; Leaf 2002; Koh 2004; Koh 2006). Whereas in China, rural industrialization took off, especially in and around coastal cities such as Guangdong, Beijing, Jiangsu, Shanghai, Zhejiang and Fujian, rural industries in Vietnam remained fragmented (O'Connor 1998), with industrialization and urbanization mainly concentrated in the Hanoi and Ho Chi Minh city regions. In Vietnam, there was a marked increase of the urban informal sector, including street vending and small-scale household businesses, whereas in China, rural townships, especially along the east coast, were able to promote greater-scale industrialization with rapid population and housing growth (O'Connor 1998; Leaf 1999; Leaf 2002; McGee 2009). Despite these initial divergences, both states have since the 1990s promoted urbanization and global integration, encouraging foreign investment and adopting a modernizing vision of development (Zhang 2001; McGee 2009; Hsing 2010; Harms 2012). The urban populations in both countries have increased significantly in the last three decades and continue to rise (Table 1), albeit at a faster pace in China.

Even with the rural industrialization program that draws more than 80 million rural laborers, China has been witnessing the rural-urban movement of between 100 and 200 million people at any time, making up 80 percent of the construction and 50 percent of the service workforce (Nielsen and Smyth 2008; van der Ploeg and Jingzhong 2010; Whyte 2010). On a smaller scale but accelerating, internal migrants in Vietnam numbered more than 6 million as of 2009, excluding the potentially huge number of unregistered migrants that were not counted by the census (United Nations 2010; GSO 2011). The great population mobility is foregrounded by an enduring rural-urban disparity. In both countries, urban income remains double that of rural areas (Đặng 2008; Sicular, Ximing et al. 2010) and urban populations have visibly better access to infrastructure, social services and social protection (Nielsen and Smyth 2008; Whyte 2010; Le et al. 2011). According to a conventional narrative, social inequality was deeply entrenched in pre-revolutionary Vietnam and China; then it was lessened during state socialism, only to reemerge under the market economy. While socialist policies were aimed at erasing the difference between social groups and regions, the rural-urban division was in fact a product of the socialist modernizing project (Fforde and De Vylder 1996; Fesselmeyer and Le 2010; Whyte 2010). This 'hierarchical ordering' of geographical locations was indeed central to the 'reproduction of the social relations of production' (Lefebvre 1991) under state socialism and remains essential to the functioning of market socialism.

Urban and industrial development was (and still is) prioritized over agriculture and the rural sector (Đặng 2008; Davis and Feng 2009; United Nations 2010; Whyte 2010) in both countries. The state used agricultural production to finance urban industries through price curbs on agricultural products and inflated prices of industrial commodities, including inputs for agriculture. While a state system of social services was set up in the countryside to provide education and healthcare, the rural populations never had access to the social securities that urban state workers and cadres were guaranteed, including provision of food, jobs, education and welfare (Kuruvilla *et al.* 2011). Finally, the household registration system, *hukou* in Chinese and *hộ khẩu* in Vietnamese, which was started in China in the early 1950s and later imported to Vietnam, practically prevented peasants from migrating elsewhere to seek alternative employment. The system ties rural and urban citizens to their household registration in a particular location, which is connected to land entitlements and access to social services (see Hardy 2001; Wang 2005 and Le *et al.* 2011 for background information on the household registration systems in China and Vietnam).

Summing up, the rural-urban disparity had existed before the revolution in both countries and to eradicate it was one of the goals of socialism. However, policies in practice made second-class citizens of the rural population, despite the socialist state's emphasis on the peasants' major role in national construction in both countries and, for Vietnam, in national defense. This has implications for understanding the class structure in both countries today. Rather than constituting a legacy of the previous 'social relations of production' (Lefebvre 1991), the post-reform rural-urban disparity has evolved in complex ways out of institutional processes in which old and new governing techniques interact with the market.

The state and the regulation of rural migrants

In both countries, the household registration (*hukou/hộ khẩu*) is no longer an instrument for the state to prevent mobility and allocate socialist provision. As an institutional boundary and commodity, it now facilitates a dualistic economy in which rural migrants are excluded from urban citizenship at different levels. While allowing for easier acquisition than before, the current household registration acts as a mechanism to control resources, limit social rights and expectation, and reinforce class and subject formation (Wang 2005; Lê and Khuất 2008; Luong 2009; Nyíri 2010; Zhao and Howden-Chapman 2010; Le *et al.* 2011). It has become an expensive commodity that can be bought, either directly from the state in highly sought-after urban centers of China (Wang 2005), or through bribes and other means (Hardy 2001; Hardy 2003). While rural people are no longer prevented from migrating, the opportunity to legally settle in the city is selectively narrow and accessible only to the most competitive migrants who are better educated, better connected or better off economically. As well, China has started to experiment with harmonizing rural and urban *hukou* in a small number of wealthy cities such as Chengdu and Beijing as a basis for reforming the local social security system. This potentially generates new spatial politics between insiders and outsiders of richer regions that exclude both rural and urban migrants from poorer areas (Shi 2012).

If the household registration previously kept rural people immobile in the service of socialist construction, it now makes them mobile for the sake of urbanization and capitalist industrialization.[4] Their 'floating' is perpetuated because they cannot easily settle in the city

---

[4]It should be noted that *hộ khẩu* rules have never been as strictly observed in Vietnam as in China (Hardy 2001; 2003).

legally, while remaining in rural areas has become a template for failure, especially for young people (Ngai and Lu Huilin 2010; Fang 2011). Their exclusion from urban social citizenship, including access to jobs and urban social services (Nielsen and Smyth 2008; Davis and Feng 2009; Le *et al.* 2011) practically separates the labor of migrant workers from its social reproduction, which must take place in their home place. This creates a low-cost and flexible labor force for urban services and industries. With the household registration, rural laborers have thus been made instrumental for the state's successive projects of socialist construction and marketization through their status as second-class citizens in both periods.

Three decades after the reform, poverty remains predominant in certain rural areas, especially in western and central China (Davis and Feng 2009; Whyte 2010) and the central coast and mountainous regions of Vietnam (Taylor 2004; World Bank 2006; Badiani *et al.* 2012). Nevertheless, the rural-urban division has been reconfigured in both countries; rural poverty has declined whereas urban poverty has become more entrenched (Badiani *et al.* 2012; Cho 2013).[5] In both countries, urban workers have borne the brunt of lay-offs and urban residents are increasingly represented amongst the urban poor while 'industrial jobs, far from being a secure way out of poverty, remain unstable and underpaid' (Porta and Beresford 2012, 78). Some rural areas have been able to capitalize on their geo-political advantage for development, especially those along the east coast of China and the Southeast of Vietnam. As well, not all urban citizens are economically better off than rural citizens or migrants, a significant proportion of whom have become upwardly mobile. In China, 11 percent of the wealthiest income decile live in rural areas (Davis and Feng 2009, 10). In Vietnam, the latest census indicates that about two thirds of registered rural-urban migrants have 'high living standards', greater than the proportion of urban residents with the same living standards (GSO 2011). The rural-urban distinction has also been blurred by the incorporation of rural villages into urban centers alongside the citification of rural villages (Guldin 2001), and the ways in which diverse groups of migrants have become part of the urban society, with or without *hukou/hộ khẩu* (Chen *et al.* 2001; Zhang 2001; Koh 2006; Anh *et al.* 2012; Zavoretti 2012).

There has also been a shift in state policies in recent years towards increased investment and social transfers to rural areas, such as grain subsidy in China or old-age allowance in both countries (World Bank 2008b; Jacka and Sargeson 2011), coupled with the removal of agricultural tax and levies. China, followed by Vietnam, has instituted a program for Building the New Countryside, with plans for land consolidation, technological investment, and improvement of rural infrastructure and services, partly in response to rural discontent from land dispossession (Hsing 2010; Gillespie 2013).[6] The actual outcomes of these programs remain unclear (Nyíri 2010; Whyte 2010), while rural people continue to be on the move and their movements remain controlled by state instruments[7] such as the household registration.

---

[5]A large part of the urban poor are migrants (c.f. Liu and Wu 2006).
[6]About 50–60 million peasants have been dispossessed of their land for industrialization and urbanization (Hsing 2010). In Vietnam, hundreds of thousands farmers have also lost their agricultural livelihoods on the same account.
[7]According to Goldstein (2006), the average single migrant needs four types of identification to work legally in Beijing: a national identity card, a temporary residence permit (issued annually at a cost), a work permit (also with yearly cost) and a health card issued by an appointed hospital. In Vietnam, the requirements are limited to the temporary residence card issued by the local police.

Not only is the *hukou/hộ khẩu* an administrative device, it is also a powerful metaphor for the discursive division between rural and urban people, a symbolic barrier that migrants have to cross to become desirable citizens. The second-class citizenship of the rural populations is indeed a product of the discursive production of categories and knowledge actively promoted by the state (Zhang 2001; Yan 2003; 2006, 2008; Harms 2011; 2012). During state socialism, the peasants were constructed both as backward and ignorant and as powerful and revolutionary, both as examples of the hard-working socialist laborers and as in need of reform (Rato 2004; Jacka 2006). In Vietnam, the mobilization of the rural populations for the American war also relied on the construction of the heroic peasant. This construction has now has given way to an image of the uncivilized peasant whose presence pollutes the urban space (Nguyen 2012; Turner and Schoenberger 2012). In China, peasants are considered 'lacking in *suzhi*' while in Vietnam they are seen as 'having a low *dân trí*'. These keywords refer to the general quality of the population, encompassing professional skills, educational level, knowledge of law and appropriate cultural and social practices.

In China there is extensive literature on the *suzhi* discourse, which reveals the emerging ways in which people construct their class subjectivities in line with the state's governing goals (Kipnis 2007; Yan 2008; Jacka 2009; Sun 2009a; 2009b). According to Andrew Kipnis (2007, 393):

> Since the early 1980s, government workers and analysts have increasingly used notions of *suzhi* to argue for all manner of policy. Any sort of 'development' project can be described in the language of raising the population's quality. Chinese educational reformers, intrigued by English-language educational theories that go under the name of 'competence education', translated this term as '*suzhi* education', and managed to get *suzhi* education inscribed as a guiding national policy for the twenty-first century (Kipnis 2006). Human resource managers in both the public and private sectors justify recruitment and salary decisions in terms of *suzhi*. Rural cadres justify their own leadership positions in terms of their *suzhi* being higher than that of the peasants around them, and, of course, urbanites discriminate against rural migrants for their lack of *suzhi*.

In Vietnam, the term '*dân trí*' or '*trình độ dân trí*' has not reached the level of usage as the term *suzhi* in China. Yet in both countries, state discourse and practices related to population quality are closely linked to a form of governmentality that emphasizes the responsibility of the individual for self-development and self-governance. Similar to China, Vietnamese 'state and non-state actors have over the past decade become increasingly interested in the projects of self-cultivation and value creation that resonate both with the needs and anxieties of the market place and with continuous socialist genealogies' (Schwenkel and Leshkowich 2012, 386). According to Zhang and Ong (2008, 11), post-reform governance in China is characterized by the coexistence of self-interest promotion and socialist control, of neo-liberal logics and national sovereignty, which they term 'socialism from afar', where 'micro-freedoms coexist with illimitable political power'. In Vietnam, a parallel form of state governance is demonstrated by Nguyen-Vo (2008) in her political economic analysis of commercial sex, and contributors to the special edition on neo-liberalism by Schwenkel and Leshkowich (2012). These studies indicate that the state has increasingly recast its interventions in social relations and citizen's life through means of technocratic guidance and scientific expertise.

In this late-socialist governing orientation, *suzhi/dân trí* is part of a biopolitics that differentiates between good and bad subjects, depending on people's ability to meet the demands of the market and their loyalty to the state. Rural migrants are cast as undesirable citizens, lacking in quality and potentially rebellious, who must reform themselves under guidance of state or market agents while providing for their own needs. Migrants' low *suzhi/dân trí* is contrasted with the higher urban quality of civility, law-abidingness and sophisticated consumption, which urban residents are able to self-cultivate with private resources (Jacka 2009; Sun 2009a; 2009b; Schwenkel and Leshkowich 2012). The assertion of this hierarchical ordering of personhood alongside the ordering of geographical locations is again not new. The peasants, deemed backward and poor, had been the target of civilizing (*wenming/văn minh*) missions by colonial governments and nationalist movements in the two countries before socialism (Gourou 1955; Han 2005). Underlined by 'a fundamental inequality between the civilizing center and the peripheral peoples on which it acts' (Friedman 2004, 714), the discourse of civility nowadays increasingly indicates that 'the privatization of space is explicitly linked to efforts to protect propertied elites from the perceived threat of an unruly, uncontrollable public' (Harms 2009, 193).

Despite its modifications, *hukou/hộ khẩu* thus remains an institutional instrument to enact the differential civil, political and social rights that are part of rural and urban citizenship (Shi 2012). According to Fong and Murphy (2006, 7), 'by focusing on personal transformations as key to citizenship rights, state leaders could blame their inability to guarantee social rights such as education and satisfactory living conditions on individual failings and dispositional inadequacies'. As an example, the urban Chinese educational system excludes migrant children from the better-quality public schools on account of *hukou*, high fees and discrimination. They then turn to migrant schools that are understaffed, lacking in facilities and providing inadequate schooling for higher education (Wang 2006), which reinforces the construction of their low *suzhi*. This valorization of citizenship is embraced by the migrants, who, however, are 'dismayed by how this valorization devalued their own citizenship status' and seek to 'capitali[ze] on the flexibility afforded by their marginality' (Fong and Murphy 2006, 6–8). Citizenship, broadly conceived as membership in a community, therefore, is also about the right to belong through recognition and responsibility, or cultural citizenship (Marshall 1964). This results not only from 'struggles between individuals and states, but also from individuals' engagement with local and global discourses and economic systems that are not necessarily controlled by the state' (Marshall 1964). However coercive, a state vision of development, or 'representation of space' in Lefebvre's terms (1991), is constantly renegotiated through the ways in which people appropriate spaces for their everyday purposes. The following section focuses on the rural migrants' negotiations with institutional and discursive structures through transnational householding, which creates social spaces that cut across different categories, imagined and real.

Translocal householding: life in between

One of the most important tropes for the migration literature in China is that of the *dagongmei* – young female and single migrant workers, who are part of the floating population (*liudong renkou*), a category that includes people living and working in the city without permanent registration. Among others (Lee 1998; Pun 2005), Fan (2002, 2003, 2004a) discusses the emergence of a migrant factory labor regime characterized by precariousness, vulnerability and exploitation that the *dagongmei* are subjected to. The *dagongmei*'s labor is necessary for urban centers, but their citizen status renders them outsiders to the urban society. Similar migrant labor regimes to those in mainland China are visible in analyses of Vietnamese factory workers (Angie 2004; Nghiem 2004; Nguyen-Vo 2006; Nghiem 2007; Bélanger and Pendakis 2009; Bélanger 2010), where a docile female labor force is constructed and deployed for capitalist production. The literature on factory daughters contributes powerful accounts of their negotiations with exploitative and alienating employment structures. However, this strand of literature pays insufficient attention to the linkage between their mobility and the translocal householding that involves migration and return, multiple trajectories of household members, and the ways in which the household's gendered reproduction spans regions and places.

Nearly as much time has passed since the reform as the time under state socialism, and rural-urban migration[8] is no longer external to either the countryside or the city in both countries. More than a generation of migrants have left the country and come back; quite a few have left for good. Studying return migration in rural Jiangxi of southern China, Murphy (2002) analyses the diverse migration trajectories of households and individuals that eventually lead to their differential standings in the village. Neither temporary nor contingent, Murphy (2002, 45) points out, these movements have been internalized in village life:

> [...] through migration and return, individuals and families use, reproduce, and reconstitute values and resources in their efforts to attain goals. This creates a continual feedback mechanism whereby migration and return become institutions internal to the village – institutions in which both migrants and non-migrants participate, and institutions that interact with the outcomes of other processes of change.

Most of the early migrants in China and Vietnam have now reached the age when they are likely to be grandparents. Many retired migrants are now caregivers for their grandchildren and the elderly in their household, as their children migrate (Judd 2010; Nguyen forthcoming). In the 1980s and 1990s, it was common for one spouse to migrate at a time while the other stayed home with the children. In China, young women often migrated for some years for work before returning to get married (Fan 2004b) whereas married women were usually pressured to return by domestic duties (Lou et al. 2004). In Vietnam, it was common for married women to migrate into urban small trade (Resurreccion and Ha 2007) on account of the greater availability of such work in urban Vietnam (McGee 2009) and the flexibility of the work that allows them to perform familial duties while migrating (Vu and Agergaard 2012a; 2012b). Since the 2000s, more young married couples migrate together in both countries, either bringing their children along or leaving them in the country with the grandparents (Fan 2011; Locke 2012; Locke et al. 2012).

In her recent works, Fan (Fan 2008; Fan and Regulska 2008; Fan and Wang 2008; Fan 2009; Fan 2011) has moved away from viewing the *hukou* as the main reason that keeps rural migrants from settling in the city. In contrast to her earlier structurally driven accounts (2002, 2003, 2004a), Fan and her colleagues are now more emphatic about the agency of migrant households. 'Split-households' featuring migrants as single persons, couples or a couple with children, they argue, are an active strategy of maintaining a combination of rural and urban livelihoods in order to deal with the insecurity of urban life and work. In straddling the city and the country, these households are able to 'make the best of both worlds' (Fan 2008, 13), namely the income opportunities in the city and the extended support system and a sort of fallback position in the country. The household thus represents

---

[8]In both countries, rural-urban migration has always taken place throughout the different historical periods, despite the state control of mobility during state socialism (Mallee and Pieke 1999; Hardy 2003); what is new after the reform is the much greater scale.

a source of security, and gendered translocal householding is a mechanism to strengthen this security.

This focus on the agency and strategies of rural household has also been central to recent works by other China scholars (Judd 2010; May 2010; van der Ploeg and Jingzhong 2010; Jacka 2012). Jacka argues for the utility of the householding concept by Douglass (2006) in examining the 'strategies and processes through which rural households create and reproduce themselves' (2012, 2). In the northwestern rural community she studied, local people have diverse strategies for combining migrant work, agriculture and care that vary according the household's labor capacity and its life cycle (see also Chen and Korinek 2010). The fluidity and flexibility of householding are premised on the cooperation between household members, but internal conflicts are equally part of the process. The conjugal bond has become more fragile due to the mobile nature of householding, despite its increasing importance for the household economy (Judd 2010; Jacka 2012). Young people's migrant work brings them closer to the family because of greater uncertainty, as it establishes their independence, which is both liberating and frightening for them. Young female migrants returning home for marriage tend to experience a clash between the perceived personal freedom enabled by migration and patriarchal relations in village life (May 2010; Ge *et al.* 2011; Zhang 2013).

The mobility of young men and women shapes householding dynamics as much as it is shaped by them. May (2010) analyses the contradictory parental motives behind the mobility of sons and daughters in a rural northeast village of China. Whereas sons are encouraged to accumulate savings for an eventual return to marry a local woman who could later take care of them in old age, daughters are actively motivated to seek marital opportunities with urban men through investing their earnings in appearance and lifestyle. The parents aim to access the social provision that urban people enjoy through their daughters' potential marriage alliance, which in fact does not occur frequently, alongside the traditional care provided by the dutiful daughter-in-laws. These contradictory aspirations, she shows, are bound to lead to disappointments among the young men in finding the right wives and among the young women seeking more urbane and less patriarchal marriages. This indicates that emerging strategies for old-age care and provision against the shortage of social security are indeed grounded in age-old notions of filial piety that are highly gendered.

Translocal householding processes are also surfacing in recent migration research on Vietnam. The emphasis has similarly shifted to the ways in which the rural household reflexively embraces migration and return as part of diversified livelihoods during its life cycle (Resurreccion 2005; Zhang *et al.* 2006; Resurreccion and Hà 2007; Pham and Hill 2008; Truong 2009; Agergaard and Thao 2011; Hoang 2011a; 2011b; Locke *et al.* 2012; Vu and Agergaard 2012a; 2012b; Vu 2013; Nguyen forthcoming). These authors likewise demonstrate that the migration of individuals is embedded in household gender and intergenerational relations while their migratory experiences are intimately connected to the well-being of their family members. Care and parenting are shown to be essential to migration, which migrants often frame as parenting rather than as a departure from it (Locke *et al.* 2012). As argued for China (Murphy 2002, 2004; Zhang 2013), translocal householding in Vietnam is fraught with emotional dislocations and adjustment and belonging issues, as shown in studies of marital dissolution among migrants (Locke *et al.* 2013) and life-work negotiations of factory workers (Bélanger and Pendakis 2009).

While the China literature pays greater attention to young female migrants, Vietnam authors focus more on the migration of married women, the accompanying household rearrangements and their implications for gender identity. The initial uneasiness created by married women's departure has been now significantly reduced; their absence from home no longer requires justification (Resurreccion and Hà 2007; Truong 2009; Hoang and Yeoh 2011; Vu and Agergaard 2012a; 2012b; Nguyen forthcoming). What emerge are the 'caring man' and the 'empowered woman' who move back and forth between city and village life, earning income and sustaining their family (Hoang and Yeoh 2011; Locke 2012; Locke *et al.* 2012; Vu and Agergaard 2012a; 2012b). Yet gender identity remains constructed around women as the dutiful and caring wives/mothers, and around and men as 'pillars of the family'. What male and female migrants do may be challenging conventional familial arrangements, yet they still seek to perform and cast their practices in ideal traditional terms of the patriarchal family (Vu and Agergaard 2012a; 2012b; Nguyen forthcoming). Similar dynamics are documented in the sexual life of couples separated by migration in which traditional norms of sexuality are invoked even when migrant men have transgressed them (Nguyen *et al.* 2011).

In contrast to the media panic about the 'left-behind' children and the elderly, abandoned and uncared for, care is central to the diverse arrangements of the translocal household. While the emotional dislocations and vulnerabilities of the children and the elderly, many of whom are caregivers themselves, have to be taken into account (Jingzhong *et al.* 2010, Mummert 2010; Hoang and Yeoh 2012), rural households in fact continually adjust migration to the care needs and caring capacity of their members (Fan and Wang 2008; Fan 2009; 2011; Nguyen forthcoming). The decision to migrate, to return or to remain in the countryside has as much to do with care as with household pursuits of livelihoods. Migration patterns, i.e. circular or permanent, over long or short distances, and the chosen type of migrant work, are closely related to local ideas regarding how care should be provided to dependents and household negotiations over the care needs of members. These needs shift over time together with changing notions of the appropriate caregivers and caring practices (Mummert 2010).

Finally, migration has facilitated changing aspirations in family life and work. Rural migrants nowadays increasingly include young people born after the reform, for whom labor migration has become an inevitable trajectory (Fang 2011; Ngai and Lu 2010). For many of these new-generation peasant-workers, return to the countryside is becoming difficult. Ngai and Lu (2010), for example, show that the attempts to go back to agricultural work by Chinese young men in Shenzhen and Dongguan fail without the support of their parents who expect them to move away from farm work. In Minh Nguyen's recent field research in rural North Vietnam, she also observed a strong aspiration among farmers for their children to be 'free from the paddy field', through urban work, education or marriage. As the population ages, young people's movement out of agriculture will affect rural social support, which is currently maintained primarily by middle-aged men and women, many of whom are former migrants (Judd 2010; Nguyen forthcoming). Their aspirations and strategies to have their children liberated from farm work and rural life inadvertently help drain care resources from the countryside, undermining intergenerational support in the future. Their children, meanwhile, continue to join the urban precariats. As the second generation of migrants, their struggles are, however, distinct from those of the previous generation, especially with an emerging class consciousness fueled by anger and resentment (Ngai and Lu 2010). These partly explain their enduring identification and connections with the home place, even when they do not plan to return (Myerson *et al.* 2010; Anh *et al.* 2012). As such, translocal householding will continue to be part of life for rural people, even though its dynamics will shift together with new patterns of population mobility. In the next section, we connect translocal householding with the production of space in the major urban centers of the two countries.

At the margin and on the edge: space, marginalization and power

So far, we have argued that both because of and despite structural conditions, translocal householding produces a relatively autonomous space through the circulation of individuals and families between the city and the country. This space bridges the rural and the urban worlds, rendering the rural-urban distinction porous and fuzzy for people involved. In this section, we show that operating at the margin of urban society, physically and metaphorically, rural migrants create other spaces that transcend the rural-urban distinction, at times precisely through practices that reproduce them.

The floating population (*liudong renkou*) remains central to scholarship on internal migration in China (Lee 1998; Ma and Xiang 1998; Solinger 1999; Zhang 2001; Zheng 2003; Zheng 2004; Jacka 2006; Pun 2005; Zheng 2007; Yan 2008; Sun 2009a; 2009b). Rather than being concentrated in factories, migrants increasingly work in the urban service sector. As domestic workers, bar hostesses or sex workers, they are subject to labor regimes that are similarly precarious and alienating while also being stigmatizing because of the association with servility and promiscuity (Zheng 2004; Zheng 2007; Yan 2008; Sun 2009a; 2009b; Otis 2012). This trend has also been observed among Vietnamese migrant laborers, the *lao động ngoại tỉnh* (Henriot 2012; Nguyen 2012; Arnold 2013). Vietnamese labor migrants, however, remain more employed in 'popular sector' activities (*nghề tự do*) such as street vending, junk trading and household businesses that are less predominant in major Chinese cities (Mc Gee 2009), except for domestic service. In Vietnam, the female street vendor carries the same analytical weight for the rural-urban relationship as the migrants who are staffing industrial zones (Higgs 2003; Resurreccion and Hà 2007; Mitchell 2008; Mitchell 2009; Truong 2009; Turner and Schoenberger 2012; Arnold 2013). The greater presence of women migrants in Vietnam, who account for more than 50 percent (GSO 2011) of migrants in contrast to 30–40 percent in China (Jacka and Sargeson 2011), might also be correlated to the greater prevalence of popular-sector activities in Vietnam that are more accessible to women of all ages, especially married women.

Depictions of rural migrants tend to reinforce an image of abjectly poor and marginalized people who move around passively, doing whatever comes their way. In fact, most migrants come to the city with specific ideas about what they are doing and a network of support that facilitates their entry and operations (Ma and Xiang 1998; Zhang 1999; Resurreccion and Hà 2007; Agergaard and Thao 2011), while many have become economically better off than urban people of the laboring class (Zhang 2001; Nguyen 2013). Rural networks guide men and women to different kinds and destinations of migrant work, depending on local norms governing male and female behavior (Hoang 2011a; 2011b). They often develop out of a particular connection to a certain trade or activity in the home place, such as junk trading in Nam Định of Vietnam (Mitchell 2008) and in Sichuan of China (Ma and Xiang 1998).

On account of these networks, migrants are commonly concentrated in certain areas of the inner city or in the urban edge of the metropolises, where about three quarters of the Chinese *liudong renkou* reside (Leaf 1999; Zhang 2001; Liu and Wu 2006; Hsing 2010; Hao *et al.* 2011; Harms 2011). They form enclaves of migrants from various native places practicing particular trades or working in the factories located in the region. Many commute daily to the inner city to work (Zhao and Chapman 2010). Migrants' life there is closely connected to that of local villagers, peasants-turned-landlords who enjoy considerable income but remain culturally inferior to urban people (Liu and Wu 2006; Siu 2007). To the urban middle class and urban governments, these 'villages in the city' (*zhengchongcun*) represent a mass of crime and poverty, which corresponds with the construction of peasants as backward and dangerous. In both countries, urban governments seek to cleanse these spaces of the perceived unruliness and disorder through eviction for construction and industrial development (Zhang 2001; Harms 2009; Hao *et al.* 2011; Harms 2011). To the migrants, these places are as much sources of livelihoods and sociality as spaces dominated by powerful people in their rural network and property-owning local residents (Zhang 2001; Siu 2007). The following accounts of the migrant communities in the urban edge of Beijing (Zhang 2001) and of Ho Chi Minh city (Harms 2011) provide further insights into these dynamics.

Zhang (2001) documents the rise and fall of a migrant enclave in the southern suburb of Beijing, the *Zhejiangcun* or Zhejiang village, a vibrant community of migrants from Zhejiang province who are renowned for business savviness and their traditional trade in garments. Informed by Lefebvre's theory of space (1991), Zhang pays special attention to the ways in which the migrants struggle to appropriate urban space to accumulate wealth and 'gain control over communal lives and economy' (2001, 9). Spatial practices in *Zhejiangcun*, however, also reveal the exploitation of migrant tenants and waged workers by the housing bosses and business owners from the same migrant network, while indicating a high level of gender differentiation. Wealthy women are removed from production to be confined to the home whilst middle- and low-income women's primary role in garment production is devalued on account of being performed at home. Young female migrant workers are subject to their migrant bosses' control, often disguised in the 'part of the family' discourse. Although deeply embedded for years in patronage ties with the local government, *Zhejiangcun* was finally demolished to make way for urban development, to reemerge in other locations that are similarly marginal but suitable for their economic operations. Zhang's account reveals the ways in which migrants, through maximizing their networks' advantages and working with the state clientele politics, are able to create autonomous social and economic spaces against the state, albeit uncivic ones. It also suggests the degree to which the Chinese state is willing to impose its visions of modernity and engineer social change, which, however, are tempered by the migrants' spatial practices.

Erik Harms analyzes the spatial and social development of Hoc Mon district on the periphery of Ho Chi Minh city, focusing on the everyday practices of different groups of residents, including the locals, rural migrants and people moving out from the inner city. He demonstrates that these practices defy the rural-urban dichotomy in a social space that is neither rural nor urban, as the city 'cuts across the rural landscape and creates new forms of space that are both rural and urban at once' (2011, 45). This is a space betwixt and between that appears both dangerous, associated with disorder and 'social evils', and a space of opportunities, where export factories and property-owning opportunities abound. Similarly to Zhang (2001), Harms stresses the production of meanings partaken by the state and the people themselves that maintain the rural-urban division in the urban edge, despite the blurring of physical boundaries. He argues that existence in such a between-category space is characterized by a combined sense of exclusion and power. This combination explains why people continue to reproduce the division through their performance and movements between the edge and the inner city, between the city and the country. Residents of the urban edge, he shows, strategically deploy rural and urban identities for social and economic purposes in different social settings, thus simultaneously contradicting and reproducing these ideal categories.

Similar dynamics of space and power are prevalent not only in the urban fringes (Leaf 2002; Siu 2007), but also among groups of migrants who mingle more with the urban populations in the inner cities, such as street vendors, sex workers and domestic workers.

Nguyen (2013)'s analysis of a migrant community specialized in junk trading in Hanoi suggests that they have capitalized on their traditional niche in the junk trade to develop a profitable recycling economy employing tens of thousands of people. Their spatial practices are similar to those described by Zhang (2001) and Harms (2011), except that they continually move around the city, making use of urban spaces that are temporarily available. These practices afford them flexibility while making them ever more transient and liminal, consolidating the urban perception of rural migrants as unruly and dangerous.

For these migrants, the performance of a rural identity is essential to their daily transactions with urban people. Like the female street vendors from the edge of Saigon who don a rural attire in order to market their goods as rurally authentic in the inner city (Harms 2011), junk traders in Hanoi highlight their inferior status to appeal to the compassion of urban people (Nguyen 2013). Performative strategies are also common among personal service workers such as bar hostesses or sex workers. As much as their 'female body is being packaged in relation to the growing consumer market' (Chen *et al.* 2001, 25), they also tailor a variety of personas to appeal to different customers to maximize their earnings (Zheng 2004; Zheng 2007; Kay Hoang 2010). Such performances are often accompanied by their consumption practices modeled on dominant cultural symbols (Zheng 2007; Otis 2012). Indeed, urban-oriented consumption is a common strategy among female migrants in China, who see it as a way to improve their *suzhi* and reduce their marginality (Sun 2009a; 2009b; Yan 2008). As Otis (2012, 150) writes:

> Through their commodity-enabled, self-imposed self-alterations women workers adapt to the tastes, preferences, and expectations of urban consumers. They spontaneously conform to their symbolic labor, the presentation of an aesthetic tied to the body, to urban gender and class expectations.

The success of this strategy is doubtful, since it is precisely through consumption that the urban middle class establishes its social distinction (Zhang 2010; Nguyen-Marshall *et al.* 2012). Migrants' spatial practices therefore work like double-edged knives (Harms 2011), both challenging their peripheral position by 'creating a parody of their constructed image and, at the same time, reinforc[ing] and reproduc[ing] hegemonic asymmetrical power relations and their marginality' (Zheng 2004, 88).

Conclusions

Rural-urban migration in Vietnam and China has been greatly transformed alongside momentous spatial and social transformations since the two countries shifted to a market economy, albeit to differing degrees and with regional variations. These transformations have reconfigured the rural and urban division. While material boundaries may be receding, symbolic boundaries are expanding, as the state increasingly seeks to govern through categories that construct rural and urban populations as occupying varied levels of development. They form 'idealized conceptions of spatial order', which social actors simultaneously internalize and play off by 'practices, strategies and socially embedded negotiations' in order to 'make space for everyday life' (Harms 2011, 238). Through their spatial practices, rural migrants have created relatively autonomous spaces to sustain families, carry out their economic activities and foster sociality and belonging across places. Since the state in both Vietnam and China continues to keep a strong hold on its citizens as it advances capitalist development, these spaces remain subject to its interventions and control, at times highly intrusive and destructive.

Translocal householding has become central to rural-urban migration. The previously rural household now operates between the city and the country, as household members migrate and return over their life course, performing gendered familial duties from both locations. Migration and return thus must be seen in the context of the rural household reshuffling and strategizing in order to reproduce itself in response to social and systemic changes. Individuals may migrate out of desires for freedom and independence, but their mobility is necessarily part of gendered household strategies to accumulate wealth and secure social provision that are continually adjusted to household members' changing care needs and capacities. This has become more imperative on account of the emphasis on the household as the primary unit of reproduction and guarantor of individual well-being. Translocal householding, however, is malleable to ruptures, instability and internal contradictions, especially when migrant household members assume identities that are not necessarily favorable to the householding project, which remains embedded in patriarchal relations and ideas.

Operating at the edge of cities and on the margin of urban society, rural migrants have also created spaces that concurrently blur the rural-urban binary and discursively reconstruct it. Local residents and rural migrants on the urban fringe are viewed as an obstacle for the state's civilizing project. Yet they effectively defy official and popular categories imposed on them by creatively carving out their own social and economic space, notably through performative strategies and consumption. Although they recognize 'how seemingly pure categories are shifty symbols, with meanings that change according to contexts' (Harms 2011, 76), their spatial practices reproduce these dominant categories while creating new patterns of social and gender differentiation.

We conclude that bringing 'class back in' to analyses of agrarian change (Bernstein 2010) enables an examination of the ways in which rural and gender differentiation feed into class formation processes through rural-urban migration in post-reform China and Vietnam. The rural-urban distinction will become less significant, as rural labor not only straddles between the city and the countryside, but also between categories of work, joining the 'classes of labour' (Bernstein 2010). However, this process will not erode the marginality that rural migrants experience, because they are most likely to occupy the lower rungs of future 'classes of labour' and, as such, precariousness is likely to remain a defining feature of their existence.

References

Agergaard, J. and V. T. Thao. 2011. Mobile, flexible, and adaptable: Female migrants in Hanoi's informal sector. *Population, Space and Place* 17, no. 5: 407–20.

Angie, T. N. 2004. What's women's work? Male negotiations and gender reproduction in the Vietnamese garment industry. In *Gender practices in contemporary Vietnam*, eds. L. Drummond and H. Rydstrom. Singapore; Copenhagen: Singapore University Press; Nordic Institute of Asian Studies: xii, 210 p.

Anh, N. T., J. Rigg, et al. 2012. Becoming and being urban in Hanoi: Rural-urban migration and relations in Viet Nam. *The Journal of Peasant Studies* 39, no. 5: 1103–31.

Arnold, D. 2013. Social margins and precarious work in Vietnam. *American Behavioral Scientist* 57, no. 4: 468–87.

Badiani, R., B. Baulch, et al. 2012. 2012 Vietnam poverty assessment – Well begun, not yet done: Vietnam's remarkable progress on poverty reduction and the emerging challenges. Hanoi: World Bank.

Bélanger, D. 2010. Vietnamese daughters in transition: Factory work and family relations. *Asia-Pacific Journal: Japan Focus* no. 5: 1–8.

Bélanger, D., and K. Pendakis. 2009. Daughters, work, and families in globalizing Vietnam. In M. Barbieri and D. Bélanger. *Reconfiguring families in contemporary Vietnam*. Stanford, Calif.: Stanford University Press: vii, 265 p.

Bernstein, H. 2010. Rural livelihoods and agrarian change: Bring class back in. In *Rural trans-formations and development – China in context: the everyday lives of policies and people*, eds. N. Long, J. Ye and Y. Wang. Cheltenham, UK; Northampton, MA: Edward Elgar: xii, 395 p.

Chan, Y.W., and T.L.T. Tran. 2011. Recycling migration and changing nationalism: The Vietnamese return diaspora and reconstruction of Vietnamese nationhood. *Journal of Ethnic and Migration Studies* 37, no. 7: 1101–17.

Chen, F., and K. Korinek. 2010. Family life course transitions and rural household economy during China's market reform. *Demography* 47, no. 4: 963–87.

Chen, N.N., C. Clark, et al. 2001. *China urban: Ethnographies of contemporary culture*. Durham, N.C.: Duke University Press.

Cho, M.Y. 2013. *The specter of 'the people': Urban poverty in Northeast China*. Ithaca: Cornell University Press.

Chu, J.Y. 2010. *Cosmologies of credit: transnational mobility and the politics of destination in China*. Durham, NC: Duke University Press.

Đặng, K.S. 2008. Rural development issues in Vietnam: Spatial disparities and some recommen-dations. In *Reshaping economic geography in East Asia*, ed. Y. Huang and A. Bocchi. Washington: World Bank: xiv, 100 p.

Dang, N.A., T.B. Tran, et al. 2010. Development on the move: measuring and optimising migration's economic and social impacts in Vietnam. Hanoi: Global Development Network and Institute for Public Policy Research.

Davis, D., and W. Feng. 2009. *Creating wealth and poverty in Postsocialist China*. Stanford, Calif.: Stanford University Press.

Douglass, M. 2006. Global householding in Pacific Asia. *International Development Planning Review* 28, no. 4: 421–46.

Douglass, M. 2012. Global householding and social reproduction: Migration research, dynamics and publlic policy in East and Southeast Asia. *ARI Working Paper Series*, Asia Research Institute. No 188.

Fan, C.C. 2002. The elite, the natives, and the outsiders: Migration and labor market segmentation in urban China. *Annals of the Association of American Geographers* 92, no. 1: 103–24.

Fan, C.C. 2003. Rural-urban migration and gender division of labor in transitional China. *International Journal of Urban and Regional Research* 27, no. 1: 24–47.

Fan, C.C. 2004a. The state, the migrant labor regime, and maiden workers in China. *Political Geography* 23, no. 3: 283–305.

Fan, C.C. 2004b. Out to the city and back to the village: The experiences and contributions of rural women migrating from Sichuan and Anhui. In *On the move: Women and rural-to-urban migration in contemporary China*, A.M. Gaetano and T. Jacka (Eds). New York: Columbia University Press: vii, 177 p.

Fan, C.C. 2008. *China on the move: Migration, the state, and the household*. London; New York: Routledge.

Fan, C.C., and J. Regulska. 2008. Gender and labor market in China and Poland. In *Urban China in transition*, ed. J.R. Logan. Malden, MA: Oxford, Blackwell Pub. Ltd., xv, 361 p.

Fan, C.C., and W. Wang. 2008. The household as security: Strategies of rural-urban migrants in China. In *Migration and social protection in China*, eds. I. Nielsen and R. Smyth. New Jersey: World Scientific, x, 265 p.

Fan, C.C. 2009. Flexible work, flexible household: Labor migration and rural families in China. *Research in the Sociology of Work* 19: 377–408.

Fan, C.C. 2011. Settlement intention and split households: Findings from a survey of migrants in Beijing's urban villages. *The China Review* 11, no. 2: 11–42

Fang, I-C. 2011. Growing up and becoming independent: An ethnographic study of new generation migrant workers in China. Thesis (PhD). The London School of Economics and Political Science.

Fesselmeyer, E., and K.T. Le. 2010. Urban-biased policies and the increasing rural-urban expenditure gap in Vietnam in the 1990s. *Asian Economic Journal* 24, no. 2: 161–78.

Fforde, A., and S. De Vylder. 1996. *From plan to market: The economic transition in Vietnam*. Boulder, Colo.: Westview Press.

Fong, V.L. 2011. *Paradise redefined: Transnational Chinese students and the quest for flexible citi-zenship in the developed world*. Stanford, California: Stanford University Press.

Fong, V.L. and R. Murphy 2006. *Chinese Citizenship: Views from the Margins*, London, New York: Routledge.

Francis, E. 2002. Gender, migration and multiple livelihoods: Cases from Eastern and Southern Africa. *Journal of Development Studies* 38, no. 5: 167–90.

Friedman, S.L. 2004. Embodying civility: Civilizing processes and symbolic citizenship in Southeastern China. *The Journal of Asian Studies* 63, no. 03: 687–718.

Ge, J., B.P. Resurreccion, et al. 2011. Return migration and the reiteration of gender norms in water management politics: Insights from a Chinese village. *Geoforum* 42, no. 2: 133–42.

Gillespie, J. 2013. Vietnam's Land Law Reforms: Radical Changes or Minor Tinkering? EastAsiaForum: Economics, Politics and Public Policy in East Asia and the Pacific.

Goldstein, J.L. 2006. The remains of the everyday: One hundred years of recycling in Beijing. In *Everyday modernity in China*, eds. M.Y. Dong and J. L. Goldstein. Seattle: University of Washington Press: 344 p.

Gourou, P. 1955. *The peasants of the Tonkin Delta: A study of human geography*. New Haven: Human Relations Area Files.

GSO. 2011. Vietnam population and housing census 2009: migration and urbanization in Vietnam – patterns, trends and differentials. Hanoi: Ministry of Planning and Investment, General Statistics Office.

Guldin, G.E. 2001. *What's a peasant to do? Village becoming town in Southern China*. Boulder, Colo.: Westview Press.

Han, X. 2005. *Chinese discourses on the peasant, 1900-1949*. Albany: State University of New York Press.

Hao, P., R. Sliuzas, et al. 2011. The development and redevelopment of urban villages in Shenzhen. *Habitat International* 35, no. 2: 214–24.

Hardy, A. 2001. Rules and resources: Negotiating the household registration system in Vietnam under reform. *Sojourn: Journal of Social Issues in Southeast Asia* 16, no. 2: 187–212.

Hardy, A. 2003. *Red hills: Migrants and the state in the highlands of Vietnam*. Copenhagen: Nordic Institute of Asian Studies.

Harms, E. 2009. Vietnam's civilizing process and the retreat from the street: A turtle's eye view from Ho Chi Minh City. *City & Society* 21, no. 2: 182–206.

Harms, E. 2011. *Saigon's edge: On the margins of Ho Chi Minh City*. Minneapolis: University of Minnesota Press.

Harms, E. 2012. Beauty as control in the new Saigon: Eviction, new urban zones, and atomized dissent in a Southeast Asian city. *American Ethnologist* 39, no. 4: 735–50.

Henriot, C. 2012. Supplying female bodies: Labor migration, sex work, and the commoditization of women in colonial Indochina and contempory Vietnam. *Journal of Vietnamese Studies* 7, no. 1: 1–9.

Hershatter, G. 2007. *Women in China's long twentieth century*. Berkeley: University of California Press.

Higgs, P. 2003. Footpath traders in a Hanoi neighbourhood. In *Consuming urban culture in contemporary Vietnam*, (eds). L. Drummond and M. Thomas. London: Routledge Curzon: xii, 75 p.

Hoang, L.A. 2011a. Gender identity and agency in migration decision-making: Evidence from Vietnam. *Journal of Ethnic and Migration Studies* 37, no. 9: 1441–57.

Hoang, L.A. 2011b. Gendered networks and migration decision-making in Northern Vietnam. *Social & Cultural Geography* 12, no. 5: 419–34.

Hoang, L.A., and B.S.A. Yeoh. 2011. Breadwinning wives and 'left-behind' husbands. *Gender & Society* 25, no. 6: 717–39.

Hoang, L.A., and B.S.A. Yeoh. 2012. Sustaining families across transnational spaces: Vietnamese migrant parents and their left-behind children. *Asian Studies Review* 36, no. 3: 307–25.

Hojman, D.E. 1989. Land reform, female migration and the market for domestic service in Chile. *Journal of Latin American Studies* 21, no. 1: 105–32.

Hsing, Y-t. 2010. *The great urban transformation: Politics of land and property in China*. New York: Oxford University Press.

Jacka, T. 2006. *Rural women in urban China: Gender, migration, and social change*. Armonk, N.Y.: M.E. Sharpe, Inc.

Jacka, T. 2009. Cultivating citizens: Suzhi (quality) discourse in the PRC. *Positions* 17, no. 3: 523–35.

Jacka, T. 2012. Migration, householding and the well-Being of left-behind women in Rural Ningxia. *The China Journal* no. 67: 1–22.

Jacka, T., and S. Sargeson. 2011. *Women, gender and rural development in China*. Cheltenham: Edward Elgar.

Jingzhong, Y., Y. Wang and K. Zhang. 2010. Rural-urban Migration and the Plight of 'Left-behind Children' Mid-west China. In *Rural Transformations and Development – China in Context: The Everyday Lives of Policies and People*, N. Long, Y. Jingzhong and Y. Wang (Eds). Cheltenham, UK; Edward Elgar: xii, 253 p.

Judd, E.R. 2010. Family strategies: Fluidities of gender, community and mobility in rural West China. *The China Quarterly* 204: 921–38.

Kay Hoang, K. 2010. Economies of emotion, familiarity, fantasy, and desire: Emotional labor in Ho Chi Minh City's sex industry. *Sexualities* 13, no. 2: 255–72.

Kerkvliet, B. 2005. *The power of everyday politics: How Vietnamese peasants transformed national policy*. New York: Cornell University Press.

Kerkvliet, B., A. Chan, et al. 1998. Comparing the Chinese and Vietnamese reforms: An introduction. *The China Journal* no. 40: 1–7.

Kerkvliet, B.J., and D.G. Marr. 2004. *Beyond Hanoi: Local government in Vietnam*. Copenhagen; Singapore: NIAS Press; Institute of Southeast Asian Studies.

Kerkvliet, B.J.T., and M. Selden. 1998. Agrarian transformations in China and Vietnam. *The China Journal*, no. 40: 37–58.

Kipnis, A. 2006. Suzhi: A Keyword Approach. *The China Quarterly*, no. 186: 295–313.

Kipnis, A. 2007. Neoliberalism reified: Suzhi discourse and tropes of neoliberalism in the People's Republic of China. *Journal of the Royal Anthropological Institute* 13, no. 2: 383–400.

Koh, D. 2004. Illegal construction in Hanoi and Hanoi's Wards. *European Journal of East Asian Studies* 3, no. 2: 337–69.

Koh, D. 2006. *Wards of Hanoi*. Pasir Panjang, Singapore: Institute of Southeast Asian Studies.

Kuruvilla, S., C.K. Lee, et al. 2011. *From iron rice bowl to informalization: Markets, workers, and the state in a changing China*. Ithaca, N.Y.: ILR Press.

Lawson, V.A. 1998. Hierarchical households and gendered migration in Latin America: Feminist extensions to migration research. *Progress in Human Geography* 22, no. 1: 39–53.

Lê, B.D. and T.H. Khuất. 2008. *Migration and social protection in Vietnam during the transition to a market economy*. Hanoi: Thế giới Publishers.

Le, B.D., G.L. Tran, et al. 2011. Social Protection for Rural-urban Migrants in Vietnam: Current Situation, Challenges and Opportunities, Center for Social Protection in Asia.

Leaf, M. 1999. Vietnam's urban edge: The administration of urban development in Hanoi. *Third World Planning Review* 21, no. 3: 297.

Leaf, M. 2002. A tale of two villages: Globalization and peri-urban change in China and Vietnam. *Cities* 19, no. 1: 23–31.

Lee, C.K. 1998. *Gender and the South China miracle: Two worlds of factory women*. Berkeley: University of California Press.

Lefebvre, H. 1991. *The production of space*. Oxford, UK; Cambridge, Mass., USA: Blackwell.

Liu, Y., and F. Wu. 2006. Urban poverty neighbourhoods: Typology and spatial concentration under China's market transition, a case study of Nanjing. *Geoforum* 37, no. 4: 610–26.

Locke, C. 2012. Visiting marriages and remote parenting: Changing strategies of rural-urban migrants to Hanoi, Vietnam. *Journal of Development Studies* 48, no. 1: 10–25.

Locke, C., N.T.N. Hoa, et al. 2012. Struggling to sustain marriages and build families: Mobile husbands/wives and mothers/fathers in Hà Nội and Hồ Chí Minh City. *Journal of Vietnamese Studies* 7, no. 4: 63–91.

Locke, C., T.T.T. Nguyen, et al. 2013. Mobile householding and marital dissolution in Vietnam: An inevitable consequence?. *Geoforum* 51: 273–283.

Long, N., J. Ye, et al. 2010. *Rural transformations and development – China in context: The everyday lives of policies and people*. Cheltenham, UK; Northampton, MA: Edward Elgar.

Lou, B., Z. Zheng, et al. 2004. The migration experiences of young women from four counties in Sichuan and Anhui. In *On the move: Women and rural-to-urban migration in contemporary China*, eds. A.M. Gaetano and T. Jacka. New York: Columbia University Press: vii, 207 p.

Luong, H.V. 2009. *Urbanization, migration, and poverty in a Vietnamese metropolis: Hồ Chí Minh City in comparative perspectives*. Pasir Panjang, Singapore: NUS Press.

Ma, L.J., and B. Xiang. 1998. Native place, migration and the emergence of peasant enclaves in Beijing. *The China Quarterly*, 155: 546–81.

Mallee, H., and F.N. Pieke. 1999. *Internal and international migration: Chinese perspectives*. London: Routledge Curzon.

Marshall, T.H. 1964. *Class, citizenship, and social development: Essays*. Westport, Con.: Greenwood Press.

May, S. 2010. Bridging divides and breaking homes: Young women's lifecycle labour mobility as a family managerial strategy. *The China Quarterly* 204: 899–920.

McGee, T.G. 2009. Interrogating the production of urban space in China and Vietnam under market socialism. *Asia Pacific Viewpoint* 50, no. 2: 228–46.

Mitchell, C.L. 2008. Altered landscapes, altered livelihoods: The shifting experience of informal waste collecting during Hanoi's urban transition. *Geoforum* 39, no. 6: 2019–29.

Mitchell, C.L. 2009. Trading trash in the transition: Economic restructuring, urban spatial transformation, and the boom and bust of Hanoi's informal waste trade. *Environment and Planning A* 41, no. 11: 2633–50.

Mummert, G. 2010. Growing up and growing old in rural Mexico and China: Care-giving for the young and the elderly at the family-state interface. In *Rural transformations and development – China in context: The everyday lives of policies and people*, N. Long, J. Ye and Y. Wang (Eds). Cheltenham, UK; Northampton, MA: Edward Elgar: xii, 215 p.

Murphy, R. 2002. *How migrant labor is changing rural China*. Cambridge, UK; New York: Cambridge University Press.

Murphy, R. 2004. The impact of labour migration on the well-being and agency of rural Chinese women: Cultural and economic contexts and the life course. In *On the move: Women and rural-to-urban migration in contemporary China*, A.M. Gaetano and T. Jacka (Eds). New York: Columbia University Press: vii, 243 p.

Myerson, R., Y. Hou, et al. 2010. Home and away: Chinese migrant workers between two worlds. *The Sociological Review* 58, no. 1: 26–44.

Ngai, P., and Lu, H. 2010. Unfinished proletarianization: Self, anger, and class action among the second generation of peasant-workers in present-day China. *Modern China* 36, no. 5: 493–519.

Nghiem, L.H. 2004. Famale garment workers: The new young volunteers in Vietnam's modernization. In *Social inequality in Vietnam and the challenges to reform*, ed. P. Taylor. Pasir Panjang, Singapore: Institute of Southeast Asian Studies: vii, 297 p.

Nghiem, L.H. 2007. Jokes in a garment workshop in Hanoi: How does humour foster the perception of community in social movements?. *International Review of Social History* 52, no. Supplement S15: 209–23.

Nguyen, H.N., M. Hardesty, et al. 2011. In full swing? How do pendulum migrant labourers in Vietnam adjust their sexual perspectives to their rural-urban lives?. *Culture, Health & Sexuality* 13, no. 10: 1193–206.

Nguyen, T.N.M. 2012. 'Doing Ô Sin': Rural migrants negotiating domestic work in Hà Nội. *Journal of Vietnamese Studies* 7, no. 4: 32–62.

Nguyen, T.N.M. 2013. 'Đi chợ' and 'Mổ bãi': Performance and Space Production in a Rural-urban Recycling Economy of Vietnam. *Working Paper Series*, no. 146. Halle/Saale: Max Planck Institute for Social Anthropology.

Nguyen, T.N.M. forthcoming. Translocal householding: Care and migrant livelihoods in a waste-trading community of Vietnam's Red River Delta. Development and Change.

Nguyen-Marshall, V., L. Drummond, et al. 2012. *The reinvention of distinction: Modernity and the middle class in urban Vietnam*. Dortrecht; Heidelberg; London; New York: Springer Verlag.

Nguyen-Vo, T-H. 2006. The body wager: Materialist resignification of Vietnamese women workers. *Gender, Place & Culture* 13, no. 3: 267–81.

Nguyen-Vo, T-H. 2008. *The ironies of freedom: Sex, culture, and neoliberal governance in Vietnam*. Seattle: University of Washington Press.

Nielsen, I., and R. Smyth. 2008. *Migration and social protection in China*. New Jersey: World Scientific.

Nyíri, P.l. 2010. *Mobility and cultural authority in contemporary China*. Seattle: University of Washington Press.

Oakes, T., and L. Schein. 2005. *Translocal China: Linkages, identities and the reimagining of space*. London; New York: Routledge.

O'Connor, D. 1998. Rural industrial development in Vietnam and China: A study in contrast. *MOST: Economic Policy in Transitional Economies* 8, no. 4: 7–43.

Ong, A. 1999. *Flexible citizenship: The cultural logics of transnationality*. Durham: Duke University Press.

Otis, E.M. 2012. *Markets and bodies: Women, service work, and the making of inequality in China*. Stanford, California: Stanford University Press.

Pearson, R. 2012. *Thailand's hidden workforce: Burmese migrant women factory workers*. London; New York: Zed Books.

Perry, E.J., and M. Selden. 2010. *Chinese society: Change, conflict and resistance*. London; New York: Routledge.

Pham, B.N., and P.S. Hill. 2008. The role of temporary migration in rural household economic strategy in a transitional period for the economy of Vietnam. *Asian Population Studies* 4, no. 1: 57–75.

Phạm, V.B. 1998. *The Vietnamese family in change: The case of the Red River delta*. London: RoutledgeCurzon.

van der Ploeg, J.D., and Y. Jingzhong. 2010. Multiple job holding in rural villages and the Chinese road to development. *The Journal of Peasant Studies* 37, no. 3: 513–30.

Porio, E. 2007. Global householding, gender, and Filipino migration: A preliminary review. *Philippine Studies* 55, no. 2: 211–42.

Porta, L., and M. Beresford. 2012. Emerging class relations in the Mekong Delta of Vietnam: A network analysis. *Journal of Agrarian Change* 12, no. 1: 60–80.

Pun, N. 2005. *Made in China: Women factory workers in a global workplace*. Durham, N.C.: Duke University Press.

Rato, M. 2004. Class, nation, and text: The representation of peasants in Vietnamese literature. In *Social inequality in Vietnam and the challenges to reform*, ed. P. Taylor. Pasir Panjang, Singapore: Institute of Southeast Asian Studies: vii, 325 p.

Resurreccion, B. 2005. Women in-between: Gender, transnational and rural-urban mobility in the mekong region. *Gender, Technology and Development* 9, no. 1: 31–56.

Resurreccion, B., and T.V.K. Hà. 2007. Able to come and go: Reproducing gender in female rural-urban migration in the Red River delta. *Population, Space and Place* 13: 211–24.

Schwenkel, C., and A. M. Leshkowich. 2012. Guest editors' introduction: How is neoliberalism good to think Vietnam? How is Vietnam good to think neoliberalism?. *Positions: East Asia Cultures Critique* 20, no. 2: 379–401.

Shi, S.J. 2012. Towards inclusive social citizenship? Rethinking China's social security in the trend towards urban–rural harmonisation. *Journal of Social Policy* 1, no. 1: 1–22.

Sicular, T., Y. Ximing, et al. 2010. How large is china's rural-urban income gap. In *One country, two societies: Rural-urban inequality in contemporary China*, ed. M.K. Whyte. Cambridge, Mass.: Harvard University Press: xii, 85 p.

Silvey, R. 2004. Transnational domestication: State power and Indonesian migrant women in Saudi Arabia. *Political Geography* 23, no. 3: 245–64.

Siu, H.F. 2007. Grounding displacement: Uncivil urban spaces in postreform South China. *American Ethnologist* 34, no. 2: 329–50.

Solinger, D.J. 1999. *Contesting citizenship in urban China: Peasant migrants, the state, and the logic of the market*. Berkeley; London: University of California Press.

Sun, W. 2009a. *Maid in China: Media, mobility, and the cultural politics of boundaries*. London: Routledge.

Sun, W. 2009b. Suzhi on the move: Body, place, and power. *Positions* 17, no. 3: 617–42.

Taylor, P., ed. 2004. *Social inequality in Vietnam and the challenges to reform*. Pasir Panjang, Singapore: Institute of Southeast Asian Studies.

Truong, C.H. 2009. A home divided: Work, body, and emotions in the post-doi moi family. In *Reconfiguring families in contemporary Vietnam*, eds. M. Barbieri and D. Bélanger. Stanford, Calif.: Stanford University Press: vii, 298 p.

Turner, S., and L. Schoenberger. 2012. Street vendor livelihoods and everyday politics in Hanoi, Vietnam: The seeds of a diverse economy?. *Urban Studies* 49, no. 5: 1027–44.

United Nations. 2010. Internal migration: opportunities and challenges for socio-economic development in Vietnam. Hanoi: United Nations in Vietnam.

Vu, T.T. 2013. Making a living in rural Vietnam from (im)mobile livelihoods: A case of women's migration. *Population, Space and Place* 19, no. 1: 87–102.

Vu, T.T., and J. Agergaard. 2012a. Doing family: Female migrants and family transition in rural Vietnam. *Asian Population Studies* 8, no. 1: 103–19.

Vu, T.T., and J. Agergaard. 2012b. 'White cranes fly over black cranes': The longue durée of rural-urban migration in Vietnam. *Geoforum* 43, no. 6: 1088–98.

Wang, F.-L. 2005. *Organizing through division and exclusion: China's hukou system*. Stanford, Calif.: Stanford University Press.

Wang, L. 2006. The urban Chinese educational system and the marginality of migrant children. In *Chinese citizenship: Views from the margins*, eds. V.L. Fong and R. Murphy. London; New York: Routledge: xii, 27 p.

Whyte, M.K. 2010. *One country, two societies: Rural-urban inequality in contemporary China*. Cambridge, Mass.: Harvard University Press.

World Bank. 2006. Vietnam Poverty Update Report 2006: Poverty and Poverty Reduction in Vietnam 1993–2004. Hanoi: World Bank.

World Bank. 2008a. Online data bank on migration worldwide. Available at: http://data.worldbank. org/indicator/SM.POP.TOTL [accessed on 21 May 2013].

World Bank. 2008b. Vietnam development report 2008: social protection. Hanoi: World Bank.

World Bank. 2010. Online data bank on urban population world wide. Available at: http://data. worldbank.org/indicator/SP.URB.TOTL.IN.ZS [accessed on 21 March 2013].

Yan, H. 2003. Specialization of the rural: Reinterpreting the labor mobility of rural young women in post-Mao China. *American Ethnologist* 30, no. 4: 578–96.

Yan, H. 2006. Rurality and labor process autonomy: The question of subsumption in the waged labor of domestic service. *Cultural Dynamics* 18, no. 1: 5–31.

Yan, H. 2008. *New masters, new servants: Migration, development, and women workers in China*. Durhan, NC: Duke University Press.

Ye, J., Wang, Y. and K. Zhang. 2010. Rural-urban migration and the plight of 'left-behind children' in Mid-west China. In *Rural transformations and development - China in context: the everyday lives of policies and people*, eds. N. Long, J. Ye and Y. Wang. Cheltenham, UK; Northampton, MA: Edward Elgar: xii, 253 p.

Zavoretti, R. 2012. An ethnographic inquiry into the limits of China's 'rural to urban migrant' paradigm. Thesis (PhD), University of London.

Zhang, H.X. 1999. Female migration and urban labour markets in Tianjin. *Development and Change* 30, no. 1: 21.

Zhang, H.X., P.M. Kelly, et al. 2006. Migration in a transitional economy: Beyond the planned and spontaneous dichotomy in Vietnam. *Geoforum* 37, no. 6: 1066–81.

Zhang, L. 2001. *Strangers in the city: Reconfigurations of space, power, and social networks within China's floating population*. Stanford, Calif.: Stanford University Press.

Zhang, L. 2010. *In search of paradise: Middle-class living in a Chinese metropolis*. Ithaca: Cornell University Press.

Zhang, L., and A. Ong. 2008. *Privatizing China: Socialism from afar*. Ithaca: Cornell University Press.

Zhang, N. 2013. Rural women migrant returnees in contemporary China. *The Journal of Peasant Studies* 40, no. 1: 171–88.

Zhao, P., and P. Howden-Chapman. 2010. Social inequalities in mobility: The impact of the 'hukou' system on migrants' job accessibility and commuting costs in Beijing. *International Development Planning Review* 32, no. 3: 363–84.

Zheng, T. 2003. Consumption, body image, and rural-urban apartheid in contemporary China. *City & Society* 15, no. 2: 143–63.

Zheng, T. 2004. From peasant women to bar hostesses: Gender and modernity in post-Mao Dalian. In *On the move: Women and rural-to-urban migration in contemporary China*, eds. A.M. Gaetano and T. Jacka. New York: Columbia University Press: vii, 80 p.

Zheng, T. 2007. Performing media-constructed images for first-class citizenship: Political struggles of rural migrant hostesses in Dalian. *Critical Asian Studies* 39, no. 1: 89–120.

Minh Nguyen is Research Fellow at the Max Planck Institute for Social Anthropology. She is currently researching *Care and migration in the Red River Delta of Vietnam*. Her book *Vietnam's socialist servants? Domesticity, class, gender, and identity* (London; New York: Routledge) is forthcoming in 2014.

Catherine Locke is Reader in Gender and Social Development at the University of East Anglia. She is a sociologist with special interest in reproduction, migration and gender. Her current research focuses on the reproductive lives of migrants in Vietnam.

Forests lost and found in tropical Latin America: the woodland 'green revolution'

Susanna B. Hecht

Forest dynamics in the Latin American tropics now take directions that no one would have predicted a decade ago. Deforestation in the Brazilian Amazon has dropped by over 80 percent, a pattern mimicked elsewhere in Amazonia, and is down by more than a third in Central America. Forest resurgence – increasing forest cover in inhabited landscapes or abandoned lands – is also expanding. In Latin America, woodland cover is increasing, at least for now, more than it is being lost. These dramatic shifts suggest quite profound and rapid transformations of agrarian worlds, and imply that previous models of understanding small-farmer dynamics merit significant review centering less on field agriculture and more on emergent forest regimes, and in many ways a new, increasingly globalized economic and policy landscape that emphasizes woodlands.

This paper analyzes changing deforestation drivers and the implications of forest recovery and wooded landscapes emerging through social pressure, social policy, new government agencies, governance, institutions, ideologies, markets, migration and 'neo-liberalization' of nature. These changes include an expanded, but still constrained, arena for new social movements and civil society. These point to significant structural changes, changes in tropical natures, and require reframing of the 'peasant question' and the functions of rurality in the twenty-first century in light of forest dynamics.

1. Introduction: the problems of paradigms – forest trend, agrarian systems

The green revolution of the twentieth century was focused on high-yielding annual crop varieties as the means of transforming peasantries and modernizing their systems of production with new institutional, economic, technological and ideological apparatuses to go with them. A different kind of 'green revolution' – one whose practices, politics and ideologies center on and produce forests – is perhaps more relevant to this next period. A century of American wilderness ideologies transposed to the Latin America tropics has shaped imagery of forests as poetic Edens, somehow outside of history and human inhabitation. We do not really see woodlands as incarnating massive structural change, because the ways that they do are hard to recognize unless they – forests – are obliterated. Yet very profound shifts are underway in markets, governments, governance, mobility and social policy that are refracted through the living mantles of these landscapes. This paper outlines some of the dynamics that are shaping forest cover in Latin America, and suggests that at the margins of the intersections of the urban, agrarian and forest transitions, different forms of rural life meanings and livelihoods are unfolding for small farmers. These diverge from the nineteenth- and early twentieth-century agrarian questions, and are especially infused with environmental/ecological politics of many kinds. Rural dynamics are affected by globalization and planetary processes occurring at multiple scales, and have changed the functions of rural systems in complex ways. Rural households are largely semi-proletarianized, semi-globalized and increasingly semi-urban, and thus reflect more hybrid politics and livelihoods than the agrarian question of earlier days.

Latin America is increasingly wooded. The paper explores some of the reasons the rate of forest clearing has declined so dramatically and woodlands have recovered so quickly. This pattern is often called the 'forest transition', which we review in more detail later in the paper. This transition reflects a number of political, institutional, social and market forces at several scales that valorized standing forests and processes that changed rural livelihoods. These dual dynamics suggest significant structural, political and livelihood shifts within a framework of new political ecologies, regionalized and globalized economic, policy and ideological contexts that produced more forest inflected political ecologies. This paper mentions REDD (Reduced Emissions from Degradation and Deforestation) processes only tangentially, largely because REDD has come to so dominate the discourses of Neotropical forests that it has obscured the number of dynamics that shape peasant livelihoods and politics that have constituted the new forested landscapes. The vast literature that engages REDD also deserves a separate treatment that is beyond the scope of this paper. The paper ends up with a discussion of the themes of agrarian change and new contexts might shape the next 40 years of JPS.

2. Forest trend in Latin America

Latin American tropical forests, covering some 11.1 km^2, with 3.3 million km^2 in open transition forests and savanna formations, are crucially important for global biodiversity, and climate change may pivot on whether they stand or fall. Global tropical deforestation is thought to generate about 13–20 percent of planetary carbon emissions – a figure roughly equal to the entire global transport sector. Thus, reducing deforestation, which reduces the forest as a carbon source, while rapidly absorbing carbon through secondary vegetation represents one of the most dramatic and efficient means of mitigation and sequestration at large scales in the global carbon economy (Galford *et al.* 2011, Fearnside 2012, Gloor *et al.* 2012). Since 60 percent of tropical forest clearing in planetary tropical forest biomes has occurred in Latin America, mostly in the countries of the Amazon Basin, declines in deforestation, increased woodland recovery in cleared lands throughout Latin America, and augmented forest cover in production landscapes are of planetary importance and critical for policy, rural livelihoods and development politics.

The forest trend in Latin America is changing profoundly and is marked by declining deforestation in every biome but those of tropical transition forests where clearing is still largely driven by agro-industrial development (Davidson *et al.* 2012, Aide *et al.* 2013). Gold mining often has localized high-impact clearing effects (Asner *et al.* 2013), livestock and speculation still stimulate clearing (Busch and Vance 2011, Carrero and Fearnside 2011, Hough 2011, Martins and Pereira 2012) and infrastructure development such as the program of integration corridors and dams of South America known as IIRSA (the acronym for Initiative for Integration of Regional Infrastructure for South America) remains a deforestation trigger whose larger impact over the next decade remains to be assessed.

Remote sensing coverage is now very good in Latin America and, while the capacity to use social data and integrate it with remote sensing data is improving rapidly, as a general rule the attention to the analytic social sciences and regional political economies in land change and climate studies has not been much of a priority and often integration lags behind the satellite monitoring (Walker *et al.* 2004, Turner *et al.* 2007, Turner and Robbins 2008, Perz *et al.* 2010, Walker *et al.* 2013). The sheer mass of geophysical and ecological research associated with climate change and land cover modeling in Latin America has considerably eclipsed the sparser volume of social science analysis, further fostering the sense of forest emptiness, a place that explodes into flame with human presence. The strong Malthusian cast within most of the scientific models and in the biological sciences also coincided with the emergence of international environmental non-governmental organizations (NGOs), which was critical for producing a globalized environmental discourse and futurology with a catastrophist emphasis (Hecht and Cockburn 2011, Medina *et al.* 2009, Borras and Franco 2010, Heise 2010, Davidov 2012, Schlosberg 2013). This was an artefact of the models, expressions of alarm and also a mechanism for fundraising. The destruction of tropical forests in its calamitous framing and imagery resembled nothing so much as nuclear annihilation with its smoldering landscapes and changed climate, albeit where nuclear winter morphed into the endless, immolating summer. The ideational parallel between nuclear annihilation and tropical deforestation perhaps fostered a sense of uni-directional destruction, without hope of landscape recovery. Faced with this prospect, and the episteme of empty wilderness, natural scientists and policy makers were largely wedded to the idea of terminal forest loss culminating in extinction or collapse of which the impacts on biodiversity, global climate and native populations would be disastrous. Hence, they argued for numerous and large conservation units (Nepstad *et al.* 2001, Peres 2005, Balch *et al.* 2008, Nepstad *et al.* 2008, Gibson *et al.* 2011, Laurance *et al.* 2012) and achieved many of their goals (Zimmerer 2006). Central American tropical biomes, where many American tropical ecologists have trained, were also seen as a Malthusian debacle (Durham 1979, Terborgh 1992). The echoes of nuclear imagery and the simplicity of the Malthus framing made tropical forest destruction legible to the average person in ways that issues of land distribution, participation, social justice, tenurial regimes, rights, violence, environmental history, globalization and agroecological debates never would be, although it can be argued (and will be later in this paper) that it was exactly the political ecologies surrounding these issues that helped change the broader deforestation dynamic. Indeed, it almost can be said that for the casual observer of Latin American development, deforestation more than any other rural feature was the hallmark of 'success' or 'failure' of rural development policy, especially in light of the lack of attention to the inhabited nature of its forests and the political economy of the drivers of forest destruction. At this juncture, though, simple Malthusian models can no longer prevail for explaining the forest trend (Aide *et al.* 2013, Hecht 2014). Even areas with exceptionally high population densities, such as Atlantic rainforests with over 75 million people and with major mega cities such as Sao Paulo and Rio de Janeiro, show significant evidence of forest resurgence (Ribeiro *et al.* 2009, Izquierdo *et al.* 2011).

2.1. *Inhabited forests*

Most of the discussion about the impacts of forest clearing over the last 40 years has centered on its biotic and climate dimensions, while agrarian issues typically focused on capitalist expansion into peasant systems with their implications for technical, structural and ecological change and their social outcomes in agricultural systems. In this discussion, the role of forests in livelihoods was often overlooked and resided in an anthropological and geographical literature, that became more salient with the movements of forest peoples such as kilombos, native populations and traditional extractors. About a fifth of Latin America's rural populations are directly forest reliant, and poorer households (and those headed by rural women) even more so (Pyhala *et al.* 2006, Gavin and Anderson 2007, Larson *et al.* 2007, Pacheco 2009, Cronkleton *et al.* 2013, Zenteno *et al.* 2013, Prado Cordova *et al.* 2013). The increasing recognition of the importance of tree systems in working landscapes for livelihoods, food security, insurance, energy, environmental services like microclimate modification, pollination and biological control, soil improvement, conservation benefits and ecosystem recuperation is drawing more analytic attention (Chazdon 2003, Chazdon 2008, Chappell *et al.* 2009, DeClerck *et al.* 2010, Vandermeer *et al.* 2010, Liere *et al.* 2012).

Although fragmented, agrarian landscapes in Latin America are often highly wooded and becoming more so in many smallholder systems (Larson *et al.* 2007, Moran-Taylor and Taylor 2010, Kandel and Cuellar 2012, Prado Cordova *et al.* 2013, Schmitt-Harsh 2013). In Meso-America for example, 98 percent of farms had more than 10 percent tree cover, 81 percent had more than 30 percent and more than half (52 percent) had 50 percent woody cover (Zomer *et al.* 2009). The transformation from cleared land to woody cover can occur quickly. For example, in El Salvador, satellite imagery from the early 1990s to 2006 showed a greater than 20 percent increase in areas with more than 30 percent tree cover, and a rise of about 7 percent in areas with more than 60 percent tree cover (Hecht and Saatchi 2007). In Guatemala, similar changes in tree density have been recorded (Holder and Chase 2012), a dynamic that has also occurred in Mexico (Mathews 2011). The Amazon estuary reveals rapid processes of forest cover intensification (Brondizio *et al.* 2014), among many other places (Farley 2007, Redo *et al.* 2012).

2.2. *Rural-urban forests*

Tropical urban dwellers often depend on transfers from rural economies, including tropical forest products, and participate in rural labor markets, often for timber, clearing, construction and artisanal mining (Chase 1999, Padoch *et al.* 2014, Reardon *et al.* 2007, Eloy 2008, Almeyda *et al.* 2010, Eloy and Lasmar 2011). Data on Latin American urban migration increasingly points to the rapid growth of rural towns and intermediate cities, settlements that are closely linked to the surrounding rural economies, and act as interfaces between these worlds. Periodic outmigration to cities for schooling, medical reasons and access to social welfare also occurs. Migrations are often circular and temporary, and involve multi-sited households, so the 'rural' and the 'urban' are relatively diffuse social, economic and spatial constructs. Local populations also move between local, regional and global product markets, as well as complex labor markets (Matos 2005, Reardon *et al.* 2007, Hecht *et al.* 2012, Limonad and Monte-Mor 2012, Brondizio *et al.* 2014). Increasingly, urban migration has been mobilized to explain declining deforestation and forest recovery, but it is only one part of the story because the natures of migration and urbanization are now so complex. Rural areas do not always depopulate and urban areas are also sites of agriculture and tree crop production.

The 'trammelled' nature of anthropogenic landscapes – whether occupied forests or peri-urban systems – produced a kind of blindness about such ecosystems: ecologists viewed them as lacking in ecological importance and unchanging, while agrarian scholars tended to focus on annual crop or commercial agriculture. Urbanists didn't see them at all. These spaces are produced by ecological processes, socio-economies, agricultural/

agroforest interventions, and by urban dynamics as well: urban land speculation on agricultural land, spontaneous occupations, urban foraging, peri-urban farming and the institutional regimes that may (or may not) accompany it. The dynamics of these systems remain relatively understudied, have been 'siloed' by disciplinary structures and still have not penetrated the development literature as well as they might have. How these systems articulate economically, ecologically, spatially, temporally and culturally with multi-scaled labor, land capital and commodity markets and with environmental governance and political economies at several levels is caught by a few case studies, but larger regional dynamics are very heterogeneous, which makes overarching narratives about anthropogenic landscapes relatively inconclusive (Brondizio 2008, Mathews 2011, Hecht *et al*. 2012). What is clear is that woodland regimes are deeply imbricated in the new tropical economies and their politics. This paper attempts to be more of a 'field guide' to such dynamics rather than a definitive statement.

3. Changing dynamics of deforestation

Over the last several years continuous satellite tracking of land use change in Latin America has recorded a sharp decline in deforestation, much to the surprise – and disbelief – of many, and documented a robust forest resurgence (INPE 2009, 2011, Schmook 2010, Mello and Alves 2011, Aide *et al*. 2013). This decline is recorded in all Amazonian countries (See Figures 1 and 2). This transformation would have been unimaginable in the 1970s and 1980s, the periods that most shaped recent Euro-American perceptions of tropical forest trends.

The next section explores both the 'whys' of earlier deforestation and the startling shift in its dynamics. We separate the analysis into two time periods because these represent the authoritarian and post-authoritarian periods whose political cultures and analytics about the dynamics of deforestation were very different.

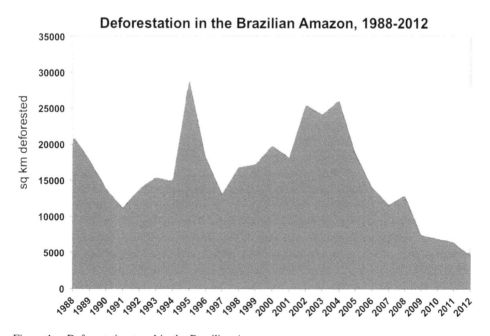

Figure 1.   Deforestation trend in the Brazilian Amazon.

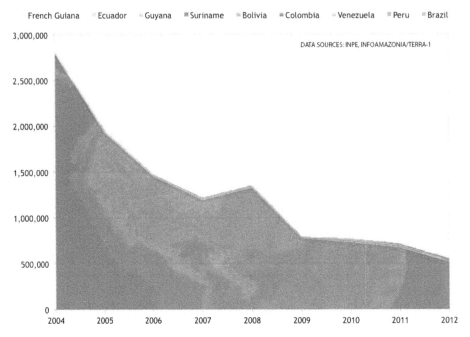

Figure 2.   Deforestation trend in non-Brazilian countries.

3.1.   *The nature of new nations: from economic growth to structural complexity*

A widespread model of economic modernization that correlates deforestation to increasing gross domestic product (GDP) is the Environmental Kuznets Curve (EKC), which asserts that as countries develop, their environmental footprints decline. Even in the 1950s, the Kuznets curve as a development framework was open to question, and later, as the EKC for deforestation moved into the limelight at the end of the 1990s, this theory took over a great deal of large-scale deforestation environmental analytics. As a quick review of carbon dynamics and/or environmental conditions in China and India might suggest, economic growth on its own is not sufficient for getting environmental improvements, an argument made by Kenneth Arrow and Robert Costanza (Arrow *et al.* 1996). In a meta-analysis of the EKC for deforestation, Choumert and colleagues reported that early studies of the EKC were largely indifferent to contrary data from the real world (Choumert *et al.* 2013). Other reviews indicated serious problems with data sources, the econometric techniques, problems in control variables, differences in geographical areas and also things like publication bias. Koop and Tole find no EKC for deforestation in Latin America – arguing that policy arrangements, institutions and distortions affecting land use differ so much as to defy useful comparisons (Koop and Tole 2001). Mills and Waite, running data on 35 countries, show no relation between country wealth and forest conservation (Mills and Waite 2009). Meyfroid *et al.* focused on trade and the transposition of deforestation from some regions to others in new patterns of spatial/economic expressions of regional and international markets, so one person's transition may rely on another's deforestation (Meyfroidt *et al.* 2010). The changes in forest dynamics require a much more nuanced review, and one that we believe is more helpfully framed by understanding transitions from authoritarian capitalism into the complex institutional, economic and political realms of the post-authoritarian period, rather than a simple GDP correlation.

For much of the post-World War II period, Latin American politics was dominated by authoritarian regimes, and many countries were rife with proxy wars, deep civil conflict and repressed civil societies. Macro-economies were largely characterized by import substitution industrialization (ISI) development policies, highly corrupt state and corporatist practices that had very negative environmental consequences, one of which was explosive deforestation. Authoritarian Latin America emphasized national projects of regional integration into the tropics, by activating old frontiers and embarking on new modernizing programs for changing rural and urban production regimes, and did so by emphasizing livestock, forest frontier colonization and road building. Land claiming in an era of questionable surveys and burned-up land offices often involved clearing for land claims. The mid-1980s to mid-1990s was a time of sharp institutional, political, policy and economic transition as the authoritarian regimes were overthrown, civil wars wound down and the 'Washington consensus' of privatization, fiscal austerity, free trade reforms, decentralization, and neoliberal policies were implemented. These new 'post-authoritarian' states developed under the discipline of two powerfully linked ideologies and practices mandated by the international community: neoliberalism and environmentalism.

Major transformations that slow deforestation have occurred in five main arenas: (1) new institutions of government at the level of the nation state such as environmental agencies, decentralized provinces and distributive social policies; (2) the rise of Latin America's civil societies and new forms of governance; (3) the evolution of markets especially in the globalizations of demand for commodities and labor, and the emergence of new environmental/ecological markets; and (4) significant changes in the paradigms of tropical science, the technologies of monitoring land-use dynamics, environmental histories, ecological economics and political ecology. These were critical in shaping new kinds of values, valuation and claims in forest dynamics; and, finally, (5) the intense processes of Latin American urbanization and migration. All had significant effects on the Latin American forest trend, and it can be argued that out of these interactive processes, major innovations in movements, policies and incentives have evolved, producing a new 'rurality' in the twenty-first century (Kay 2008, Hecht 2010). These are not necessarily landscapes of equity, nor have they simply transposed traditional rural questions into a forest register. Rather, they reflect deeper structural changes. This paper focuses on governance, markets and transfers (like conditional cash transfers and remittances) since the complexities of new scientific models and urbanization are beyond the scope of this paper.

As Latin American political systems ended their authoritarian phase the analysis of deforestation shifted from centralized state-led development approaches, corruption and destructive policies, and increasingly focused on more decentralized processes to slow clearing: the roles of forms of governance (the rise of NGOs and increasingly influential forest-based social movements like the rubber tappers, agriculture and territorial associations), resource rights and new environmental property regimes (recognition of traditional holdings such as Kilombos), new environmental institutional framings (inhabited reserves of many kinds, sustainable use areas, rather than simple conservation set asides), decentralization and participatory forms of development, as well as more monetarist dimensions of globalized land uses like market expansion and the neoliberal dynamics of expanded globalized and green markets. These provided inputs into the policy and political processes of an increasingly science-inflected, and ecologically monitored, tropical development. These were all seen as means of reducing deforestation and have had significant regional impacts to various degrees.

3.2. Social movements, new institutions and the new Amazon map

How can we explain the turnaround that defied the trends of previous decades and flouted some of the most cherished received ideas about deforestation drivers? These included strong commodity prices, drought (Amazonia's two worst droughts occurred in 2005, 2010), road building and free trade policies often identified as major clearing drivers. It is not that there wasn't any deforestation (there clearly was) but that institutional changes, social movement politics favoring human occupied reserves and conservation policies were paying off. Regulation, better markets for tree niche products, 'regional thinking' among NGOs and social movements involved in regional and large-scale planning exercises such as state-level economic and ecological zoning became more influential. The rise of what came to be known as socio-environmentalism—socio-ambientalismo— forged a 'third way' in rural development. Rather than simple agro-industrial yield efficiency or biodiversity conservation asides as the rationales for Amazonian occupation, socio-environmentalism marshaled discourses of social justice, historical resilience, political autonomy and sustainability of working landscapes. This differed from both the dominant modernization development models and classic conservation and provided the ideological framing which dovetailed with regional planning processes that were relatively participatory. Brazil provides the case that has been most extensively analyzed. The factors affecting deforestation turnaround or forest transition in Brazil are summarized below in Table 1.

The upshot of these dynamics is manifold: first, as the spatial map below shows, there has been considerable expansion in a wide variety of conservation settings, and the reality is that the institutional and legislative advances represent the triumph of social movements, traditional peoples, indigenous populations and environmentalists in shaping landscape at the macro scale in relatively intact forests within the broader frameworks of 'green governance' (Figure 3).

Social movements had significant effects on the Latin American forest trend, and it can be argued that out of these interactive processes, major innovations in policies and incentives have evolved that produced unusual environmentalisms, whether elite, popular or mercantile, that made the 'new' 'Amazonian map'. Conservation in native and biodiversity set-asides exploded in greater Amazonia to about 22 percent of these types of set-asides in Brazil. When inhabited environments and sustainability use categories are included, more than 60 percent of Brazilian Amazonia is now in some form of conservation or sustainable use landscapes. For all of greater Amazonia, about 45 percent is in some form of protection (Red Amazonica de informacion 2012). These increasingly encompass emergent forms of state, science, NGO and producer associations allied in pacts over natural resources management such as occurred in Acre, Amazonas and Amapa – the so-called 'green development' places whose dynamics were largely shaped by state policy and social movements.

In different modalities, Municipio Verdes (Green Counties) emerged as part of a focused regional policy, and a kind of combination of 'shaming politics', fiscal sanctions and new recuperation credit lines coupled to continuous monitoring. Mato Grosso as a state effectively decoupled agro-industrial soy and corn production from local deforestation as it intensified land uses in a switch from pasture to soy rotations. Paragominas, an early site of ranch development, later followed by timber, mining and charcoal, focused much more on patterns of commercial organizations, firms, markets, regulation, sanctions and new forms of credit lines. Municipios Verdes projects were means of slowing deforestation in areas of high clearing, recuperation of highly degraded landscapes and funding for intensifications in the older areas of regional

Table 1.   Governance, institutions and policy in slowing deforestation in Amazonia.

National commitments and formal institutions and processes
Development of IBAMA (The Brazilian Institute of the Environment)
Development of a national system of conservation areas (complete preservation to inhabited
   landscapes) (SNUP National System of Conservation areas), expansion of conservation areas
Forest code (in contest)
Recognition and demarcation of native reserves
Development of legal mechanisms for inhabited, non-native reserves
Enabling of sub-national agreements and autonomy
ZEE: econo-ecological zoning exercises
Placing parks and reserves in active development zones
INCRA (National institute for Colonization and Agrarian Reform) reform (Art. 88 which recognizes
   traditional rights)
Decentralization: regional- and state-level policies
'Territorialization' of conservation in policies: Acre, Amazonas and Amapa
Substantial state community forests
Ability to implement local PES (Payment for Environmental Services) REDD+ (Acre/Amazonas)
Investment in socio-environmental based sciences (agroforestry forestry and NTFP other resources
   management)
Mato Grosso 'panopticon': TNC (The Nature Conservancy)/IBAMA monitor clearing with GIS
   (Geographic Information System) cadastral overlay to control clearing
Municipio Verdes (Green municipalities) credits for recuperation
Deforestation zone credit blacklisting
Community autonomy in resource management
Historical communities (Kilombos, extractive reserves, traditional peoples)
Other forms of community management of forests
Regional planning through larger-scale organizations in civil society
Transfer of forest rights to communities
Better enforcement of legal sanctions
Real time monitoring
Better cadastral systems with GIS; reform of land institute (INCRA)
Green markets, commodity chains and their management
NTFP (Non Timber Forest Products) expansion in national markets (Brazil nuts, açai, coffee, fruits,
   oils and natural rubber)
NTFP expansion in international markets
Debate on GMOs (Genetically Modified Organisms)
Timber certification (quite fraudulent)
International boycotts of Amazon beef: Fribor, Carrefour, Wal-mart
International boycott of Amazon soy
Credit blackout zones in speculative deforestation expansion zones
National multi-layered climate policy
Signatories to UNFCC, Copenhagen, Cancun
Development of National Climate Policy
PES: Payments for lower deforestation, watershed and national CO_2 offsets
Elaboration of REDD policies (and pilot projects)
Other processes and spillovers
Social policy spillovers: 'Bolsa Familia' food subsidies to rural inhabitants, remittances
Ideological spillovers (desmatemento zero)
Political will

development, and can be seen as a kind of 'junker' model of land recuperation. The
evolution of state policies, institutions and monitoring within the framework of
NGOs and commodity round tables were a significant part of the story, but one of
an array of practices known as 'green governance'.

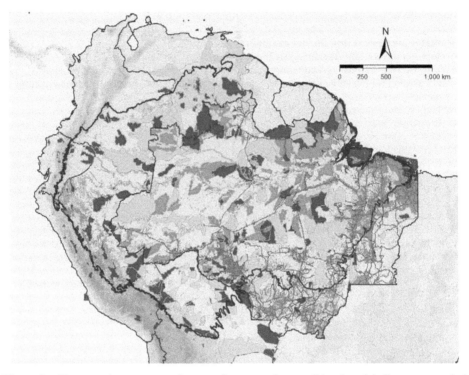

Figure 3.   The new Amazon map: impact of conservation, traditional and indigenous peoples' movements.

3.3.   *Green governance*

'Green governances' can be roughly divided into seven general categories with their related institutions: (1) inhabited conservation areas like extractive reserves, Kilombo lands, native territories, sustainable development zones and communal tenure areas managed more or less autonomously; (2) local governance of set-asides or regional park buffers where management is organized and directed by international conservation NGOs (da Cunha and de Almeida 2000, Brosius 2004, Linhares 2004, Bray *et al.* 2005, Goeschl and Igliori 2006, Nepstad *et al.* 2006, Garcia-Frapolli *et al.* 2007, Gray *et al.* 2008, Sletto 2008, Zarin *et al.* 2010); (3) large-scale producer associations like 'sustainable soy' which evolved to address environmental critiques and elaborated monitoring regimes, regulatory norms, and sanctions and the emergent 'Municipio Verdes' projects in conjunction with monitoring NGOs like TNC (The Nature Conservancy), IPAM (The Amazon Research Institute) and the provincial state (Nepstad *et al.* 2013); and (4) small-farm producer associations have also emerged as the institutional mechanisms for state and NGO transfers. This has expanded to (5) associations with a more territorial cast like those along the Xingu, or ejidos or Kilombo lands, and watershed resources councils, and (6) regional and forest peasant movements deploying environmental discourses to validate practices, to engage allies in regions of land struggle and to gain greater control over territory and territorial development processes such as Via campesina and MST (Land less People's Movement) (Bebbington 1999, Padoch 1999, Brondizio 2004, Edelman 2005, Bray *et al.* 2006, Page 2010, Klepek 2012). Finally, (7) private landowners engaged in the markets for environmental services organizing their activities along commercial land management carbon dioxide (CO_2) guidelines. These all suggest substantive institutional transformations unimaginable in the authoritarian period, dynamics that were reinforced by new kinds of markets.

3.4. *From green hell to green markets: production, consumption and services*

Green markets developed in several arenas: first through the labeling and segmented marketing of the production process in social and organic/biodiversity friendly traditional crops like coffee since the 1970s, and later through an emphasis on unusual non-timber forest products in national and international markets including Açai and Guayusa. During the 1990s, certification – a market-based system of regulation overseen by NGOs – was increasingly required to enter elite markets. This emphasis on 'branded' elite or non-traditional niche crops, NTFPs, timber and tree-based crops became an important small-farmer strategy in light of the collapsing prices of grains and the conservation gloss that adhered to tree commodities, although costly for smaller farmers and inhibited many kinds of agroforest uses and traditional practices (McAfee 1999, Brondizio 2008, Avalos-Sartorio and Blackman 2010, Bacon 2010, Beuchelt and Zeller 2011, Abbots 2012, Bitzer *et al.* 2013, Zenteno *et al.* 2013). In a world where social identity is increasingly signified by consumption patterns, the simultaneous extension of 'distinction' with ubiquity lies at the heart of these business models and contributes to much of the externally perceived cultural content of specialized woodland products. The practical result is that these systems help generate what Meyfroidt and Lambin (2011) call the 'smallholder tree-based intensification' pathway of forest transition. These affected forest recovery as well as deforestation. Next, markets expanded for local urban construction inputs and other products from intensifying successional plots or agroforests for popular building and regional uses as local urbanization triggered by roads and state policy expanded (Sears *et al.* 2007, Medina *et al.* 2009, Cossio *et al.* 2011). Finally, second-generation frontier families increasingly intensified tree holdings as a consequence of labor shortages with outmigration, better marketing and better infrastructure (Browder and Godfrey 1997, Bilsborrow *et al.* 2004).

In the case of larger-scale holdings, commodity chain analysis began to also play a role in the marketing dynamics as part of consumption politics managed by firms, NGOs and consumers (Bakker 2010, Millard 2011, Hecht 2012, Peluso 2012, Hejkrlik *et al.* 2013, Newton *et al.* 2013). Firm and consumer boycotts of products associated with deforestation like the Greenpeace boycott against Amazon soy and the 'deforestation livestock' embargo by big supermarkets like Carrefour Walmart and the Brazilian supermarket chain Pão de Azucar used their market leverage. This was also accompanied by the 'stick': as 'Municipios Verdes' unfolded 'Environmental crimes' could include sourcing beef from deforestation areas, with hefty fines for buyers and rural credit blacklisting for sellers.

Another evolving market engaged the expansion of ecotourism. Advanced by conservationists as a means of monetizing the hedonic value of forests, these enterprises make state or indigenous territories (like the Cuyabeno or Yasuni reserves in Ecuador) as well as national parks intelligible to outsiders through the experience of the smaller ecotourism area adjacent to the larger park or native reserve. The total global amounts of revenue generated by ecotourism are estimated at US 29 billion a year: a miniscule portion of the largest global industry, tourism (Kirkby *et al.* 2010). Ecotourist holdings are also eligible for REDD funding. Kirkby, analyzing ecotourism in Peru, suggested returns of USD 1100 per ha on 55,000 ha of ecotourism enterprise per year compared to unsustainable gold mining and timber extraction, and especially low-value grain agriculture and extensive livestock. Ecotourism is an elite niche, and does generate some local jobs as well as producing a kind of conservation episteme among both visitors and local populations, but it also comes in for criticism about the forms of employment and other kinds of local impacts (Wunder 2000, Garcia-Frapolli *et al.* 2007, Horton 2009, Almeyda *et al.* 2010, Kirkby *et al.* 2010, Ojeda 2012, Peralta 2012, Doan 2013).

Finally, the most publicized form of green markets is emergent payment for environmental or ecological services – markets that monetize previously unvalued, or at least unsystematically valued, ecological or environmental landscapes, including the conservation value of inhabited landscapes, plantations and conservation forests especially for carbon offsets. These have become controversial as questions of local sovereignty and management as value and values have become contentious (Borner and Wunder 2007, Daniels *et al.* 2010, McAfee and Shapiro 2010, Danielsen *et al.* 2011, de Koning *et al.* 2011, Ibarra *et al.* 2011, Barbier and Tesfaw 2012, McAfee 2012, Shapiro-Garza 2013). However, REDD+ (the 'plus' refers to social additionalities) is also viewed as the most likely mechanism for supporting commodity and non-commodity forests in working landscapes at least in terms of conservation funding. There are now close to a thousand REDD- or REDD +-ready projects in Latin America such as the well-known Juma program in Amazonas and the indigenous Paiter-Surui carbon offset (the first REDD project for an indigenous Amazonian group); Acre and Chiapas are offset venues as parts of California's AB 32 sustainable energy programs, and these do not include the projects developed in the private offset or 'non-REDD' offset markets like Ecuador's 'socio bosques'. The financial mediator fund Forest Trend for many private investors' has mobilized USD 598,600,000 to support REDD activities just in Brazil, but mostly for plantations. The debates on REDD+ engage an enormous literature that is beyond the scope of this paper but must be considered in how they articulate in the larger dynamic of forest cover.

3.5. *Social policies with deforestation impacts: cheap food; conditional cash transfers*

Cheap food policies were key to urban stability throughout the authoritarian transition period, and remain so in the current day since food price hikes in Latin America correlate with urban unrest. These policies can be seen as a form of urban stabilization, poverty alleviation, general consumer support and state legitimation within a globalized, free trade system and political economies that produce high levels of inequality (Lavinas 2013). These policies have injected a fundamental instability into small-farmer grain economies (Fitting 2006, Gravel 2007, McAfee 2008, Keleman 2010, Radel *et al.* 2010). Peasant wage food production had been seen as the linchpin in earlier twentieth-century rural development and modernization programs, with credits and technology development focused on small-farm systems. Small farmers, after all, were the whole point of the green revolution and meant as a green counterweight to the 'red' marxist revolutions (McAfee 2008, Cullather 2010, Patel 2012). The role of the small farmer as basic wage food provider was mostly eroded as prices for basic food grains declined throughout the Latin American tropics, especially as free trade blocs like NAFTA (North America Free Trade agreement) and CAFTA (Central America Free Trade Agreement) expanded. Small farms lacked the ability to compete with national or international agro-industry due to the dynamics of imports, credit lines, subsidies, political position and national political economies of food (Wilcox 1992, Hecht 1993, Murphy 2001, Van Ausdal 2009, Siegmund-Schultze *et al.* 2010, Del Angel-Perez and Villagomez-Cortes 2011, Martins and Pereira 2012). Small-farm annual cropping systems began to contract in areas of southern Mexico, Central America and Amazonia (Fitting 2006, Neeff *et al.* 2006, Pascual and Barbier

2007, Brondizio 2008, Keleman 2010, Hecht *et al.* 2012, Browning 2013). This, along with widespread semi-proletarianization within both rural and urban (and international) labor markets, meant that by the new millennium, what was meant by a peasant in its classic definition as an autonomous *agricultural* producer who retained the means of production was transformed. Households' 'answer' to the agrarian question – how peasants would articulate, or transition to, modern globalized capitalism – stretched into multiple forms of livelihood, spatially complex living arrangements and from tropical forests to urban centers in the First World and the Third. While peasants still continue to clear and intervene in tropical ecosystems, production activities with multifunctionalities, multiple use land management with flexible labor requirements, products that can move between markets and subsistence like small-scale cattle, home compounds, small-scale agro forests and unfarmed or fallowed holdings increasingly hold sway. Thus, what had been a driver for about 30 percent of clearing in tropical biomes – annual crops – has gradually contracted even as, in many cases, woodlands expanded (Mathews 2011, Robson and Berkes 2011a, Hecht *et al.* 2012).

Conditional cash transfer policies emerged as one response to increasing patterns of inequality that resulted from authoritarian regimes and neoliberal reforms. As part of the 'left turn' in the forms of Latin American states as commodity booms and foreign investment enhanced growth but exacerbated socio- economic inequalities and increasing social instabilities in the transition periods, elected governments with left-leaning politics began to explore redistributive programs such as the 'Bolsa Familia' in Brazil, 'Oportunidades' in Mexico and similar programs in Honduras, Nicaragua, Bolivia, Ecuador, Chile, Uruguay, El Salvador and elsewhere. These programs provide women in poor households with a cash supplement on the condition that kids study and get vaccinated for childhood diseases. As a poverty alleviation program, the justifications include the elimination of hunger (President Lula da Silva's 'Fome Zero'), human capital development, public health and reduction in the use of child labor. In Brazil, 'Bolsa Familia' and the social security program 'Beneficia de Prestaçao Continuada', as well as retirement pensions, accrue to the impoverished elderly or disabled. There is a considerable policy literature about the welfare impacts of these programs, the basic Keynesian redistributive approach of which has improved food access, reduced infant mortality and enhanced schooling (Hall 2008). Bolsa Familia covers some 12 million families, affecting 44 million Brazilians. Mexico's Oportunidades covers about 20 percent of Mexican households (Rawlings and Rubio 2003) and some 44 percent of Mexican rural households. Duarte and colleagues, studying the impact of Bolsa Familia in rural Pernambuco, found that 88 percent of the income from these transfers was spent on food, which meant rather than growing, people bought food (Duarte *et al.* 2009). This by no means suggests that annual cropping completely stops, but rather that annual cropping becomes more a supplement and security than a subsistence strategy. This has many implications for agrodiversity, regional landscapes, indigenous knowledge systems and nutritional profiles for those who move from their own production to more processed food.

When looking at Amazonia, where deforestation has declined so much, it is useful to note that regional income source shares for Amazonia include three percent from retirements, 7.2 percent from Bolsa Familia and 10.3 percent from the 'Benficia Continuada'. That is, 20 percent of regional income source shares come from these transfer programs (Silveira-Neto and Azzoni 2012). These data highlight the importance of targeted poverty alleviation subsidies for reducing deforestation pressures to a certain degree and, with less clearing, allowing more potential forest recovery.

The impact of conditional cash transfers on deforestation is not known precisely because it hasn't been specifically studied, but the limited data available suggest that these transfers in Brazil have stabilized and reduced clearing pressure. Piperata and colleagues (2011), with longitudinal data from 2003–2009, describe a 37 percent decline in manioc gardens in households with Bolsa Familia. Padoch and colleagues (Sears *et al.* 2007) and Brondizio *et al.* (2014) report agricultural contraction associated with pensions, and that people switched from more agricultural activities to working their açai holdings. In the Atlantic Rainforests, Kilombo dwellers also reduced clearing with transfers and more waged activities associated with periodic outmigration (Adams *et al.* 2013). Van Vliet and colleagues found that Bolsa Familia income stabilized clearing on new road systems in central Amazonia (van Vliet *et al.* 2013). The impact of these domestic transfers, like remittances, have complex outcomes on forest dependence – they increased it in the Piperata, Padoch and Brondizio studies, but do not always. The impact is further compounded by the rise in wage labor and the expansion of urban and rural labor markets with multi-sited households, but where transfers provide a baseline household subsidy in the context of increasing informality of labor markets and volatility in commodities (Lavinas 2013). In this light, it is useful to look at the effect of remittances, also a cash transfer (one we discuss more further on), on agriculture. As authors from Central America point out, even households that get subsidies (remittances, transfers, etc.) still often keep some agriculture going, but as the next table based on data from Central America shows, remittances generally were associated with retraction in agriculture, although small-scale livestock investment (because of its flexibility and low labor requirements) was part of the mix. The situation is complex, however. For Oportunidades, Alix-Garcia found that transfers increased the probability of deforestation especially for cattle investment with distance from infrastructure (Alix-Garcia *et al.* 2013) (Table 2).

Current social policy is virtually invisible in the deforestation literature, but it serves analytically as a proxy for thinking about other forms of domestic transfers: remittances, and other state welfare, retirement and transfer payments. They can also serve as proxies for understanding the effects of REDD more generally. Prior to REDD initiatives, and for many areas outside of REDD financing, various kinds of transfers, including migration, remittances and state transfers, have had a substantive impact on forest trends. In the case of Puerto Rico, these were instrumental in changing the countryside to a 'post-agricultural landscape' (Aide *et al.* 1996, Rudel *et al.* 2000, Grau *et al.* 2003, Lugo 2009). This dynamic of declining clearing and enhanced forest recovery is well documented for El Salvador (Hecht and Saatchi 2007) and exemplified in Figure 4, which leads us to the question of forest resurgence (Figure 4).

4. Forest recovery and the forest transition

Forest regeneration and reforestation – what is described in the literature as the 'forest transition' – is very widespread in the Latin American tropics, which gained 360,000 km^2 of forest cover over the last decade (Aide *et al.* 2013, Clark *et al.* 2013). This trend is shown in Figure 5. Famously deforested places like El Salvador have experienced remarkable forest recovery (Hecht *et al.* 2006, Hecht and Saatchi 2007), as has much of the Caribbean, which was largely deforested for hundreds of years as slave-based sugarcane, tobacco and coffee production took its ecological as well as social toll. Andean zones, often derided for vistas of despair and deforestation, are now sites of rapid woodland regrowth. Xeric forests in northern Mexico and the Brazilian northeast have also expanded significantly (Aide *et al.* 1996).

Table 2. Changes in land uses in migrant communities in Southern Mexico and Central America.

Agricultural retraction: annual cropping systems

Veracruz (MX)	90 percent retraction in migrants' households, and significant retraction in households engaged in wage labor or urban commerce; (high levels of 'Oportunidades' as well as remittances)
Sierra Juarez (MX)	52 percent annual crop retraction
Coastal Oaxaca (MX)	10 percent annual crop retraction
Sierra Sul (MX)	10 percent annual crop retraction
Olancho (HO)	43 percent decline in cultivated areas in migrant households 8 percent complete retraction
Las Vueltas (ES)	45 percent households
Nueva Concepción (ES)	Cattle to houses
El Salvador	50 percent of migrants still cultivated some land as opposed to the 88 percent of non-migrant households

Coffee/tree crops expansion

Veracruz (MX)	Coffee
Coastal Oaxaca (MX)	Commercial coffee, fruit trees
Sierra Juarez (MX)	Commercial timber
Olancho (HO)	35 percent increase in coffee area (65 percent of land purchased goes to coffee); coffee can claim land in conservation areas hence its expansion at the frontier; also, Honduras has a national policy of coffee promotion
El Salvador	Coffee expansion

Pasture

Olancho (HO)	24,000 ha conversion between 2004–2009 in households with immigrants (household area in pasture increased 60 percent); non-migrant households less than 14 percent of area of migrants in cattle)
Nuevo Concepción (ES)	Honduran extension of cattle frontier
Vera Cruz (MX)	Long history of cattle

Secondary vegetation expansion
Olancho (HO)
El Salvador
Vera Cruz (MX)
Sierra Juarez (MX)

New conservation areas

Olancho (HO)	
Coastal Oaxaca (MX)	Payment for environmental services (PES)
Sierra Sul (MX)	

National programs affecting land use

Olancho (HO)	Land claiming through coffee okay in conservation areas: livestock credits
Oportunidades	Schooling for kids #2 source of household income in Sierra Sul; Veracruz
Credit Lines	Cattle – Olancho, southern Mexico; coffee – El Salvador, Honduras, Mexico

Source: Hecht *et al.* (2012).

Forest resurgence is a function of natural regeneration of agricultural lands and pastures, but also reflects agroforestry expansion, densifying silvopastoral systems, forest plantations, regeneration projects and intensification on existing holdings and in home compounds. The drivers and the biotics of this forest resurgence vary widely, as we will discuss later. The expansion of monocultural forests is described in the literature as a

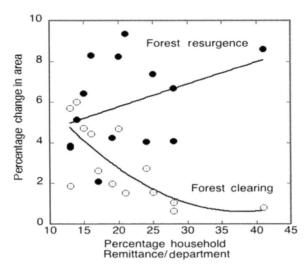

Figure 4.   Decline in deforestation and rise in forest recovery in El Salvador.

type of forest transition in which the definition of transition reflects observable expansion of *all* kinds of woodlands, but clearly such forests have different ecological characteristics. This expansion has also brought into focus the complexity of anthropogenic landscapes as *forested* ecosystems, rather than only cropping systems.

There is currently considerable debate about the 'value' of resurgent forests compared to old growth in terms of ecological and environmental services (Chazdon 2008, Lugo 2009, DeClerck *et al.* 2010, Robson and Berkes 2011b, Blinn *et al.* 2013, Hecht 2014). Forests are dynamic and their successional parameters can be and are manipulated over time. Their biophysical, cultural and social functionalities depend on basic ecosystem parameters (like soil type, rainfall, slope, aspect), previous systems of land use (for example, swidden versus pasture), length of time in different land uses, the kind of ecological matrix the land use was embedded in, the forms and kinds of the connecting corridors and the ecological characteristics of the new forests (from novel ecosystems, intensified forest gardens to monocrop clonal tree plantations). These resurgent landscapes are often novel ecosystems with high levels of ecological function including carbon storage, biomass and diversity (Hobbs *et al.* 2006, Chazdon *et al.* 2009, Lugo 2009, Lugo *et al.* 2012), although significant forest areas are now in monocultural systems in the Andes (Farley 2007) and parts of the Atlantic rainforest and cerrado (Hylander and Nemomissa 2009, Rocha *et al.* 2013). Oil palm, though much smaller in area compared to Asia, is expanding (Gutierrez-Velez *et al.* 2011, Gutierrez-Velez and DeFries 2013, Castiblanco *et al.* 2013). That said, Latin America's forest transitions are more ecologically complex compared to Asia's exploding monocultures of rubber, oil palm and monospecific plantations (Danielsen *et al.* 2009, Zhai *et al.* 2012, Fox and Castella 2013, Mertz *et al.* 2013).

4.1.   *'The forest transition' in the North and South: Latin American qualifications of Euro-American transitions*

The rewooding of the US and Europe was widely associated with urbanization and industrialization and the structural shift implied with European land enclosures and regional

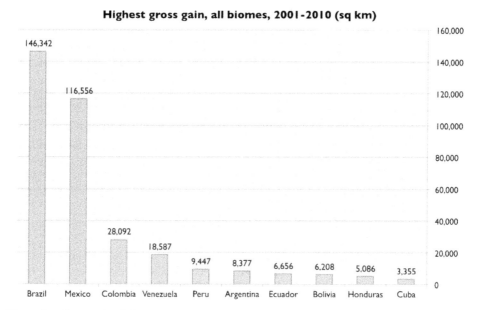

Figure 5.   Gross gain of biomass in all biomes.

dislocations of agriculture through production efficiencies and technical change within agricultural landscapes (Mather 1992, Mather and Needle 1998). Trees regrew and populations declined in the countryside as urbanization, industrialization and agro-industrialization expanded. Some authors have sought to universalize this Euro-American model of the 'forest transition' as part of the environmental Kuznets curve, and invoke the modernization narrative that developing countries will more or less repeat the patterns of the First World.

Half a century of development analytics have urged caution about translating First-World experiences and histories to the developing world. There are the many differences between them, not the least of which are the ecologies of the tropics and the structural conditions under which tropical economies developed (Hecht 2004, Perz 2007). Current transitions in the developing world differ from the Euro-American pattern: (1) they have occurred extremely quickly – over decades – as opposed to northern processes that often took centuries; (2) they reflect strong exogenous pressures at least as much as endogenous dynamics; (3) rural areas, even though forested, often have high population densities so the 'demographic scouring' of Euro-American transitions has not consistently occurred in absolute terms; (4) forest recovery processes are often driven by more globalized pressures, responding to commodities, financial flows and, often, labor outmigration into regional or global markets (Asner et al. 2002, Rudel et al. 2005, Hecht and Saatchi 2007, Garcia-Barrios et al. 2009, Barbier et al. 2010, Hecht et al. 2012). These transformations involved significant recasting of the livelihood possibilities, meaning, the 'nature' and politics of these Latin American landscapes. These general features are made more complex by varying historical roots, diverse cultural-ecological matrices, an array of tenurial systems, emerging regional, global and environmental markets, policy interventions, war, institutional rivalries, new ideologies (including environmental ideologies), territorial identities and competing authorities that vie for the political spaces that forests have become. What drives forest resurgence in Latin America? We outline some of the major influences: catastrophes, socio-ecological landscape abandonment, forest policy and remittances.

4.2. *Catastrophic events and forest resurgence*

Events that produce massive demographic changes or severe economic disruptions have historically been associated with forest resurgence. Emblematic in this regard is the extensive forest recovery after the population collapse in the New World associated with the arrival of Europeans and their diseases (Mann 2005, Mann 2008, Nevle and Bird 2008, Nevle *et al.* 2011). Climate-related collapses that ended the high classic Maya also produced rewooding (Sheets 2002, Turner 2006, Mueller *et al.* 2010, Ross and Rangel 2011). Problems like nuclear waste dumps and testing sites in the US induced a 'rewilding' of hyper-contaminated sites (Mascoe 2007). Increasing severity of weather events or other geophysical catastrophes, like hurricanes, tsunamis, earthquakes, etc., may contribute to land use change (1) by making areas uninhabitable, (2) through efforts of biotic stabilization (and new tree-based production systems) to reduce vulnerabilities, or (3) through resettlement (Ensor 2009, Timms 2011).

War and militarization have catastrophic impacts that often produce forest resurgence when agricultural frontiers are inhibited, productive infrastructure is ruined, local populations are displaced and remnant populations become both targets and collateral damage in proxy wars and civil conflict. Emblematic in this regard are El Salvador in the 1980s and Colombia over the last few decades. While conflicts can lead to forest recovery, a kind of gun-barrel capitalism can produce land uses that claim lands of others or close off future land occupation as in the development of pastures in conflict zones in Guatemala (Ybarra 2009, Moran-Taylor and Taylor 2010, Ybarra 2012) in parts of Colombia (Sanchez-Cuervo *et al.* 2012, Thoumi 2012, Sanchez-Cuervo and Aide 2013) and Nicaragua (Mascaro *et al.* 2005, Stevens *et al.* 2011).

Economic collapse can also trigger profound landscape change as occurred in the Purus and much of the Upper Amazon after the collapse of the rubber economy (Coomes and Barham 1994, Hecht 2013). The decline in Mexico and Central America's corn prices might also be considered an economic collapse or catastrophe. Climatic, conflict and market volatilities increasingly may interact with each other. Catastrophic events are driven by exogenous forces, part of planetary and human history, and should be incorporated into an understanding of forest recovery through land abandonment, migration and tree-based adaptive strategies.

4.3. *Socio-ecological abandonment*

Land degradation with soil nutrient declines, erosion, resource overuse, catastrophic pest outbreaks in agriculture and weather events, etc. can produce forest recovery as land is taken out of production. Lambin and Meyfroidt refer to this as a 'socio-ecological feedback', where ecosystem services and utility are degraded by past land uses; human use is withdrawn and the endogenous processes of ecosystem recovery occur. Such huge, derelict landscapes are widespread in Latin America where the abandoned or extensive pasture is the iconic countryside in many regions. About 20 percent of former pastures in Amazonia are now in some form regrowth (INPE 2009, Mello and Alves 2011). The empty pasture is indeed an ecological story of soil nutrient decline and brush invasion (Hecht 1985, Fearnside 2008), but it is also expressive of more complex dynamics, often as much an institutional as an ecological narrative: land claiming, speculation, money laundering, institutional rents, a low labor enterprise, a failed strategy of economic diversification, collapsed markets and regional land conflicts can all be manifested in wasted grassland. The abandoned upland milpa in El Salvador speaks to a history of dispossession that drove farmers out of traditional holdings into erodible steep lands, but the real coup de grace came with the civil war. Soil nutrients and erodibility are scientifically legible as proximate causes; the social underpinnings far less so. 'Ecological abandonment' perhaps needs more comprehensive framing, one underpinned by political ecologies as well as edaphic or biotic stories. Degraded landscapes often become targets of state- or NGO-led recuperation which is often justified in terms of ecosystem services; such landscapes are now increasingly sites of carbon offset recuperations which include the aims of livelihood improvement, and often revalorize land holdings. These policies often become part of what are called 'forest scarcity' state approaches.

4.4. *Forest scarcity*

Rudel *et al.* (2005) and Meyfroidt and Lambin (2011), among others, point to 'forest loss' as a pressure that stimulates state policies and subsidies for reforestation as woodland resources and services become scarcer. These models focus on how decline in forests enhances demand for marketable forest goods and changes land values of woodlands, while loss of forest environmental services, especially erosion control, water resources and conservation, becomes an alternative stimuli for recuperative state intervention (Barbier 2009, Meyfroidt and Lambin 2011). While these are often seen stimulating wood product markets, these forest programs are typically highly subsidized and often subject to institutional rather than production rents. States often use forest policy to change tenure regimes, forest access and regional economies (Ribot 2001, Brown 2003, Kant 2009, Larson *et al.* 2010, Peluso and Vandergeest 2011). Recently, however, new forest rights policies have been helpful in forest expansion as 'effective use' – that is, cleared land – is no longer a necessary requirement for claiming, and traditional forest uses are legitimated in emergent legal systems. Complex landscapes with successional features increasingly feature as part of 'forest' rather than agrarian reform, and this rights-based approach has been instrumental in keeping forests in place even when under pressure, such as with extractive reserves, and other kinds of traditional peoples' holdings, (Pacheco *et al.* 2012).

4.5. *'Unconditional' transfers: remittances*

Migration has emerged as one of the most salient features of modern globalization. The World Bank estimated that there are some 215 million international and some 700 internal migrants worldwide, which, given the scale of undocumented movement, is surely an underestimate (World Bank 2013). More than USD 550 billion in formal remittances was projected for 2013, with a rise to some USD 700 billion expected by 2015. At the macro level, migration patterns vary widely by regions, by histories and by current drivers. At micro levels, migratory processes differ by class, by gender, culture, by occupations and receiving countries, and by the networks and institutions that mediate and drive migration. Latin America receives about USD 65 billion per annum. The magnitudes of these transfers rival direct foreign investment and often exceed it in small economies. Remittances are more than triple the multilateral development assistance worldwide. For many countries in Central America, remittances are now a major portion of their GDP. Remittances count for 20 percent of the GDP of Honduras, 17 percent of El Salvador and Nicaragua's economies, and 12 percent of Guatamala's GDP. Remittances as a percentage of the value of exports are now worth about half for Nicaragua, 60 percent for Guatemala and 75 percent for Honduras eclipsing any national agricultural or industrial export product. This mobility and its revenues in all its forms affects households and communities profoundly and has multiplier effects on wellbeing and livelihoods. Migration affects urban and peri-urban forms, but also – due to labor changes and remittances – rural strategies, investments, local cultures, labor markets, governance, institutions and, far less noted but no less consequential, forest landscapes. Given the vast attention to forest policy in terms of deforestation, climate change, the micro analytics of forest science and a surging literature on conservation ecology, it is surprising how relatively little forests are addressed as part of the migration-development nexus and most especially how invisible forests are as investments in urban livelihoods and complex 'matrix landscapes'. Virtually every rural study in Central America and Mexico makes references to the role of remittances in the dynamics of rural landscapes, even if they do not specifically reference forests. Relatively little attention has focused on the remittance impact for forests in either the migration or the forestry literature, but the emerging data show that remittances have clearly affected forest patterns (Klooster 2003, Hecht and Saatchi 2007, Schmook and Radel 2008, Mathews 2011, Vaca *et al.* 2012).

Forests hold potential for illuminating some of the 'deagrarianization' as well as 'repeasantization' debates by expanding our understanding of the rural beyond the crop field and into the landscape, and by placing rural livelihoods in a more dynamic framework than the small-scale peasant agricultural community that long prevailed in rural analytics. In the same way that institutional analysis has augmented our understanding of forest dynamics, so too have forests expanded our knowledge of institutional repertoires, and it is likely that forests can help broaden our understanding of some of the socio-environmental features of migration by lifting our eyes from the field to the forests that are in many ways the defining feature of the Neotropics (Bray *et al.* 2006, Mathews 2011, Robson and Berkes 2011a).

The impact of remittances on forests depends on a number of factors, but how investment is occurring can be difficult to assess. Intensification of agricultural production – irrigation, machinery, agricultural inputs – is more visible and widely documented. Tree-based intensification is often not easily seen for several reasons: (1) tree intensification is based usually on knowledge systems rather than equipment, irrigation or fertilizer, and this may not show up in national accounts and thus is difficult to track; (2) agroforestry and tree intensification structures can range from fairly casual management to intensive production, a feature that fits well with labor availability in migrant households but may be invisible to outsiders; (3) the systems can be quite spatially diffuse, occurring in highly managed to almost unmanaged landscapes; (4) the multi-use features of these systems, their 'hidden harvests', may obscure forms of intensification for non-marketed or local niche markets that may not be obvious to outsiders (Brondizio 2004); and (5) the habit of collapsing complex forests into one product with trackable commodity chains (like firewood) obscures the range of possible managements and intensifications and the inherent flexibility of these systems which are especially important in households with periodic migration and diverse livelihoods. This invisibility, however, makes them potentially vulnerable to forms of extra-legal and legal expropriations.

The expansion of forests reflects other effects of migration. These include the demographics: many communities in Central America have lost more than 60 percent of their men. This has several significant impacts along with the other vast and difficult socio-psychological and economic problems that attend migration. First, it changes community demographics, contributing to the aging and the 'feminization' of the campo (Van Wey *et al.* 2005, Bates 2007, McSweeney and Jokisch 2007, Ibanez and Velez 2008, Radel *et al.* 2010, Koster *et al.* 2013). Then, migration creates a labor deficit. This is precisely the strong labor needed for forest clearing, field work and many forms of natural resources management. This labor shortage cannot be easily compensated for by the traditional labor exchanges and collective labor that prevailed in many communities in earlier times, since former or potential participants may be residing out of the country. This can produce destabilization of the traditional institutions of governance since 'cargas' – symbolic and community social obligations – may be difficult to meet, especially with enhanced border security worldwide that limits what had been earlier transnational flows from destination countries for festivals and family events. While remittances may substitute for labor and provide celebratory goods, the bonds of reciprocity and identity can be weakened. There are also other elements of the intergenerational and intergroup transfers that develop knowledge and experience about the physical management and manipulation of forest and successional systems as well as the knowledge and ritual systems that underpin them. These impacts at the locality affect communities as well as households, and ultimately forests as well. Lack of management can increase vulnerability of forests to fire, change levels of availability of natural resources of many kinds (such as forage, wildlife and fruits), and increase the vulnerability of such areas to incursions or resource theft (Putzel *et al.* 2008, Cardozo 2011, Duchelle *et al.* 2011, Fule *et al.* 2011, Shanley *et al.* 2012, Zenteno *et al.* 2013).

Finally, the processes of rural differentiation are enhanced by remittances. International remittances roughly double rural household incomes (Hecht *et al.* 2012). Investments in rural production may continue; other activities, including building spec houses, urban real estate, new land acquisition, animal acquisition, expanding tree crops and intensifying home compounds, form part of the new rurality that shapes investment and forest dynamics. Indeed, in much of Latin America, it is almost impossible to understand forest dynamics, especially those in small-farmer contexts, without placing migration in its many modalities at the heart of the question. Why was land abandoned, why did forests expand, who bought cattle, why was some agriculture intensified, etc.? These are but a few of the many questions that inhere in the Latin American diaspora, and that place the forest question at the very heart of rural studies.

5. Final thoughts: forests, farmers, 'green revolutions' and 'green globalizations'

This paper has reviewed some of the more salient processes underpinning the transformations in forest cover in Latin America, to suggest what realms of analysis will be necessary to understand 'the rural' over the next 40 years of JPS. The previous decades of JPS focused on how peasants as agrarian actors integrated into capitalist production systems, the new dynamics of proletarianization, technical change, knowledge systems and technical choice in agriculture. How did peasants shape issues of revolution and resistance, and the rise of identity politics, especially gender and indigeneity, out of the morass of the term 'peasantry' as a means of maintaining autonomy? These efforts largely centered on peasants as *agricultural* producers and rural agricultural laborers. What this paper has tried to show is that several key parameters will be structuring peasant debates in the future and many of them revolve around a 'green axis' of forests, planetary instabilities, migration, globalization and environmentalism whose politics are likely to be structured by more globalized 'disciplines of nature' (in both the Foucauldian and academic senses) via green markets of commodity production, consumption, environmental service markets and new forms of governance including those framed within international scientific concerns – like climate change and conservation. These will probably emphasize social and environmental policy in lieu of agrarian or agricultural policy. This configuration generates new forms of rurality that function as economic and security 'platforms' in a context of increasing informality in labor and volatility in commodity and service markets and greater planetary uncertainty. The emergence of "neo-liberal nature" takes several forms, whether triggered by markets for environmental services and offsets like REDD, new kinds management regimes whether mandated by environmental NGOs or resulting out of migration, regional management regimes using market incentives as in the Municipio verdes and the Central American Biological corridor, as well as emergent "green" markets of various types continue as sites of contention. These will not be 'easy' sites of governmentality. Resistance to forms of expropriation by environmental enclosures embracing the 'wilderness' of the 'Ur wald' – the imaginary primeval Ur nature – has as its analog the emergence of new monocultural 'neo-natures' of genetically modified and highly manipulated genomes of soy, clonal eucalyptus and oil palm, among others, that are harbingers of structural discontinuities. These affect rural areas and impose different pressures on rural peoples – different at least in form. The questions of the 'reproduction of the peasantry' are now quite variant in the extent and forms of their solutions from earlier periods. These are also, obviously, significant sites of contestation over access, resources, roles, reproduction, power and meaning in these landscapes.

Data from modeling exercises at fairly macro scales are increasingly central to rural development policy and are likely to become more so as technical expertise becomes more developed and the legitimacy accorded this information in political debates and land-use politics becomes greater within the emerging technocratic capacity within Latin America. Increasingly, surveillance of rural land use tied to systems of monitoring and sanctions frames a new kind of tropical 'panopticon' as well as inputs into models. These also invoke a kind of 'moral' landscape. Rural analysts and activists need to get more comfortable with such models both to understand what they are saying epistemically and politically, and to develop useful critiques, new politics and policies.

Within this broad framework, many old themes still remain in a new globalized context:

(1) The continuing questions about the farmer and access to land, but with the understanding that the agrarian question actually exists (and always has) with complex relations to forests and their politics. Most forests in the tropics are under the stewardship of rural populations, but their autonomy may be rapidly undermined through environmental markets, the power relations in emergent forms of governance mediated by non-state actors such as NGOs, and the global dimensions of new attributes of 'nature' that may limit traditional uses such as occurs now with REDD. This interplay between environmental services and underlying contested property or forest autonomy claims raises questions about future access to forested land, regional sovereignty and self-rule of use regardless of property regime.

(2) The larger questions of rural livelihoods under conditions of increased volatility triggered by planetary processes and dynamic and unpredictable processes within globalization will become more salient. What will 'rural' mean, and how will emerging ideologies, economies and technologies shape it? The idiocy of rural life or the stewards of the planet? Globalization and migration, whether regional or international, move peasant analysis away from the livelihood dynamics of a single locality, which had been the approach earlier in the life of JPS, into complex scales and nodal flows rather than just an analysis of the dynamics of place. Globalization has facilitated many small-scale rural economies, but undermined many as well. Political ecology and environmental history remain powerful tools for understanding changes over time, and for evaluating the forces unfolding outside the locality.

(3) Knowledge systems require the practices of everyday life and they remain among the least visible and most vulnerable of the processes in the reproduction of rural lives, meaning and livelihoods. This aspect of structural change is given short shrift all around. In the larger love affair of hard technologies and modernist landscapes, the software that has been running many rural systems is in danger of being lost due to migration, lack of documentation and the more general lack of prestige with which knowledge that emerges from non-western epistemes is held. The agro- and biodiversity produced by such systems is also menaced, but critical to planetary futures whether rural or urban. Especially with more forested landscapes, and the weak capacity of western science to manage complex biotic systems, indigenous knowledge becomes a true linchpin of natural resources management.

(4) The institutions that mediate rural life are undergoing very strong transformations, some of which consolidate gains (such as new tenurial forms) while others are eroded, due to migration, to NGOs overriding social movements, to political indifference. New forms of globalization, governance and markets that support forests are far more contested than it appears in this review of the 'new green revolution'. The structural changes that inhered in these emergent dynamics will require constant assessment and critical analysis of the power relations that underpin them simply because the forest 'result' is considered benign. These political ecologies are more complicated than the simple triumph of woodlands as a certificate of good policy or practices.

(5) The equity issues in rural development have been moderated by 'development funds from below' with remittances and the state-mediated poverty alleviation with cash transfers, coupled to a degree by robust commodity markets in timber, coffee, cacao, Açai and others. These have made the poverty profile of Latin America more optimistic than it otherwise might have been as millions left extreme poverty behind. Any shift in the troika of remittances, global resource booms and national social transfers can easily destabilize this configuration. It can be argued that rural populations – because of the 'subsidy from nature' – are better buffered against volatility, human or planetary, but this is actually an empirical question.

(6) It is worth keeping in mind that 'non-forest' and 'non-environmental' processes had huge effects on forest recuperation and peasant livelihoods. Cheap food policies, poverty alleviation programs of conditional cash transfers, and migration coupled to the transformations in tenurial regimes that legalized traditional holdings (and not just of natives) were substantive as drivers of forest maintenance and forest recovery than specifically environmental policies simply because the number of households affected by migration, remittances, cash transfers and tenure changes was so significant.

What does this mean about the future of peasantries? As they were described by Shanin in an earlier time and an earlier JPS, with a more agricultural and European focus, 'peasants' seem quite distant from the insurgent and economically innovative forest dwellers and creators who increasingly characterize Latin American small-scale farmers. We don't even really have a name for what these sorts of rural dwellers are since the categories of urban-rural, agricultural-forest, local-regional and national-international are so intertwined. What we can say is peoples and communities have shaped woodlands, and been shaped by woodlands, and these infuse their economies, institutions, forms of knowledge, meaning and history. There is a lot of empirical work ahead, and probably no master narratives for explaining what goes on in the verdant heart of the tropics other than the large pulse of the planet.

References

Abbots, E.-J. 2012. Coffee and community: Maya farmers and fair-trade markets. *Journal of the Royal Anthropological Institute*, 18, 702–703.

Adams, C., L.C. Munari, N. Van Vliet, R.S.S. Murrieta, B.A. Piperata, C. Futemma, N.N. Pedroso, C.S. Taqueda, M.A. Crevelaro and V.L. Spressola-Prado. 2013. Diversifying incomes and losing landscape complexity in Quilombola shifting cultivation communities of the Atlantic rainforest (Brazil). *Human Ecology*, 41, 119–137.

Aide, T.M., M.L. Clark, H.R. Grau, D. Lopez-Carr, M.A. Levy, D. Redo, M. Bonilla-Moheno, G. Riner, M.J. Andrade-Nunez and M. Muniz. 2013. Deforestation and reforestation of Latin America and the Caribbean 2001–2010. *Biotropica*, 45, 262–271.

Aide, T.M., J.K. Zimmerman, M. Rosario and H. Marcano. 1996. Forest recovery in abandoned cattle pastures along an elevational gradient in northeastern Puerto Rico. *Biotropica*, 28, 537–548.

Alix-Garcia, J., C. McIntosh, K.R.E. Sims and J.R. Welch. 2013. The ecological footprint of poverty alleviation: Evidence from Mexico's Oportunidades program. *Review of Economics and Statistics*, 95, 417–435.

Almeyda, A.M., E.N. Broadbent, M.S. Wyman and W.H. Durham. 2010. Ecotourism impacts in the Nicoya Peninsula, Costa Rica. *International Journal of Tourism Research*, 12, 803–819.

Arrow, K., B. Bolin, R. Costanza, P. Dasgupta, C. Folke, C.S. Holling, B.O. Jansson, S. Levin, K.G. Maler, C. Perrings and D. Pimentel. 1996. Economic growth, carrying capacity, and the environment. *Ecological Applications*, 6, 13–15.

Asner, G.P., M. Keller, R. Pereira and J.C. Zweede. 2002. Remote sensing of selective logging in Amazonia – Assessing limitations based on detailed field observations, Landsat ETM+, and textural analysis. *Remote Sensing of Environment*, 80, 483–496.

Asner, G.P., W. Llactayo, R. Tupayachi and E.R. Luna. 2013. Elevated rates of gold mining in the Amazon revealed through high-resolution monitoring. *Proceedings of the National Academy of Sciences*, 110(46), 18454–18459.

Avalos-Sartorio, B. and A. Blackman. 2010. Agroforestry price supports as a conservation tool: Mexican shade coffee. *Agroforestry Systems*, 78, 169–183.

Bacon, C.M. 2010. Who decides what is fair in fair trade? The agri-environmental governance of standards, access, and price. *Journal of Peasant Studies*, 37, 111–147.

Bakker, K. 2010. The limits of 'neoliberal natures': Debating green neoliberalism. *Progress in Human Geography*, 34, 715–735.

Balch, J.K., D.C. Nepstad, P.M. Brando, L.M. Curran, O. Portela, O. de Carvalho and P. Lefebvre. 2008. Negative fire feedback in a transitional forest of southeastern Amazonia. *Global Change Biology*, 14, 2276–2287.

Barbier, E.B., J.C. Burgess and A. Grainger. 2010. The forest transition: Towards a more comprehensive theoretical framework. *Land Use Policy*, 27, 98–107.

Barbier, E.B. and A.T. Tesfaw. 2012. Can REDD+ Save the Forest? The role of payments and tenure. *Forests*, 3, 881–895.

Bates, D.C. 2007. The Barbecho crisis, La Plaga del Banco, and international migration: Structural adjustment in Ecuador's Southern Amazon. *Latin American Perspectives*, 34, 108–122.

Bebbington, A. 1999. Capitals and capabilities: A framework for analyzing peasant viability, rural livelihoods and poverty. *World Development*, 27, 2021–2044.

Beuchelt, T.D. and M. Zeller. 2011. Profits and poverty: Certification's troubled link for Nicaragua's organic and fairtrade coffee producers. *Ecological Economics*, 70, 1316–1324.

Bilsborrow, R.E., A.F. Barbieri and W. Pan. 2004. Changes in population and land use over time in the Ecuadorian Amazon. *Acta Amazonica*, 34, 635–647.

Bitzer, V., P. Glasbergen and B. Arts. 2013. Exploring the potential of intersectoral partnerships to improve the position of farmers in global agrifood chains: Findings from the coffee sector in Peru. *Agriculture and Human Values*, 30, 5–20.

Blinn, C.E., J.O. Browder, M.A. Pedlowski and R.H. Wynne. 2013. Rebuilding the Brazilian rainforest: Agroforestry strategies for secondary forest succession. *Applied Geography*, 43, 171–181.

Borner, J. and S. Wunder. 2007. *Divergent opportunity costs of REDD on private lands in the Brazilian Amazon. ed. CIFOR*. Belem: CIFOR.

Borras, S.M. and J.C. Franco. 2010. Contemporary discourses and contestations around pro-poor land policies and land governance. *Journal of Agrarian Change*, 10, 1–32.

Bray, D.B., C. Antinori and J.M. Torres-Rojo. 2006. The Mexican model of community forest management: The role of agrarian policy, forest policy and entrepreneurial organization. *Forest Policy and Economics*, 8, 470–484.

Bray, D.B., L.M. Perez and D. Barry. 2005. *The community forests of Mexico: Managing for sustainable landscapes*. Austin: University of Texas Press.

Brondizio, E.S. 2004. Agricultural intensification, economic identity, and shared invisibility in Amazonian peasantry. *Culture and Agriculture*, 26, 1–24.

Brondizio, E.S. 2008. *The Amazon Caboclo and the Acai Palm: Forest farmers in the global market*. The Bronx: The New York Botanical Garden.

Brondizio, E.S., A.D. Siqueira and N. Vogt. 2014. Forest resources, city services: Globalization, Household networks, and urbanization in the amazon estuary. In: Susanna B. Hecht, K. Morrison Christine Padoch, ed. *The social lives of forests*. Chicago: University of Chicago, pp. 348–361.

Brosius, J.P. 2004. Indigenous peoples and protected areas at the World Parks Congress. *Conservation Biology*, 18, 609–612.

Browder, J.O. and B.J. Godfrey. 1997. *Rainforest cities: urbanization, development, and globalization of the Brazilian Amazon*. New York; Chichester, England: Columbia University Press.

Brown, K. 2003. 'Trees, forests and communities': Some historiographical approaches to environmental history on Africa. *Area*, 35, 343–356.

Browning, A. 2013. Corn, Tomatoes, and a dead dog: Mexican agricultural restructuring after NAFTA and rural responses to declining maize production in Oaxaca, Mexico. *Mexican Studies-Estudios Mexicanos*, 29, 85–119.

Busch, C.B. and C. Vance. 2011. The diffusion of cattle ranching and deforestation: Prospects for a hollow frontier in Mexico's Yucatan. *Land Economics*, 87, 682–698.

Cardozo, M. 2011. Economic displacement and local attitude towards protected area establishment in the Peruvian Amazon. *Geoforum*, 42, 603–614.

Carrero, G.C. and P.M. Fearnside. 2011. Forest clearing dynamics and the expansion of landholdings in Apui, a deforestation hotspot on Brazil's Transamazon Highway. *Ecology and Society*, 16(2), 1–18.

Castiblanco, C., A. Etter and T.M. Aide. 2013. Oil palm plantations in Colombia: a model of future expansion. *Environmental Science and Policy*, 27, 172–183.

Chappell, M.J., J. Vandermeer, C. Badgley and I. Perfecto. 2009. Wildlife-friendly farming vs land sparing. *Frontiers in Ecology and the Environment*, 7, 83–84.

Chase, J. 1999. Exodus revisited: The politics and experience of rural loss in central Brazil. *Sociologia Ruralis*, 39, 165–185.

Chazdon, R.L. 2003. Tropical forest recovery: Legacies of human impact and natural disturbances. *Perspectives in Plant Ecology Evolution and Systematics*, 6, 51–71.

Chazdon, R.L. 2008. Beyond deforestation: Restoring forests and ecosystem services on degraded lands. *Science*, 320, 1458–1460.

Chazdon, R.L., C.A. Peres, D. Dent, D. Sheil, A.E. Lugo, D. Lamb, N.E. Stork and S.E. Miller. 2009. The potential for species conservation in tropical secondary forests. *Conservation Biology*, 23, 1406–1417.

Choumert, J., P.C. Motel and H.K. Dakpo. 2013. Is the environmental Kuznets curve for deforestation a threatened theory? A meta-analysis of the literature. *Ecological Economics*, 90, 19–28.

Clark, M.L., T.M. Aide and G. Riner. 2013. Land change for all municipalities in Latin America and the Caribbean assessed from 250-m MODIS imagery (2001–2010). *Remote Sensing of Environment*, 126, 84–103.

Coomes, O.T. and B.L. Barham. 1994. The Amazon Rubber Boom – Labor control, resistance and failed plantation development revisited. *Hispanic American Historical Review*, 74, 231–257.

Cossio, R.E., S. Perz and K. Kainer. 2011. Capacity for timber management in small and medium forest enterprises: A case study from the Peruvian Amazon. *Small-Scale Forestry*, 10, 489–507.

Cronkleton, P., A.M. Larson, L. Feintrenie, C. Garcia and P. Levang. 2013. Reframing community forestry to manage the forest-farm interface. *Small-Scale Forestry*, 12, 5–13.

Cullather, N. 2010. *The hungry world*. Cambridge: Harvard University Press.

da Cunha, M.C. and M.W.B. de Almeida. 2000. Indigenous people, traditional people, and conservation in the Amazon. *Daedalus*, 129, 315–338.

Daniels, A.E., K. Bagstad, V. Esposito, A. Moulaert and C.M. Rodriguez. 2010. Understanding the impacts of Costa Rica's PES: Are we asking the right questions? *Ecological Economics* 69, 2116–2126.

Danielsen, F., H. Beukema, N.D. Burgess, F. Parish, C.A. Bruhl, P.F. Donald, D. Murdiyarso, B. Phalan, L. Reijnders, M. Struebig and E.B. Fitzherbert. 2009. Biofuel plantations on forested lands: Double jeopardy for biodiversity and climate. *Conservation Biology*, 23, 348–358.

Danielsen, F., M. Skutsch, N.D. Burgess, P.M. Jensen, H. Andrianandrasana, B. Karky, R. Lewis, J.C. Lovett, J. Massao, Y. Ngaga, P. Phartiyal, M.K. Poulsen, S.P. Singh, S. Solis, M. Sorensen, A. Tewari, R. Young and E. Zahabu. 2011. At the heart of REDD+: a role for local people in monitoring forests? *Conservation Letters*, 4, 158–167.

Davidov, V. 2012. From colonial primitivism to ecoprimitivism: Constructing the indigenous 'savage' in South America. *Arcadia*, 46, 467–487.

Davidson, E.A., A.C. de Araujo, P. Artaxo, J.K. Balch, I.F. Brown, M.M.C. Bustamante, M.T. Coe, R.S. DeFries, M. Keller, M. Longo, J.W. Munger, W. Schroeder, B.S. Soares, C.M. Souza and S.C. Wofsy. 2012. The Amazon basin in transition. *Nature*, 481, 321–328.

DeClerck, F.A.J., R. Chazdon, K.D. Holl, J.C. Milder, B. Finegan, A. Martinez-Salinas, P. Imbach, L. Canet and Z. Ramos. 2010. Biodiversity conservation in human-modified landscapes of Mesoamerica: Past, present and future. *Biological Conservation*, 143, 2301–2313.

Del Angel-Perez, A.L. and J.A. Villagomez-Cortes. 2011. Public demands, environmental perceptions, and natural resource management in Mexico's tropical lowlands. *African Journal of Business Management*, 5, 2083–2092.

Doan, T.M. 2013. Sustainable ecotourism in Amazonia: Evaluation of six sites in Southeastern Peru. *International Journal of Tourism Research*, 15, 261–271.

Duarte, G.B., B. Sampaio and Y. Sampaio. 2009. Programa Bolsa Família: Impacto das transferências sobre os gastos com alimentos em famílias rurais. *Revista de Economia e Sociologia Rural*, 47, 903–918.

Duchelle, A.E., P. Cronkleton, K.A. Kainer, G. Guanacoma and S. Gezan. 2011. Resource theft in tropical forest communities: Implications for non-timber management, livelihoods, and conservation. *Ecology and Society* 16(1): 1–20.

Durham, W.H. 1979. *Scarcity and survival in Central America: Ecological origins of the Soccer Wars*. Palo Alto: Stanford University Press.

Edelman, M. 2005. Bringing the Moral Economy back in … to the study of 21st-century transnational peasant movements. *American Anthropologist*, 107, 331–345.

Eloy, L. 2008. Food diversity and urbanization, agrarian transformations and food recomposition in the north-western Amazon Brazil. *Anthropology of Food*, 2882–2882.

Eloy, L. and C. Lasmar. 2011. Urbanization and transformation of indigenous resource management: The case of upper Rio Negro Brazil. *Acta Amazonica*, 41, 91–101.

Ensor, M. 2009. *Legacy of Hurricane Mitch: Lessons from post-disaster reconstruction in Honduras*. Phoenix: University of Arizona Press.

Farley, K.A. 2007. Grasslands to tree plantations: Forest transition in the Andes of Ecuador. *Annals of the Association of American Geographers*, 97, 755–771.

Fearnside, P.M. 2008. The roles and movements of actors in the deforestation of Brazilian Amazonia. *Ecology and Society*, 13(1), 23, http://www.ecologyandsociety.org/vol13/iss1/art23/ (accessed April 1, 2014).

Fearnside, P.M. 2012. Brazil's Amazon forest in mitigating global warming: unresolved controversies. *Climate Policy*, 12, 70–81.

Fitting, E. 2006. Importing corn, exporting labor: The neoliberal corn regime, GMOs, and the erosion of Mexican biodiversity. *Agriculture and Human Values*, 23, 15–26.

Fox, J. and J.-C. Castella. 2013. Expansion of rubber Hevea brasiliensis in Mainland Southeast Asia: What are the prospects for smallholders? *The Journal of Peasant Studies*, 40, 155–170.

Fule, P.Z., M. Ramos-Gomez, C. Cortes-Montano and A.M. Miller. 2011. Fire regime in a Mexican forest under indigenous resource management. *Ecological Applications*, 21, 764–775.

Galford, G.L., J.M. Melillo, D.W. Kicklighter, J.F. Mustard, T.W. Cronin, C.E.P. Cerri and C.C. Cerri. 2011. Historical carbon emissions and uptake from the agricultural frontier of the Brazilian Amazon. *Ecological Applications*, 21, 750–763.

Garcia-Barrios, L., Y.M. Galvan-Miyoshi, I.A. Valdivieso-Perez, O.R. Masera, G. Bocco and J. Vandermeer. 2009. Neotropical forest conservation, agricultural intensification, and rural out-migration: The Mexican experience. *Bioscience*, 59, 863–873.

Garcia-Frapolli, E., B. Ayala-Orozco, M. Bonilla-Moheno, C. Espadas-Manrique and G. Ramos-Fernandez. 2007. Biodiversity conservation, traditional agriculture and ecotourism: Land cover/land use change projections for a natural protected area in the northeastern Yucatan Peninsula, Mexico. *Landscape and Urban Planning*, 83, 137–153.

Gavin, M.C. and G.J. Anderson. 2007. Socioeconomic predictors of forest use values in the Peruvian Amazon: A potential tool for biodiversity conservation. *Ecological Economics*, 60, 752–762.

Gibson, L., T.M. Lee, L.P. Koh, B.W. Brook, T.A. Gardner, J. Barlow, C.A. Peres, C.J.A. Bradshaw, W.F. Laurance, T.E. Lovejoy and N.S. Sodhi. 2011. Primary forests are irreplaceable for sustaining tropical biodiversity. *Nature*, 478, 378–381.

Gloor, M., L. Gatti, R. Brienen, T.R. Feldpausch, O.L. Phillips, J. Miller, J.P. Ometto, H. Rocha, T. Baker, B. de Jong, R.A. Houghton, Y. Malhi, L.E.O.C. Aragão, J.L. Guyot, K. Zhao, R. Jackson, P. Peylin, S. Sitch, B. Poulter, M. Lomas, S. Zaehle, C. Huntingford, P. Levy and J. Lloyd. 2012. The carbon balance of South America: A review of the status, decadal trends and main determinants. *Biogeosciences*, 9, 5407–5430.

Goeschl, T. and D.C. Igliori. 2006. Property rights for biodiversity conservation and development: Extractive reserves in the Brazilian Amazon. *Development and Change*, 37, 427–451.

Grau, H.R., T.M. Aide, J.K. Zimmerman, J.R. Thomlinson, E. Helmer and X.M. Zou. 2003. The ecological consequences of socioeconomic and land-use changes in postagriculture Puerto Rico. *Bioscience*, 53, 1159–1168.

Gravel, N. 2007. Mexican smallholders adrift: The urgent need for a new social contract in rural Mexico. *Journal of Latin American Geography*, 6, 77–98.

Gray, C.L., R.E. Bilsborrow, J.L. Bremner and F. Lu. 2008. Indigenous land use in the Ecuadorian Amazon: A cross-cultural and multilevel analysis. *Human Ecology*, 36, 97–109.

Gutierrez-Velez, V.H. and R. DeFries. 2013. Annual multi-resolution detection of land cover conversion to oil palm in the Peruvian Amazon. *Remote Sensing of Environment*, 129, 154–167.

Gutierrez-Velez, V.H., R. DeFries, M. Pinedo-Vasquez, M. Uriarte, C. Padoch, W. Baethgen, K. Fernandes and Y.L. Lim. 2011. High-yield oil palm expansion spares land at the expense of forests in the Peruvian Amazon. *Environmental Research Letters*, 6.

Hall, A. 2008. Brazil's Bolsa Familia: A double-edged sword? *Development and Change*, 39, 799–822.

Hecht, S.B. 1985. Environment, Development and Politics – Capital Accumulation and the Livestock Sector in Eastern Amazonia. *World Development*, 13, 663–684.

Hecht, S.B. 1993. The logic of livestock and deforestation in Amazonia. *Bioscience*, 43, 687–695.

Hecht, S.B. 2004. Invisible Forests: The Political Ecology of Forest Resurgence in El Salvador. In: M. Watts and R. Peet, ed. *Liberation ecologies*. London: Routledge, pp. 64–104.

Hecht, S.B. 2010. The new rurality: Globalization, peasants and the paradoxes of landscapes. *Land Use Policy*, 27, 161–169.

Hecht, S.B. 2012. From eco-catastrophe to zero deforestation? Interdisciplinarities, politics, environmentalisms and reduced clearing in Amazonia. *Environmental Conservation*, 39, 4–19.

Hecht, S.B. 2013. *The scramble for the Amazon, and the lost paradise of Euclides da Cunha*. Chicago: University of Chicago Press.

Hecht, S. 2014. The social lives of forests: Successions and forest transitions: Theories forest resurgence. In: S.B. Hecht, C. Padoch and K. Morrison, ed. *Social lives of forests*. Chicago: University of Chicago Press, pp. 97–113.

Hecht, S.B. and A. Cockburn. 2011. *Fate of the forest: Developers, destroyers and defenders of the Amazon*. London: Verso.

Hecht, S.B., S. Kandel, I. Gomes, N. Cuellar and H. Rosa. 2006. Globalization, forest resurgence, and environmental politics in El Salvador. *World Development*, 34, 308–323.

Hecht, S., S. Kandel and A. Morales. 2012. *Migration, livelihoods and natural resources*. San Salvador: IDRC PRISMA.

Hecht, S.B. and S.S. Saatchi. 2007. Globalization and forest resurgence: Changes in forest cover in El Salvador. *Bioscience*, 57, 663–672.

Heise, U. 2010. *Sense of Place, Sense of Planet*. New York: Oxford.

Hejkrlik, J., J. Mazancova and K. Forejtova. 2013. How effective is Fair Trade as a tool for the stabilization of agricultural commodity markets? Case of coffee in the Czech Republic. *Agricultural Economics-Zemedelska Ekonomika*, 59, 8–18.

Hobbs, R.J., S. Arico, J. Aronson, J.S. Baron, P. Bridgewater, V.A. Cramer, P.R. Epstein, J.J. Ewel, C.A. Klink, A.E. Lugo, D. Norton, D. Ojima, D.M. Richardson, E.W. Sanderson, F. Valladares,

M. Vila, R. Zamora and M. Zobel. 2006. Novel ecosystems: theoretical and management aspects of the new ecological world order. *Global Ecology and Biogeography*, 15, 1–7.

Holder, C.D. and G. Chase. 2012. The role of remittances and decentralization of forest management in the sustainability of a municipal-communal pine forest in eastern Guatemala. *Environment Development and Sustainability*, 14, 25–43.

Horton, L.R. 2009. Buying up nature economic and social impacts of Costa Rica's ecotourism boom. *Latin American Perspectives*, 36, 93–107.

Hough, P.A. 2011. Disarticulations and commodity chains: Cattle, coca, and capital accumulation along Colombia's agricultural frontier. *Environment and Planning A*, 43, 1016–1034.

Hylander, K. and S. Nemomissa. 2009. Complementary roles of home gardens and exotic tree plantations as alternative habitats for plants of the Ethiopian Montane Rainforest. *Conservation Biology*, 23, 400–409.

Ibarra, J.T., A. Barreau, C. Del Campo, C.I. Camacho, G.J. Martin and S.R. McCandless. 2011. When formal and market-based conservation mechanisms disrupt food sovereignty: Impacts of community conservation and payments for environmental services on an indigenous community of Oaxaca, Mexico. *International Forestry Review*, 13, 318–337.

Ibanez, A.M. and C.E. Velez. 2008. Civil conflict and forced migration: The micro determinants and welfare losses of displacement in Colombia. *World Development*, 36, 659–676.

INPE. 2009. INPE maps secondary vegetation of Para, Mato Grosso and Amapa. www.inpe.br/ingles/index.php

INPE. 2011. *PRODES. Taxas Annuais de Desmatemento*. Sao Jose dos Campos: INPE.

Izquierdo, A.E., H.R. Grau and T.M. Aide. 2011. Implications of rural-urban migration for conservation of the Atlantic Forest and urban growth in Misiones, Argentina 1970–2030. *Ambio*, 40, 298–309.

Kandel, S. and N. Cuellar. 2012. Migration dynamics, rural livelihoods and their challenges. In: S. Hecht, S. Kandel and A. Morales, ed. *Migration, Licvelohoods and Natural Resources Management*. San Salvador: IDRC, pp. 125–146.

Kant, S. 2009. Recent global trends in forest tenures. *Forestry Chronicle*, 85, 849–858.

Kay, C. 2008. Reflections on Latin American rural studies in the neoliberal globalization period: A New Rurality? *Development and Change*, 39, 915–943.

Keleman, A. 2010. Institutional support and in situ conservation in Mexico: Biases against small-scale maize farmers in post-NAFTA agricultural policy. *Agriculture and Human Values*, 27, 13–28.

Kirkby, C.A., R. Giudice-Granados, B. Day, K. Turner, L.M. Velarde-Andrade, A. Duenas-Duenas, J.C. Lara-Rivas and D.W. Yu. 2010. The market triumph of ecotourism: An economic investigation of the private and social benefits of competing land uses in the Peruvian Amazon. *PLOS ONE*, 5(9), http://www.plosone.org/article/info%3Adoi%2F10.1371%2Fjournal.pone.0013015#pone-0013015-g003 (accessed April 1, 2014), DOI: 10.1371/journal.pone.0013015.

Klepek, J. 2012. Selling Guatemala's next green revolution: Agricultural modernization and the politics of GM maize regulation. *International Journal of Agricultural Sustainability*, 10, 117–134.

Klooster, D. 2003. Forest transitions in Mexico: Institutions and forests in a globalized countryside. *Professional Geographer*, 55, 227–237.

de Koning, F., M. Aguinaga, M. Bravo, M. Chiu, M. Lascano, T. Lozada and L. Suarez. 2011. Bridging the gap between forest conservation and poverty alleviation: The Ecuadorian Socio Bosque program. *Environmental Science and Policy*, 14, 531–542.

Koop, G. and L. Tole. 1999. Is there an environmental Kuznets curve for deforestation? *Journal of Development Economics*, 58, 231–244.

Koop, G. and L. Tole. 2001. Deforestation, distribution and development. *Global Environmental Change*, 11, 193–202.

Koster, J.M., M.N. Grote and B. Winterhalder. 2013. Effects on household labor of temporary out-migration by male household heads in Nicaragua and Peru: An analysis of spot-check time allocation data using mixed-effects models. *Human Ecology*, 41, 221–237.

Larson, A.M., D. Barry and G.R. Dahal. 2010. New rights for forest-based communities? Understanding processes of forest tenure reform. *International Forestry Review*, 12, 78–96.

Larson, A.M., P. Pacheco, F. Toni and M. Vallejo. 2007. Trends in Latin American forestry decentralisations: Legal frameworks, municipal governments and forest dependent groups. *International Forestry Review*, 9, 734–747.

Laurance, W.F., D.C. Useche, J. Rendeiro, M. Kalka, C.J.A. Bradshaw, S.P. Sloan, S.G. Laurance, M. Campbell, K. Abernethy, P. Alvarez, V. Arroyo-Rodriguez, P. Ashton, J. Benitez-Malvido,

A. Blom, K.S. Bobo, C.H. Cannon, M. Cao, R. Carroll, C. Chapman, R. Coates, M. Cords, F. Danielsen, B. De Dijn, E. Dinerstein, M.A. Donnelly, D. Edwards, F. Edwards, N. Farwig, P. Fashing, P.-M. Forget, M. Foster, G. Gale, D. Harris, R. Harrison, J. Hart, S. Karpanty, W.J. Kress, J. Krishnaswamy, W. Logsdon, J. Lovett, W. Magnusson, F. Maisels, A.R. Marshall, D. McClearn, D. Mudappa, M.R. Nielsen, R. Pearson, N. Pitman, J. van der Ploeg, A. Plumptre, J. Poulsen, M. Quesada, H. Rainey, D. Robinson, C. Roetgers, F. Rovero, F. Scatena, C. Schulze, D. Sheil, T. Struhsaker, J. Terborgh, D. Thomas, R. Timm, J.N. Urbina-Cardona, K. Vasudevan, S.J. Wright, J.C. Arias-G, L. Arroyo, M. Ashton, P. Auzel, D. Babaasa, F. Babweteera, P. Baker, O. Banki, M. Bass, I. Bila-Isia, S. Blake, W. Brockelman, N. Brokaw, C.A. Bruehl, S. Bunyavejchewin, J.-T. Chao, J. Chave, R. Chellam, C.J. Clark, J. Clavijo, R. Congdon, R. Corlett, H.S. Dattaraja, C. Dave, G. Davies, B.d.M. Beisiegel, R.d.N. da Silva, A. Di Fiore, A. Diesmos, R. Dirzo, D. Doran-Sheehy, M. Eaton, L. Emmons, A. Estrada, *et al.* 2012. Averting biodiversity collapse in tropical forest protected areas. *Nature*, 489, 290–294.

Lavinas, L. 2013. 21st century welfare. *New Left Review*, 84, November, December, 1–25.

Liere, H., D. Jackson and J. Vandermeer. 2012. Ecological complexity in a coffee agroecosystem: Spatial heterogeneity, population persistence and biological control. *PLOS ONE*, 7(9), http://www.plosone.org/article/info%3Adoi%2F10.1371%2Fjournal.pone.0045508#pone-0045508-g005 (accessed April 1, 2014), DOI:10.1371/journal.pone.0045508.

Limonad, E. and L.R. Monte-Mor. 2012. The Right to the City, between rural and urban. *Scripta Nova-Revista Electronica De Geografia Y Ciencias Sociales*, 16(418). http://www.ub.es/geocrit/sn-418/sn-418–25.htm (accessed April 1, 2014).

Linhares, L.F.D. 2004. Kilombos of Brazil – Identity and land entitlement. *Journal of Black Studies*, 34, 817–837.

Lugo, A.E. 2009. The emerging era of novel tropical forests. *Biotropica*, 41, 589–591.

Lugo, A.E., T.A. Carlo and J.M. Wunderle, Jr. 2012. Natural mixing of species: Novel plant-animal communities on Caribbean Islands. *Animal Conservation*, 15, 233–241.

Mann, C. 2005. *1491: New revelations about the Americas before Columbus*. New York: Knopf.

Mann, C.C. 2008. Ancient earthmovers of the Amazon. *Science*, 321, 1148–1152.

Martins, P.F.D. and T.Z.D. Pereira. 2012. Cattle-raising and public credit in rural settlements in Eastern Amazon. *Ecological Indicators*, 20, 316–323.

Mascaro, J., I. Perfecto, O. Barros, D.H. Boucher, I.G. de la Cerda, J. Ruiz and J. Vandermeer. 2005. Aboveground biomass accumulation in a tropical wet forest in Nicaragua following a catastrophic hurricane disturbance. *Biotropica*, 37, 600–608.

Mascoe, J. 2007. *Nuclear borderlands*. Princeton: Princeton University Press.

Mather, A.S. 1992. The forest transition. *Area*, 24, 367–379.

Mather, A.S. and C.L. Needle. 1998. The forest transition: A theoretical basis. *Area*, 30, 117–124.

Mathews, A. 2011. *Instituting nature: Authority, experitise and power in Mexican forests*. Cambridge: MIT Press.

Matos, R. 2005. *Espacialidades en Rede: Populacao, urbanizacao e migracao no Brasil*. Belo Horizonte: C/Arte.

McAfee, K. 1999. Selling nature to save it? Biodiversity and green developmentalism. *Environment and Planning D-Society and Space*, 17, 133–154.

McAfee, K. 2008. Beyond techno-science: Transgenic maize in the fight over Mexico's future. *Geoforum*, 39, 148–160.

McAfee, K. 2012. The Contradictory Logic of Global Ecosystem Services Markets. *Development and Change*, 43, 105–131.

McAfee, K. and E.N. Shapiro. 2010. Payments for ecosystem services in Mexico: Nature, neoliberalism, social movements, and the state. *Annals of the Association of American Geographers*, 100, 579–599.

McSweeney, K. and B. Jokisch. 2007. Beyond rainforests: Urbanisation and Emigration among Lowland Indigenous Societies in Latin America. *Bulletin of Latin American Research*, 26, 159–180.

Medina, G., B. Pokorny and J. Weigelt. 2009. The power of discourse: Hard lessons for traditional forest communities in the Amazon. *Forest Policy and Economics*, 11, 392–397.

Mello, A.Y.I. and D.S. Alves. 2011. Secondary vegetation dynamics in the Brazilian Amazon based on thematic mapper imagery. *Remote Sensing Letters*, 2, 189–194.

Mertz, O., K. Egay, T.B. Bruun and T.S. Colding. 2013. The last swiddens of Sarawak, Malaysia. *Human Ecology*, 41, 109–118.

Meyfroidt, P. and E.F. Lambin. 2011. Global forest transition: Prospects for an end to deforestation. In: A. Gadgil and D.M. Liverman, ed. *Annual review of environment and resources*, Vol. 36, pp. 343–371.

Meyfroidt, P., T.K. Rudel and E.F. Lambin. 2010. Forest transitions, trade, and the global displacement of land use. *Proceedings of the National Academy of Sciences of the United States of America*, 107, 20917–20922.

Millard, E. 2011. Incorporating agroforestry approaches into commodity value chains. *Environmental Management*, 48, 365–377.

Mills, J.H. and T.A. Waite. 2009. Economic prosperity, biodiversity conservation, and the environmental Kuznets curve. *Ecological Economics*, 68, 2087–2095.

Moran-Taylor, M.J. and M.J. Taylor. 2010. Land and leña: Linking transnational migration, natural resources, and the environment in Guatemala. *Population and Environment*, 32, 198–215.

Mueller, A.D., G.A. Islebe, F.S. Anselmetti, D. Ariztegui, M. Brenner, D.A. Hodell, I. Hajdas, Y. Hamann, G.H. Haug and D.J. Kennett. 2010. Recovery of the forest ecosystem in the tropical lowlands of northern Guatemala after disintegration of classic maya polities. *Geology*, 38, 523–526.

Murphy, L.L. 2001. Colonist farm income, off-farm work, cattle, and differentiation in Ecuador's northern Amazon. *Human Organization*, 60, 67–79.

Neeff, T., R.M. Lucas, J.R. dos Santos, E.S. Brondizio and C.C. Freitas. 2006. Area and age of secondary forests in Brazilian Amazonia 1978–2002: An empirical estimate. *Ecosystems*, 9, 609–623.

Nepstad, D.C., W. Boyd, C.M. Stickler, T. Bezerra and A.A. Azevedo. 2013. Responding to climate change and the global land crisis: REDD+, market transformation and low-emissions rural development. *Philosophical Transactions of the Royal Society B-Biological Sciences*, http://rstb. royalsocietypublishing.org/content/368/1619/20120167.full (accessed April 1, 2014), DOI:10. 1098/rstb.2012.0167.

Nepstad, D., G. Carvalho, A.C. Barros, A. Alencar, J.P. Capobianco, J. Bishop, P. Moutinho, P. Lefebvre, U.L. Silva and E. Prins. 2001. Road paving, fire regime feedbacks, and the future of Amazon forests. *Forest Ecology and Management*, 154, 395–407.

Nepstad, D., S. Schwartzman, B. Bamberger, M. Santilli, D. Ray, P. Schlesinger, P. Lefebvre, A. Alencar, E. Prinz, G. Fiske and A. Rolla. 2006. Inhibition of Amazon deforestation and fire by parks and indigenous lands. *Conservation Biology*, 20, 65–73.

Nepstad, D.C., C.M. Stickler, B. Soares and F. Merry. 2008. Interactions among Amazon land use, forests and climate: Prospects for a near-term forest tipping point. *Philosophical Transactions of the Royal Society B-Biological Sciences*, 363, 1737–1746.

Nevle, R.J. and D.K. Bird. 2008. Effects of syn-pandemic fire reduction and reforestation in the tropical Americas on atmospheric CO_2 during European conquest. *Palaeogeography Palaeoclimatology Palaeoecology*, 264, 25–38.

Nevle, R.J., D.K. Bird, W.F. Ruddiman and R.A. Dull. 2011. Neotropical human-landscape interactions, fire, and atmospheric CO2 during European conquest. *Holocene*, 21, 853–864.

Newton, P., A. Agrawal and L. Wollenberg. 2013. Enhancing the sustainability of commodity supply chains in tropical forest and agricultural landscapes. *Global Environmental Change-Human and Policy Dimensions*, 23, 1761–1772.

Ojeda, D. 2012. Green pretexts: Ecotourism, neoliberal conservation and land grabbing in Tayrona National Natural Park, Colombia. *The Journal of Peasant Studies*, 39, 357–375.

Pacheco, P. 2009. Smallholder livelihoods, wealth and deforestation in the Eastern Amazon. *Human Ecology*, 37, 27–41.

Pacheco, P., D. Barry, P. Cronkleton and A. Larson. 2012. The recognition of forest rights in Latin America: Progress and shortcomings of forest tenure reforms. *Society and Natural Resources*, 25, 556–571.

Padoch, C., A. Steward, M. Pinedo-Vasquez, L. Putzel and M. Ruiz. 2014. Urban residence, rural employment and the future of Amazonian forests. In: S. Hecht, K. Morrison, C. Padoch, eds. *The social Lives of forests*. Chicago: Univeristy of Chicago Press.

Page, T.L. 2010. Can the state create campesinos? A comparative analysis of the Venezuelan and Cuban Repeasantization Programmes. *Journal of Agrarian Change*, 10, 251–272.

Pascual, U. and E.B. Barbier. 2007. On price liberalization, poverty, and shifting cultivation: An example from Mexico. *Land Economics*, 83, 192–216.

Patel, R. 2012. The Long Green Revolution. *The Journal of Peasant Studies*, 40, 1–63.

Peluso, N.L. 2012. What's nature got to do with it? A situated historical perspective on socio-natural commodities. *Development and Change*, 43, 79–104.

Peluso, N.L. and P. Vandergeest. 2011. Political ecologies of war and forests: Counterinsurgencies and the making of national natures. *Annals of the Association of American Geographers*, 101, 587–608.

Peralta, N. 2012. Ecotourism as an incentive to biodiversity conservation: The case of Uakari Lodge, Amazonas, Brazil. *Uakari*, 8, 75–93.

Peres, C.A. 2005. Why we need megareserves in Amazonia. *Conservation Biology*, 19, 728–733.

Perz, S.G. 2007. Grand theory and context-specificity in the study of forest dynamics: Forest transition theory and other directions. *Professional Geographer*, 59, 105–114.

Perz, S.G., F. Leite, C. Simmons, R. Walker, S. Aldrich and M. Caldas. 2010. Intraregional migration, direct action land reform, and new land settlements in the Brazilian Amazon. *Bulletin of Latin American Research*, 29, 459–476.

Piperata, B.A., J.E. Spence, P. Da-Gloria and M. Hubbe. 2011. The Nutrition transition in Amazonia: Rapid economic change and its impact on growth and development in Ribeirinhos. *American Journal of Physical Anthropology*, 146, 1–13.

Prado Cordova, J.P., S. Wunder, C. Smith-Hall and J. Boerner. 2013. Rural income and forest reliance in Highland Guatemala. *Environmental Management*, 51, 1034–1043.

Putzel, L., C. Padoch and M. Pinedo-Vasquez. 2008. The Chinese timber trade and the logging of Peruvian Amazonia. *Conservation Biology*, 22, 1659–1661.

Pyhala, A., K. Brown and W.N. Adger. 2006. Implications of livelihood dependence on non-timber products in Peruvian Amazonia. *Ecosystems*, 9, 1328–1341.

Radel, C., B. Schmook and S. McCandless. 2010. Environment, transnational labor migration, and gender: Case studies from southern Yucatan, Mexico and Vermont, USA. *Population and Environment*, 32, 177–197.

Rawlings, L. and G. Rubio. 2003. *Evaluating the impact of conditional cash transfers. 25.* Washington: World Bank.

Reardon, T., K. Stamoulis and P. Pingali. 2007. Rural nonfarm employment in developing countries in an era of globalization. *Agricultural Economics*, 37, 173–184.

Redo, D.J., H. Ricardo Grau, T.M. Aide and M.L. Clark. 2012. Asymmetric forest transition driven by the interaction of socioeconomic development and environmental heterogeneity in Central America. *Proceedings of the National Academy of Sciences of the United States of America*, 109, 8839–8844.

Ribeiro, M.C., J.P. Metzger, A.C. Martensen, F.J. Ponzoni and M.M. Hirota. 2009. The Brazilian Atlantic Forest: How much is left, and how is the remaining forest distributed? Implications for conservation. *Biological Conservation*, 142, 1141–1153.

Ribot, J.C. 2001. Reframing deforestation. Global analysis and local realities – Studies in West Africa. *Development and Change*, 32, 181–182.

Robson, J. and F. Berkes. 2011a. How does out-migration affect community institutions? A study of two indigenous municipalities in Oaxaca, Mexico. *Human Ecology*, 39, 179–190.

Robson, J.P. and F. Berkes. 2011b. Exploring some of the myths of land use change: Can rural to urban migration drive declines in biodiversity? *Global Environmental Change-Human and Policy Dimensions*, 21, 844–854.

Rocha, P., B. Viana, M. Cardoso, A. Melo, M. Costa, R. Vasconcelos and T. Dantas. 2013. What is the value of eucalyptus monocultures for the biodiversity of the Atlantic forest? A multitaxa study in southern Bahia, Brazil. *Journal of Forestry Research*, 24, 263–272.

Ross, N.J. and T.F. Rangel. 2011. Ancient Maya agroforestry echoing through spatial relationships in the extant forest of NW Belize. *Biotropica*, 43, 141–148.

Rudel, T.K., O.T. Coomes, E. Moran, F. Achard, A. Angelsen, J.C. Xu and E. Lambin. 2005. Forest transitions: Towards a global understanding of land use change. *Global Environmental Change-Human and Policy Dimensions*, 15, 23–31.

Rudel, T.K., M. Perez-Lugo and H. Zichal. 2000. When fields revert to forest: Development and spontaneous reforestation in post-war Puerto Rico. *Professional Geographer*, 52, 386–397.

Sanchez-Cuervo, A.M. and T.M. Aide. 2013. Consequences of the armed conflict, forced human displacement, and land abandonment on forest cover change in Colombia: A multi-scaled analysis. *Ecosystems*, 16, 1052–1070.

Sanchez-Cuervo, A.M., T.M. Aide, M.L. Clark and A. Etter. 2012. Land cover change in Colombia: Surprising forest recovery trends between 2001 and 2010. *PLOS ONE*, 7(8), e43943.

Schlosberg, D. 2013. Theorising environmental justice: The expanding sphere of a discourse. *Environmental Politics*, 22, 37–55.

Schmitt-Harsh, M. 2013. Landscape change in Guatemala: Driving forces of forest and coffee agro-forest expansion and contraction from 1990 to 2010. *Applied Geography*, 40, 40–50.

Schmook, B. 2010. Shifting maize cultivation and secondary vegetation in the Southern Yucatan: Successional forest impacts of temporal intensification. *Regional Environmental Change*, 10, 233–246.

Schmook, B. and C. Radel. 2008. International labor migration from a tropical development frontier: Globalizing households and an incipient forest transition. *Human Ecology*, 36, 891–908.

Sears, R.R., C. Padoch and M. Pinedo-Vasquez. 2007. Amazon forestry transformed: Integrating knowledge for smallholder timber management in eastern Brazil. *Human Ecology*, 35, 697–707.

Shanley, P., M.D. Silva, T. Melo, R. Carmenta and R. Nasi. 2012. From conflict of use to multiple use: Forest management innovations by small holders in Amazonian logging frontiers. *Forest Ecology and Management*, 268, 70–80.

Shapiro-Garza, E. 2013. Contesting the market-based nature of Mexico's national payments for eco-system services programs: Four sites of articulation and hybridization. *Geoforum*, 46, 5–15.

Sheets, P.D. 2002. *Before the volcano erupted : The ancient Cerén village in Central America*. Austin: University of Texas Press.

Siegmund-Schultze, M., B. Rischkowsky, J.B. da Veiga and J.M. King. 2010. Valuing cattle on mixed smallholdings in the Eastern Amazon. *Ecological Economics*, 69, 857–867.

Silveira-Neto, R.M. and C.R. Azzoni. 2012. Social policy as regional policy: Market and nonmarket factors determining regional inequality. *Journal of Regional Science*, 52, 433–450.

Sletto, B. 2008. The knowledge that counts: Institutional identities, policy science, and the conflict over fire management in the Gran Sabana, Venezuela. *World Development*, 36, 1938–1955.

Stevens, K., L. Campbell, G. Urquhart, D. Kramer and J.G. Qi. 2011. Examining complexities of forest cover change during armed conflict on Nicaragua's Atlantic Coast. *Biodiversity and Conservation*, 20, 2597–2613.

Terborgh, J. 1992. *Diversity and the tropical rain forest*. New York: Scientific American Library: Distributed by W.H. Freeman.

Thoumi, F.E. 2012. Illegal drugs, anti-drug policy failure, and the need for institutional reforms in Colombia. *Substance Use and Misuse*, 47, 972–1004.

Timms, B.F. 2011. The misuse of disaster as opportunity: Coerced relocation from Celaque National Park, Honduras. *Antipode*, 43, 1357–1379.

Turner, B.L. 2006. Culture, ecology, and the classic Maya collapse. *Geographical Review*, 96, 490–493.

Turner, B.L., E.F. Lambin and A. Reenberg. 2007. The emergence of land change science for global environmental change and sustainability. *Proceedings of the National Academy of Sciences of the United States of America*, 104, 20666–20671.

Turner, B.L. and P. Robbins. 2008. Land-change science and political ecology: Similarities, differences, and implications for sustainability science. *Annual Review of Environment and Resources*, 33, 295–316.

Vaca, R.A., D.J. Golicher, L. Cayuela, J. Hewson and M. Steininger. 2012. Evidence of incipient forest transition in Southern Mexico. *PLOS ONE*, 7(8), http://www.plosone.org/article/info%3Adoi%2F10.1371%2Fjournal.pone.0042309 (accessed April 1, 2014), DOI:10.1371/journal.pone.0042309.

Van Ausdal, S. 2009. Pasture, profit, and power: An environmental history of cattle ranching in Colombia, 1850–1950. *Geoforum*, 40, 707–719.

Vandermeer, J., I. Perfecto and N. Schellhorn. 2010. Propagating sinks, ephemeral sources and per-colating mosaics: Conservation in landscapes. *Landscape Ecology*, 25, 509–518.

VanWey, L.K., C.M. Tucker and E.D. McConnell. 2005. Community organization migration, and remittances in Oaxaca. *Latin American Research Review*, 40, 83–107.

van Vliet, N., C. Adams, I.C. Guimaraes Vieira and O. Mertz. 2013. 'Slash and burn' and 'shifting' cultivation systems in forest agriculture frontiers from the Brazilian Amazon. *Society and Natural Resources*, 26, 1454–1467.

Walker, R., E. Arima, J. Messina, B. Soares, S. Perz, D. Vergara, M. Sales, R. Pereira and W. Castro. 2013. Modeling spatial decisions with graph theory: Logging roads and forest fragmentation in the Brazilian Amazon. *Ecological Applications*, 23, 239–254.

Walker, R., S.A. Drzyzga, Y.L. Li, J.G. Qi, M. Caldas, E. Arima and D. Vergara. 2004. A behavioral model of landscape change in the Amazon Basin: The colonist case. *Ecological Applications*, 14, S299–S312.

Wilcox, R. 1992. Cattle and environment in the Pantanal of Mato-Grosso, Brazil, 1870–1970. *Agricultural History*, 66, 232–256.

World Bank 2013. *Migration and Development Brief 21*. Washington, D.C.: World Bank.

Wunder, S. 2000. Ecotourism and economic incentives – an empirical approach. *Ecological Economics*, 32, 465–479.

Ybarra, M. 2009. Violent visions of an ownership society: The land administration project in Peten, Guatemala. *Land Use Policy*, 26, 44–54.

Ybarra, M. 2012. Taming the jungle, saving the Maya Forest: Sedimented counterinsurgency practices in contemporary Guatemalan conservation. *Journal of Peasant Studies*, 39, 479–502.

Zarin, D., J. Alavalpari, F. Putz and M. Schmink. 2010. *Working forests in the Neotropics*. New York: Columbia University Press.

Zenteno, M., P.A. Zuidema, W. de Jong and R.G.A. Boot. 2013. Livelihood strategies and forest dependence: New insights from Bolivian forest communities. *Forest Policy and Economics*, 26, 12–21.

Zhai, D.L., C.H. Cannon, J.W.F. Slik, C.P. Zhang and Z.C. Dai. 2012. Rubber and pulp plantations represent a double threat to Hainan's natural tropical forests. *Journal of Environmental Management*, 96, 64–73.

Zimmerer, K.S. 2006. *Globalization and new geographies of conservation*. Chicago: University of Chicago Press.

Zomer, R.J., A. Trabucco, R. Coe and F. Place. 2009. Trees on Farm: Analysis of Global Extent and Geographical Patterns of Agroforestry. ICRAF Working Paper, 89. Nairobi, Kenya: World Agroforestry Centre.

Susanna Hecht is Professor in the Luskin School of Public Affairs in Regional and International Development, and Environmental Analysis and Policy at University of California, Los Angelos (UCLA), with appointments in UCLA's Institute of the Environment and in the Department of Geography. She is the author of *Scramble for the Amazon and the lost paradise of Euclides da Cunha* (2013), the editor with Kathy Morrison and Christine Padoch of *The social lives of forests* (2014), the award winning *Fate of the forest: developers, destroyers and defenders of the Amazon* (2011, 4th ed.) and *Migration, livelihoods and natural resources* (2012) with Susan Kandel and Abelardo Morales, among more than 14 books and numerous articles. Dr. Hecht specializes in the development and Political Ecology of the Latin American tropics.

Index

Note: Page numbers followed by 'f' refer to figures and followed by 't' refer to tables.

For Product Safety Concerns and Information please contact our EU
representative GPSR@taylorandfrancis.com
Taylor & Francis Verlag GmbH, Kaufingerstraße 24, 80331 München, Germany